THE ORCHIDS OF THE PHILIPPINES

Jim Cootes

FEATURING PHOTOGRAPHS BY
DAVID P. BANKS AND DAVID TITMUSS

TIMBER PRESS
Portland, Oregon

This book is dedicated to Jack Jannese,
who introduced me to species orchids, and
Elias and Alicia Javier, who introduced me to
the native orchids of the Philippines.

Opposite: ***Dendrochilum cymbiforme***, *in situ*, Mountain Province (photograph by Dr. Henrik Pedersen)

CONTENTS

Opposite: ***Dendrochilum cootesii***, *in situ*, Neuva Ecija (photograph by Jim Cootes)

PREFACE

Southeast Asia is a rich area orchid-wise. Whilst many species are found in several countries in the region, many are native to a particular, and at times, very small locality. The Philippines is a country with a high ratio of endemic species.

This book has an excellent blend of information for both the serious orchid grower and for orchid enthusiasts with a few plants of Philippines origin in their collection. It covers both the showier species and those of botanical interest.

The information is presented in a unique and concise format to enable and assist in the identification of some of the "unknown" species inevitably found in specialist collections. The descriptions have been made from living specimens, which the author has studied, and the quality photographs accompanying the descriptions are from these plants. Details are provided on the habitats where the particular species are found in the wild and the conditions to which they have accordingly become adapted.

The author has been to the Philippines on more than 25 occasions and has seen a considerable number of the species growing *in situ.* This includes many that are seldom found in collections. Some may prove to be new species.

Jim has been a close family friend for more years than I really care to remember. His fascination for orchids of the Philippines (particularly the miniatures) has never waned over the years. His knowledge, and his dedication to passing such knowledge on, is widely acknowledged in orchid circles.

This book is a milestone in modern orchid literature covering the Philippines—and a tribute to the efforts and enthusiasm of a fine human being.

Graeme Banks
Past President
(1994–1997)
Australian Orchid Council
May 1999

ACKNOWLEDGEMENTS

When a work such as this reaches fruition it is frequently because of the assistance of many individuals. I wish to thank the following people from the bottom of my heart.

David Banks helped considerably with many aspects of this work, along with photographing many of the illustrated plants, often at short notice.

David Titmuss photographed many of the plants used to illustrate this book, frequently at short notice and often, inopportune times.

Slides were also provided by **Lloyd Bradford**, **Dr. Guido Braem** (Germany), **Dr. Mark Clements**, **Jim Comber** (United Kingdom), **Hans Fessel** (Germany), **Gary Yong Gee**, **Andres Golamco Jr.** (Philippines), **Olaf Gruß** (Germany), **Ray Hill**, **Jack Jannese**, **Elias Javier** (Philippines), **Kelvin Manubay** (Philippines), **Charles Nishihira** (Hawaii), **Peter O'Byrne** (Singapore), **Dr. Henrik Æ. Pedersen** (Denmark), **Anton Sieder** (Austria), **Danny Tiu** (Philippines) and **Dr. Howard Wood** (USA).

Much of the information contained in this book would not have been available without the assistance of **Dr. Mark Clements**, **David Jones** and the staff of the Centre for Plant Biodiversity Research, Australian National Herbarium, Canberra, Australia. I am forever grateful for their interest and encouragement in the writing of this book.

Miguel Garcia and **Anna Hallett**, librarians at the Royal Botanic Gardens, Sydney, N.S.W., were extremely helpful in locating the correct names of sometimes very obscure literature and for other assistance so freely given.

Paul Ormerod provided much information I could not have obtained elsewhere.

Lianna McCann's knowledge of computers helped improve the appearance of the manuscript immeasurably.

Graeme Banks, past president of the Australian Orchid Council, wrote the kind preface.

Dr. Eric Christenson (USA) willingly answered my various questions about ***Aerides*** and ***Ascocentrum*** species.

Dr. Peter Weston, senior research scientist, Royal Botanic Gardens, Sydney, N.S.W., assisted in locating the names of the type species of some genera.

Dr. Henrik Æ. Pedersen (Denmark) kindly sent me copies of his magnificent drawings of Philippine ***Dendrochilum*** species. These illustrations made the identification of many of my previously 'unnamed' species much easier. He also provided slides of other genera that were photographed during a field trip to the Philippines in 1994.

Jaap Vermeulen (the Netherlands) willingly identified ***Bulbophyllum*** species and provided the sections for them, plus other useful information.

Dr. Howard Wood (USA) provided assistance with the ***Dendrobium*** alliance.

Danny Tiu (Philippines) assisted with ***Dendrobium*** and other aspects of the flora of the Philippines. Danny was also my guide on many trips into the mountains of the Philippines to see the orchids growing in their natural habitat.

Darryl Smedley offered many useful suggestions regarding the manuscript.

Dr. Karlheinz Senghas (Germany) provided useful information on ***Vandaceous*** species.

Dr. Ed de Vogel (the Netherlands) assisted with ***Coelogyne***.

Dr. Leslie Garay (USA) assisted with the ***Bulbophyllum*** alliance and the etymology of numerous specific names.

Dudley Clayton (UK) provided much information on ***Coelogyne*** species.

Peter O'Byrne (Singapore) assisted with literature and distribution of some species. We also had a number of interesting e-mail discussions.

Rudolf Jenny (Switzerland) provided literature I could not find elsewhere.

Dr. Guido Braem (Germany) helped with sections and etymology of some ***Paphiopedilum*** species.

Shova Loh, publishing manager, and **Sue Sismondo,** editor, for their thoroughness in the final stages of the publication of this book; you both are an inspiration. **Tuck Loong** for his patience and professionalism in designing this book. **Enrico Sismondo** for the photograph of the author on the front jacket flap.

Rolando 'Lando' Alamo, Daniel 'Boboy' Escalante, **Bhoy** and **Zaldy Bulliunan** accompanied me on many of my forays into the forests of the central mountains of Luzon, never complaining and always willing to assist. **Bienvenido 'Atorne' Asturias Jr.** and **Eddie Martinez** joined me in exploring the forests of eastern Mindoro. **Orlando 'Toto' Adamos** accompanied me to many of the above destinations and willingly carried many plants. Their assistance is gratefully acknowledged here.

My thanks to the native people of the mountains of the Philippines who accompanied me into the forests so I could observe many species growing in their natural habitat—their knowledge of their local areas is astounding, and they were always ready to assist in any way possible.

If I have forgotten anyone I humbly apologise—just put it down to old age, forgetfulness and maybe a few too many drinks.

INTRODUCTION

The orchids of the Philippines are amongst the most beautiful in the world. Many horticulturally attractive orchid species occur within the more than 7,000 islands that make up the Philippine archipelago. Approximately 141 genera, which have over 1,100 species, varieties or forms, are represented in the Philippines. Many of my favourites are described here.

The most important of all vandaceous species (in my opinion), the magnificent ***Euanthe sanderiana*** (Rchb. *f.*) Schltr., is endemic to the second largest island, Mindanao.

The genus ***Phalaenopsis*** has its centre of distribution in the Philippines with at least 20 species, varieties and forms.

Many fine ***Paphiopedilum***, ***Aerides***, ***Renanthera*** and ***Vanda*** species grow there and are eagerly sought by orchid collectors worldwide.

The dendrobiums of the section ***Formosae***, whose pristine white flowers are unequalled in purity of colour, are very popular show-bench specimens at orchid society meetings.

The large, showy blooms are not the only gems that occur in the Philippines. In the mountains, there are the botanical species, whose numbers far outweigh those of their more widely grown cousins.

The genus ***Bulbophyllum*** is represented by over 120 species. I am certain that many more await description.

Dendrochilum has its centre of distribution in the high mountains, and there are over 90 recorded species and varieties.

Eria is depicted by over 70 species and ***Dendrobium*** has more than 80 members. These four genera alone constitute over one-third of the recorded orchid species in the Philippines.

On the other hand there are genera with only one or two species growing in the archipelago, for example, ***Acanthephippium mantinianum*** Linden and Cogniaux, ***Amesiella monticola*** Cootes and Banks, ***Amesiella philippinensis*** (Ames) Garay, ***Sarcophyton pachyphyllus*** (Ames) Garay and ***Megalotus bifidus*** (Lindley) Garay.

Then there are the widely spread species that are found in many countries throughout the world. The epiphytic ***Polystachya concreta*** (Jacq.) Garay and Sweet is found in Africa, most of Asia, Central and South America and the West Indies.

Some widely cultivated terrestrial species, such as ***Phaius tancarvilleae*** (Banks *ex* l'Heritier) Bl. and ***Spathoglottis plicata*** Bl., grow throughout tropical Asia, down to Australia and across to the Pacific Islands.

These are but some of the wonderful orchids that I wish to introduce in this book, *The Orchids of the Philippines.*

Scale 1:8,000,000
0
100
200
300 Kilometres
Metres
5,000
3,000
2,000
1,000
500
200
0
PACIFIC
OCEAN
South
China
Sea
Philippine
Sea
Sibuyan
Sea
Visayan
Sea
Bohol
Sea
Sulu
Sea
Babuyan Channel
Lingayen
Gulf
Manila
Bay
Lamon Bay
Babuyan
Calayan
Fuga
Camiguin
ILOCOS
NORTE
KALINGA
APAYAO
CAGAYAN
ABRA
Cordillera Central
Cagayan
Sierra Madre
MOUNTAIN
PROVINCE
ILOCOS
SUR
ISABELA
IFUGAO
LA UNION
Mt Pulog
(2929 m)
BENGUET
Caraballo
Mts
LUZON
NUEVA
VIZCAYA
QUIRINO
PANGASINAN
Zambales Mts
NUEVA
ECIJA
AURORA
TARLAC
Mt Arayat
(1030 m)
ZAMBALES
PAMPANGA
BULACAN
RIZAL
Polillo
Patnanongan
Jomalig
BATAAN
MANILA
QUEZON
LAGUNA
CAVITE
BATANGAS
CAMARINES NORTE
CATANDUANES
Lubang
Mt Halcon
(2587 m)
MARINDUQUE
CAMARINES SUR
MINDORO
Batan
Burias
ALBAY
Mayon Volcano (2421 m)
OCCIDENTAL
MINDORO
ORIENTAL
MINDORO
SORSOGON
Romblon
Tablas
SIBUYAN
Ticao
NORTHERN SAMAR
Busuanga
MASBATE
WESTERN
SAMAR
SAMAR
Culion
Coron
Biliran
Linapacan
AKLAN
EASTERN
SAMAR
CAPIZ
PANAY
LEYTE
ANTIQUE
ILOILO
CEBU
Dumaran
Guimaras
Cebu
SOUTHERN
LEYTE
Dinagat
NEGROS
OCCIDENTAL
Siargao
PALAWAN
BOHOL
Panaon
NEGROS
SURIGAO
DEL
NORTE
NEGROS
ORIENTAL
CAMIGUIN
Mt Victoria (1727 m)
SIQUIJOR
AGUSAN
DEL
NORTE
SURIGAO
DEL SUR
Mt Mantalingajan
(2054 m)
MISAMIS
ORIENTAL
MISAMIS
OCCIDENTAL
AGUSAN
DEL SUR
Bugsuk
LANAO
DEL
NORTE
BUKIDNON
Agusan
Balabac
Lake Lanao
DAVAO
DEL
NORTE
MINDANAO
ZAMBOANGA
DEL SUR
LANAO DEL
SUR
ZAMBOANGA
DEL NORTE
Olutanga
NORTH
COTABATO
Davao
DAVAO
ORIENTAL
N
MAGUINDANAO
Mt Apo
(2954 m)
Samal
BASILAN
Sulu Archipelago
SULTAN
KUDARAT
DAVAO
DEL SUR
Pangutaran
SOUTH
COTABATO
JOLO
SABAH
(EAST MALAYSIA)
Siasi
TAWITAWI
Sarangani

GEOGRAPHY

The archipelago of the Philippines consists of over 7,100 islands situated to the southeast of mainland Asia between the latitudes of 5 degrees North to 21 degrees North. The archipelago separates the South China Sea from the Pacific Ocean and extends for over 1,800 kilometres from north to south and almost 1,000 kilometres from east to west.

Geological faults and major earth movements have played their part in the formation of the archipelago, giving it a staggering array of topographical features, from rugged mountain ranges to jagged coastlines.

Volcanoes form a notable feature of the Philippine archipelago; there are 19 active and 180 dormant volcanoes. The eruption of Mount Pinatubo in June 1991 wreaked great havoc to the surrounding countryside and to the people who lived on and around the mountain. Problems from this eruption will continue for many years.

The Philippines is generally mountainous, but large rivers and wide valleys run between the mountain ranges, creating fertile plains that adjoin the coastlines. The longest river in the Philippines is the Cagayan, which stretches for over 325 kilometres in northern Luzon. The Rio Grande de Mindanao, in the province of North Cotabato is the second longest river in the Philippines. The Agusan River, also in Mindanao, is the third longest river in the archipelago.

The physical characteristics are best discussed in terms of four principal geographic regions: Luzon, the Visayan Islands, Mindanao and Palawan.

LUZON

Luzon is the largest island and has about 35 percent of the archipelago's total land area (approximately 106,000 square kilometres). At the northernmost tip, adjoining the China Sea, is the narrow strip of the Ilocos coastal plain. Moving south, one comes to the mountain range of the Cordillera Central with its rich mining regions.

The Cagayan valley is a north-south depression in northern Luzon bounded by the Babuyan Channel in the north, the Caraballo Mountains on the south, the Sierra Madre Mountains on the east and the Cordillera Central to the west. The valley is about 240 kilometres long by 85 kilometres wide and is drained by the Cagayan River, the longest and widest in the Philippines.

The Cordillera Central stretches from the Ilocos highlands to the northern flank of the Caraballo Mountains. It is about 320 kilometres long by about 85 kilometres wide with rugged topography. Luzon's highest peak, Mount Pulog (2,929 metres), forms the common boundary of the provinces of Benguet, Ifugao and Nueva Vizcaya.

The Caraballo Mountains are the southeastern extension of the Cordillera Central and separate the Cagayan valley from the Luzon central plain.

The Luzon central plain, between the Zambales Mountains and the Sierra Madre Mountains, is the largest in the country. The plain is about 150 kilometres long with an average width of 60 kilometres. It opens into Manila Bay in the south and the Lingayen Gulf in the north. Mount Arayat (1,030 metres), a lone volcanic peak, stands prominently in the centre of the plain.

The Sierra Madre Mountains, in northeast Luzon, are the longest mountain system of the Philippines. The northern Sierra Madre Mountains follow the Pacific Ocean shoreline from Cagayan province in the north to Quezon province in the south. There are several peaks in the system with elevations over 1,100 metres. Mount Anacuano at 1,852 metres is the highest. The southern Sierra Madre Mountains extend from Aurora province to Lamon Bay in the south of Quezon province. The peaks here are fewer and smaller than their neighbours in the north. The tallest peak is Mount Irid at 1,448 metres.

Further south lies the volcanic region of southwest Luzon and the irregularly shaped peninsula of Bicol in southeast Luzon. This area is dominated by Mount Mayon (2,421 metres high), an active volcano that erupts at fairly frequent intervals. A number of dormant volcanoes are also found in this region. Notable among these are Isarog (1,976 metres), Malinao (1,553 metres), Bulusan (1,553 metres), Iriga (1,159 metres) and Labo (942 metres).

The islands of Mindoro, Marinduque and Masbate, together with smaller nearby islands, complete the picture of this region.

Mindoro is approximately 10,000 square kilometres in area. Its landscape is characterised by some very rugged mountains, intervening valleys and elongated plateaus, with rolling land along the coast. Among the mountain peaks that dot the countryside are Mount Halcon (2,587 metres), Mount Baco (2,487 metres), Mount Calavite (1,312 metres), Mount Exline (1,170 metres) and Mount Pamucuban (790 metres). Only Mount Halcon has been explored botanically. One can only wonder what botanical delights await to be found on the unexplored peaks.

Marinduque is about 900 square kilometres in area with several mountain peaks ranging in elevation from 558 to 1,157 metres (Mount San Antonio). The western slopes of the mountains are steeper than the eastern.

Masbate is a two-pronged island with the eastern prong consisting of discontinuous highlands and broad bands of lowlands. The highest elevation on Masbate is 718 metres.

THE VISAYAN ISLANDS

The Visayan Islands lie between Luzon and Mindanao. Samar, Negros, Panay, Leyte, Cebu and Bohol are the main islands of this region. Mountains dominate the central regions of these islands, yielding to narrow strips of arable land.

Samar is the third largest island in the Philippines with a land area of approximately 13,429 square kilometres. The island consists mainly of low, extremely rugged mountains; the highest is Mount Capotoan at 896 metres above sea level.

Negros is about 190 kilometres long by 36 kilometres wide with an area of about 12,541 square kilometres. The Canlaon volcano (2,450 metres high), which is still active, dominates the northern landscape. In the southeast, volcanic peaks dot the countryside; Cuernos de Negros (1,303 metres) is the tallest.

Panay has a land area of approximately 12,000 square kilometres. Mountains on this island run in a north-south direction along the west coast. The highest peak is Mount Nangtud at 2,049 metres.

Leyte is generally mountainous, particularly in the west and northeast. The western range is rugged, with at least three mountains attaining elevations over 1,150 metres: Mount Sacripante 1,150 metres, Mount Lumas 1,252 metres and Mount Lobi 1,349 metres. Leyte is one of the few islands in the Visayas that have been well explored botanically. In fact, three new ***Dendrobium*** species and one new ***Flickingeria*** species have been named from this island in 1996, 1997 and 1998.

Cebu is about 200 kilometres long by 40 kilometres wide at its widest point. The island has a mountainous backbone. The highest mountain in the north is Mount Cabalasan at 1,013 metres. In the south, the highest peak is Mount Uling at 659 metres.

Bohol is approximately 4,117 square kilometres in size, making it the 10th largest island in the Philippine archipelago. The central region of Bohol is a cultivated plateau with the unique 'Chocolate Hills' a feature. These hills generally rise about 60 metres above an almost flat landscape.

MINDANAO

Mindanao is the second largest island at approximately 96,000 square kilometres and has an extremely varied landscape. High mountain ranges, volcanic peaks, plateaus and swampy plains all lie within the much twisted coastline.

The Diwata Mountains extend down the east coast of Mindanao from north to south. The mountains are very rugged, and there are numerous peaks with elevations from 900 to 2,196 metres. Mount Kampalili, Mount Hilonghilong and Mount Tadagalit are some of the higher peaks.

The Central Range runs the whole length of the island, from the provinces of Agusan del Norte, Agusan del Sur and Davao del Norte to North and South Cotabato. Mount Apo (2,954 metres), the Katanglad mountains (reaching elevations of 2,379 metres), Mount Matutum (2,293 metres) and the Kalatungan mountains are some of the larger peaks.

A third range forms the spine of the Zamboanga Peninsula and Misamis Occidental province. Mount Silingan (899 metres) and Sharp Peak (753 metres) are the tallest mountains on the Zamboangan peninsula. In Misamis Occidental, Mount Malindang at 2,425 metres towers over the surrounding countryside.

The Lanao-Bukidnon plateau is rolling lava upland at around 700 metres with its dominant feature, Lake Lanao, draining into the Agusan river, which passes through a steep precipice, forming the Maria Cristina Falls.

The two large swampy plains of Mindanao, the Agusan-Davao plain and the Cotabato valley, are the key to Mindanao's agricultural wealth.

The Sulu Archipelago, with 2,600 islands, stretches southwest from the tip of the Zamboangan peninsula. The southern-most islands are visible from Borneo. The biggest island in the group is Jolo, which is of volcanic origin, as are most of the larger islands in the Sulu Archipelago.

PALAWAN

Palawan includes the islands of Busuanga, Coron and Culion. These three islands are generally mountainous with elevations over 650 metres. Palawan, at about 12,000 square kilometres, is the fifth largest island and the largest province in the Philippines and is mountainous along its entire length. North Palawan has rugged mountain ridges with prominent peaks; Cleopatra's Needle (1,593 metres), Mount Kapoas (1,021 metres) and the Pagdanan Range reach elevations of up to 644 metres. Central Palawan has Mount Victoria at 1,727 metres and a broad coastal plain about 80 kilometres long. Southern Palawan has two, very rugged mountain ranges that culminate in Mount Mantalingajan at 2,054 metres.

CLIMATE

There are four major climatic zones in the Philippine archipelago.

The western sides of northern Luzon, Mindoro, Panay, Negros and Palawan islands have a distinct wet and dry season.

The eastern coast of northern Luzon, the west coast of southern Luzon, eastern Mindoro, western Samar, western Leyte, Bohol and central Mindanao have no pronounced dry season, and it can rain at any time during the year.

The Cagayan valley of northern Luzon, eastern Panay, eastern Negros, western Cebu, western Mindanao and eastern Palawan have a rainy season of about three months each year.

The eastern coast of southern Luzon, Catanduanes island, eastern Samar, eastern Leyte and eastern Mindanao have no distinct dry season, but it rains heavily during the cooler months (December and January).

FLORAL ZONES

Vegetation formations in the Philippines can be grouped into the following entities.

In the LOWLAND RAIN FOREST, which is found from sea level to 1,000 metres or more, the typical growth is of three tree layers over shrub and ground cover.

Most widespread is the dipterocarp forest, accounting for some three-quarters of the lowland rain forest. The uppermost trees, mostly dipterocarps, reach heights of about 40 metres, the middle layer trees heights of up to 20 metres, the lowermost layer up to 10 metres. Below the trees are the shrubs, normally less than 6 metres high. The dipterocarp forest is the region where the majority of the most spectacular and horticulturally attractive orchid species occur.

Some of the better known species from this zone are: ***Aerides lawrenciae*** Rchb. *f.*, ***Ascoglossum calopterum*** (Rchb. *f.*) Schltr., ***Bulbophyllum dearei*** Rchb. *f.*, ***Cymbidium finlaysonianum*** Lindl., ***Dendrobium aloifolium*** (Bl.) Rchb. *f.*, ***Dendrobium anosmum*** Lindl., ***Dendrobium bullenianum*** Rchb. *f.*, ***Dendrobium crumenatum*** Sw., ***Dendrobium spurium*** (Bl.) J.J. Sm., ***Dipodium paludosum*** (Griff.) Rchb. *f.*, ***Eria ornata*** (Bl.) Lindl., ***Euanthe sanderiana*** (Rchb. *f.*) Schltr., ***Grammatophyllum multiflorum*** var. ***tigrinum*** Lindley, ***Phalaenopsis aphrodite*** Rchb. *f.*, ***Phalaenopsis equestris*** (Schauer) Rchb. *f.*, ***Phalaenopsis sanderiana*** Rchb. *f.*, ***Renanthera philippinensis*** Ames and Quis., ***Trichoglottis philippinensis*** Lindl., ***Vanda lamellata*** Lindl.

SEASONAL FOREST once covered about two percent of the Philippine lowlands but is now virtually cleared, as is the mixed swamp forest. I have only observed ***Dendrobium*** species of the section ***Crumenata***, members of the genus ***Grastidium*** and ***Luisia*** species in this type of forest.

BEACH FOREST is relatively uniform throughout the Pacific. In the Philippines this coastal forest is well developed, forming a salt-spray-tolerant fringe along the seashore. I have never observed any orchid species growing in this type of forest.

MANGROVE SWAMP FORESTS are restricted to the tidal mud flats at the mouths of rivers where the water is brackish to salty. A number of orchid species occur within these forests; for example, ***Bulbophyllum cheiri*** Lindley, ***Cleisostoma weberi*** (Ames) Garay, ***Dendrobium taurinum***

Lindley, ***Eria aporoides*** Lindley, ***Renanthera philippinensis*** Ames and Quis., and ***Renanthera storiei*** Rchb. *f.* have been recorded from this habitat.

The LOWER MONTANE FOREST starts at about 1,000 metres where the humid lowlands give way to a cooler, more temperate, mid-mountain region where oak-laurel forests predominate. Where the forest reaches optimum development, the structure is of two tree layers. The first canopy, at 25 to 30 metres, is lower than that of typical lowland forest. The subcanopy attains heights of about 16 metres, and usually the trees are of less dense growth in this region. This type of forest is home to some very spectacular orchids. ***Bulbophyllum facetum*** Garay et al., ***Dendrobium amethystoglossum*** Rchb. *f.*, ***Dendrobium fairchildiae*** Ames and Quis., ***Dendrobium victoriae-reginae*** Loher, ***Phalaenopsis lindenii*** Loher, ***Phalaenopsis philippinensis*** Golamco *ex* Fowlie and C.Z. Tang, ***Vanda javierae*** Tiu *ex* Fessel and Lückel and ***Vanda roeblingiana*** Rolfe are found here. The majority of the species from the genus ***Dendrochilum*** grow here, as do many ***Bulbophyllum*** species. Members of the ***Dendrobium*** sections ***Pedilonum*** and ***Calcarifera*** inhabit the trees. It is also the abode of a multitude of extremely beautiful, small-flowered species commonly known as 'botanicals'.

I am of the opinion that there are many orchids in the lower montane forests that are still unknown to science. During a trip to the Philippines in February 1996 I found several specimens of the genus ***Epicrianthes***. This genus has not been previously recorded from the Philippines.

A series of articles published between June and October 1995 by David Banks and myself in *Orchids Australia* on the genus ***Dendrochilum*** has also brought to light what may well be two species previously not described. One of these species (species 'K') has been named in honour of the author of this book as ***Dendrochilum cootesii*** by Dr. Henrik Æ. Pedersen, a Danish taxonomist. Dr. Pedersen spent several years studying Philippine dendrochilums. The second species (species 'J') is currently (March 2000) being studied by Dr. Pedersen.

The lower montane forest is a species orchid-grower's paradise and the source of many of the plants described within these pages.

PINE FORESTS, which occupy much of the mountains of Luzon, are considered to be an intruding wedge. Millions of years ago, when climatic conditions were cooler and the islands were in physical contact with the mainland of the northwest, the pine trees and other temperate species may have migrated down from the Himalayas. The Benguet Pine (***Pinus insularis*** Endl.) remains as a relict species of this past invasion, still occupying the dryer, cooler mid-mountain levels in the provinces of central Luzon. *Tapu'lau* or Mindoro Pine (***Pinus merkusii*** Jungh.) is a species that is restricted to the province of Zambales on the west coast of Luzon and the island of Mindoro. Many orchid species are also found here. ***Dendrobium sanderae*** Rolfe and its varieties (except var. ***surigaense***), ***Dendrobium auriculatum*** Ames and Quis., ***Dendrochilum*** and ***Vanda*** species are inhabitants of this region. Without a doubt many other genera utilise the pine trees as their host tree.

LOWER MONTANE MIST FORESTS are found along the main mountain range of Luzon, the Cordillera Central. The middle slopes are typically covered with open, sunny, pine forests extending to about 2,500 metres, or the lower edge of the persistent daily mists.

Within the misted levels the vegetation change is both abrupt and startling. Here one enters a dimly lit, dense, wet, mossy forest of mixed broadleaf trees. These trees form a single layer about 8 metres tall.

Mount Apo in Mindanao is the highest mountain in the Philippines and has a similar mist-stunted, lower montane forest starting at 2,500 metres. This is a single-layered forest dominated by ***Myrica javanica***.

Other parts of the Philippines at the same altitude, but mist-free, have a normal two-layered, lower montane forest. The so-called mossy montane forest on the summit of Mount Makiling (1,009 metres) in Laguna province is only a misted and stunted lower montane forest.

True montane forest is not found in the Philippines, even at elevations over 2,700 metres.

Dendrobium amethystoglossum (left) and ***Ceratostylis incognita*** (centre and right), *in situ*, Neuva Vizcaya (photograph by Jim Cootes)

HOW TO USE THIS BOOK

The purpose of this book is for the identification of orchid species found in the Philippines. It is not within the realms of this book to try and describe all the orchid species found in the Philippines; many of the species are only known from single collections made in the early 1900s. I have, though, included a checklist of all recorded Philippine orchid species, as currently understood, as an appendix. Also included as an appendix are the ***Bulbophyllinae*** and ***Dendrobiinae*** divided into their respective genera and sections.

Please remember some species are widely distributed throughout Asia and the Pacific Islands (and in the case of ***Polystachya concreta*** in Africa and Central and South America) and sometimes show a wide range of variation. The old, well-known adage that a picture is worth a thousand words has been applied here. Each described species has a colour photograph of the flower of the plant concerned. Many of the species are illustrated for the first time.

The description of the plant, for the majority of the species, is from specimens grown by the author or plants to which I have ready access. It should be noted that a number of the plants included have not been grown by the author. The plant and flower dimensions of these species have been taken from the available literature or supplied by the grower(s) of the described plants.

My description of the plant is from cultivated specimens and will, possibly, vary from previously published descriptions. The reason for this is that plants in cultivation often grow much larger than specimens from the forest. This is because of regular fertilising, fewer insect pests other natural climatic calamities and far more congenial conditions in which the plant can grow.

All genera and species listed are in alphabetical order for ease of location.

A number of the species included in the book are of wide distribution, so an explanation of the geographical terms used in the following pages is included here:

- **Indochina** refers to the countries of Cambodia, Laos, and Vietnam.
- **Indonesia** includes the islands of Sumatra, Java and the smaller islands of the archipelago. It does not include the Indonesian parts of Borneo or New Guinea, which are listed separately, as is Sulawesi.
- **Borneo** is meant as the whole island.
- **New Guinea** includes Irian Jaya.

It should be noted here that I have had a great deal of assistance from botanists and taxonomists world-wide in the compilation of this book and they have been acknowledged earlier. If there are any incorrect identifications of the plants, or other mistakes, they are mine and do not reflect on the previously mentioned assistance. The author, alone, has made the final decisions on taxonomy.

GENERA AND SPECIES

Opposite: ***Dendrochilum stenophyllum***, *in situ*, Neuva Vizcaya (photograph by Jim Cootes)

Acanthephippium Blume

Carl Blume established this genus of terrestrial orchids in 1825. The name was published in *Bijdragen tot de Flora van Nederlandsch Indie.* ***Acanthephippium*** is a combination of two Greek words, *akantha,* meaning thorn, and *ephippion,* saddle, which alludes to the shape of the labellum.

The type species for this genus is ***Acanthephippium javanicum*** Blume.

There are approximately 15 species in the genus occurring from the Himalayas, China, Japan, Thailand, Malaysia, Indonesia, New Guinea, the Pacific Islands and the Philippines. There is (currently) only one ***Acanthephippium*** species described from the archipelago, though in February 1996 I saw slides of what may prove to be an undescribed species.

Acanthephippium mantinianum

Linden and Cogniaux

Jean Linden and Alfred Cogniaux named this plant in the *Journal des Orchidees*, published in 1896. It was named after Georges Mantin of Olivet, France, who was an avid orchid grower and developer of hybrids.

Growth habit: upright, sympodial. **Pseudobulbs:** tapered, up to 15 cm long; base up to 5 cm in diameter. **Leaves:** very large, up to 60 cm long, 15 cm wide and of a thin texture. **Inflorescence:** upright, to 15 cm long, bearing up to six flowers about 2.5 cm in diameter. **Flower colour:** bright yellow with red striping. **Dorsal sepal:** concave, linear, up to 3.3 cm long, 1.5 cm wide, unequally bilobed at the tip. **Petals:** sub-rhombic, up to 2.8 cm long, 1.5 cm wide, rounded at the tips. **Lateral sepals:** obliquely oblong and recurving slightly at tips; basal two-thirds fused to form an obscurely bilobed mentum, up to 3.5 cm long, 1.6 cm wide. **Labellum:** three-lobed; lateral lobes erect, enclosing column; mid-lobe rounded at tip, up to 2.9 cm long, 1.1 cm wide (flattened), hinged halfway down its length.

Habitat and Distribution: ***Acanthephippium mantinianum*** is found in the mountainous regions of Cavite, Laguna, Mountain Province, Quezon and Sorsogon provinces on Luzon where it grows in shaded locations at elevations from 500 to 1,500 metres. It is endemic to the Philippines.

Notes: The flowers of this species do not open widely and are somewhat tubular in shape. The dorsal and lateral sepals are fused for most of their length. The floral segments are extremely fleshy.

This species has an interesting feature: the inflorescence elongates after pollination to ensure wider dispersal of the seed.

A most unusual species that is worthy of a place in any comprehensive orchid collection.

Photo: David Titmuss

Acanthephippium mantinianum

Acriopsis Reinwardt *ex* Blume

Carl Blume established this genus of six species and four varieties in 1825 in *Bijdragen tot de Flora van Nederlandsch Indie.* The generic name refers to the flower's resemblance to a locust.

The type species for this genus is ***Acriopsis javanica*** Reinwardt *ex* Blume.

Acriopsis have been recorded from Sikkim, Myanmar (Burma), Thailand, Laos, Cambodia, Vietnam, Malaysia, throughout Indonesia, Sulawesi (Celebes), the Philippines, New Guinea, the Solomon Islands and northern Australia. The Philippines has two species and one variety.

Acriopsis javanica Reinwardt *ex* Blume

Carl Blume established this species in *Bijdragen tot de Flora van Nederlandsch Indie* in 1825. The specific name refers to Java where the original specimens were collected.

Growth habit: upright, sympodial. **Pseudobulbs:** up to 4 cm tall by 2.5 cm wide, carry up to three leaves. **Leaves:** linear, up to 20 cm long, 1.5 cm wide. **Roots:** Apart from securing the small ***Cymbidium***-like plant to its host, they form litter traps to catch falling leaves. **Inflorescences:** upright, to 25 cm tall, branching, bearing up to 200 blooms about 8 mm in diameter. **Flower colour:** pale green, sometimes spotted with purple; labellum white. **Dorsal sepal:** concave, linear, up to 4 mm long, 1.5 mm wide. **Petals:** oblong, up to 5 mm long, 2 mm wide. **Lateral sepals:** fused to form a synsepalum up to 4 mm long, 2 mm wide. **Labellum:** three-lobed, forming a tube with base of column.

Habitat and Distribution: ***Acriopsis javanica*** is a widely distributed species found in Thailand, Indochina, Malaysia, throughout Indonesia, Borneo, Sulawesi (Celebes), New Guinea, the Solomon Islands and northern Australia. In the Philippines it has been found in Sorsogon on Luzon; Leyte in the Visayas and Agusan del Norte Agusan del Sur, Davao del Norte, Davao Oriental, Davao del Sur, Lanao del Norte, Lanao del Sur, Misamis Occidental and Misamis Oriental on Mindanao. It grows as an epiphyte, occasionally a lithophyte, at any elevation to about 1,600 metres and frequently inhabits exposed sites.

Photo: Jim Cootes

Acriopsis javanica

Aerides Loureiro

Juan Loureiro established this beautiful monopodial genus in 1790. The generic name was published in *Flora Cochinchinensis* and is derived from the Greek *aer* for air and *eides,* resembling. The name means literally "children of the air".

The type species for this genus of beautiful orchids is ***Aerides odorata*** Lour.

The genus is found in Japan, China, the Himalayas, Myanmar (Burma), Thailand, Vietnam, Indonesia, New Guinea and the Philippines. Seven ***Aerides*** species are found in the Philippines.

Aerides lawrenciae Reichenbach *f.*

Professor Heinrich G. Reichenbach named this magnificent species in the *Gardeners' Chronicle* of 1883. He named it after Lady Lawrence, wife of the president of the British Royal Horticulture Society at that time.

Growth habit: upright, monopodial; tall plants can become semi-pendulous. The plant is of typical ***Vanda***-like growth habit and can reach a height of more than one metre. **Leaves:** leathery, curved, up to 30 cm long by 5 cm wide, tips unequally lobed. **Inflorescences:** pendulous; several on a large, healthy plant bearing up to 30 flowers, blooms about 4 cm in diameter. **Flower colour:** white with purple tips to each segment; spur is green. **Dorsal sepal:** oblong, up to 20 mm long by 12 mm wide. **Petals:** ovate, up to 15 mm long by 10 mm wide. **Lateral sepals:** elliptic, up to 18 mm long by 15 mm wide. **Labellum:** erect lateral lobes not overlapping; mid-lobe rectangular with toothed edges.

Habitat and Distribution: ***Aerides lawrenciae*** is endemic to the provinces of North and South Cotabato, Davao Norte and Davao Sur on the island of Mindanao. It grows as an epiphyte at elevations up to 500 metres in brightly lit situations.

Photo: Jim Cootes

Aerides lawrenciae

Aerides lawrenciae
var. ***fortichii*** Ames and Quisumbing

Professor Oakes Ames and Dr. Eduardo Quisumbing named this lovely species in the *Philippine Journal of Science* in 1935. The specific name honours the Honourable Manuel Fortich on whose property the type specimens were collected.

This is the albino form of the species. Plant and flowers are the same size as the species, except the flowers are pure white with a greenish spur.

Habitat and Distribution: ***Aerides lawrenciae*** var. ***fortichii*** is endemic to Bukidnon province on Mindanao. It grows as an epiphyte at elevations up to 500 metres.

Notes: This species is extremely rare.

Photo: David Banks

Aerides lawrenciae var. ***fortichii***

Aerides leeana Reichenbach *f.*

Synonym: ***Aerides jarckiana*** Schltr.

Professor Heinrich G. Reichenbach named this plant in the *Gardeners' Chronicle* in 1881 and named it after a Mr. Lee of Leatherhead, England.

Growth habit: upright, monopodial, up to 35 cm tall. **Leaves:** leathery, up to 25 cm long by 3 cm wide and do not curve, tips unequally lobed. **Inflorescences:** erect for the first two-thirds of their length, then pendulous for approximately 15 cm and occasionally branching: this portion of the inflorescence carries about 30 flowers 1 cm in diameter. Large, well-grown specimens can bear up to four inflorescences. **Flower colour:** purple; pure white clones also exist but are extremely rare. **Dorsal sepal:** elliptic, up to 7 mm long by 6 mm wide. **Petals:** elliptic, up to 7 mm long by 5 mm wide. **Lateral sepals:** elliptic, up to 7 mm long by 8 mm wide. **Labellum:** rectangular, 10 mm long by 4 mm wide with incurved lateral lobes that totally cover the entrance to the spur, which points down.

Habitat and Distribution: ***Aerides leeana*** is endemic to the provinces of Bataan, Camarines Norte, Camarines Sur, Cavite, Quezon and Rizal on Luzon and the island of Samar in the Visayas. It grows as an epiphyte at elevations of around 800 metres.

Notes: ***Aerides leeana*** will grow and flower successfully in shaded situations, which is unusual for members of the genus ***Aerides***, as they generally require bright light to grow and flower to their best potential.

Photo: David Banks

Aerides leeana

Photo: Jim Cootes

Aerides odorata

Aerides odorata Loureiro

Juan Loureiro named this species in *Flora Cochinchinensis* in 1790. The specific name refers to the lovely perfume of the blooms.

Growth habit: upright, monopodial; large plants become semi-pendulous. This plant can grow 1.5 metres tall and readily branches. **Leaves:** very leathery, up to 30 cm long by 3 cm wide. **Inflorescences:** arching, longer than the leaves, bear up to 40 flowers about 2.5 cm in diameter. **Flower colour:** variable. Some clones have white segments tipped with pink and a pink labellum, other specimens have cream sepals and petals with a golden-yellow lip. There are also albino clones, but I have never seen any in the Philippines. **Dorsal sepal:** elliptic, up to 1.3 cm long by 1 cm wide. **Petals:** elliptic, up to 1.3 cm long by 8 mm wide. **Lateral sepals:** elliptic, up to 1.5 cm long by 1.2 cm wide. **Labellum:** three-lobed; lateral lobes upright and not overlapping; mid-lobe covers lateral lobes and is toothed along the edges; spur curves forward and is usually green or yellowish.

Habitat and Distribution: ***Aerides odorata*** is a widely distributed species found in China, Nepal, the Himalayas, Myanmar (Burma), Thailand, Vietnam, Peninsular Malaysia, Indonesia and New Guinea. In the Philippines this orchid is only recorded from the province of Oriental Mindoro on the island of Mindoro and on Palawan, but undoubtedly has a wider distribution. This species is also found at a wide range of elevations—from 300 to 2,000 metres. It grows as an epiphyte high in the branches of trees where it receives plenty of light, often full sun for the whole day.

Aerides quinquevulnera Lindley

Dr. John Lindley named this species in 1838 in *Sertum Orchidaceum*. The specific name refers to the coloured tips of the floral segments, which resemble five wounds.

Growth habit: upright, monopodial. **Leaves:** leathery, curved, up to 30 cm long by 4 cm wide, unequally bilobed at their tips. **Inflorescences:** pendulous, carry up to 30 flowers 2.5 cm in diameter. **Flower colour:** white to cream with bright purple tips and spots to the segments; labellum spotted with purple. **Dorsal sepal:** elliptic, up to 1 cm long by 6 mm wide. **Petals:** elliptic, up to 8 mm long by 5 mm wide. **Lateral sepals:** ovate, up to 1.5 cm long by 1 cm wide. **Labellum:** three-lobed; lateral lobes upright and sometimes overlapping; mid-lobe rectangular.

Habitat and Distribution: ***Aerides quinquevulnera*** is found throughout the Philippines and has been recorded from the provinces of Aurora, Bataan, Cavite, Laguna, Mountain Province, Quezon, Rizal and Zambales on Luzon, the province of Oriental Mindoro on the island of Mindoro and the province of Negros Oriental on the island of Negros in the Visayas. It has recently been recorded from Papua New Guinea. This plant grows at elevations between 300 and 2,000 metres as an epiphyte in brightly lit situations.

Photo: David Titmuss

Aerides quinquevulnera

Photo: Andres Golamco Jr.

Aerides quinquevulnera var. ***farmeri*** 'Dalaginding'

A number of varieties are recognised within this species and all have similar dimensions and flower shape as the species:

Aerides quinquevulnera var. ***farmeri*** (Boxall) Stein is the albino form with pure white flowers.

Endemic to the islands of Babuyan and Calayan, to the north of Luzon, where it grows as an epiphyte at elevations up to 300 metres.

Aerides quinquevulnera var. ***flava*** Valmayor and Tiu has bright yellow segments spotted with purple.

Only known from the province of Quezon on the east coast of Luzon where it grows as a epiphyte at elevations up to 300 metres.

Aerides quinquevulnera var. ***purpurata*** Rchb. *f.* has almost solid purple blooms (the spotting has merged).

It has been recorded from the provinces of Quezon, Rizal and Zambales on Luzon, the province of Oriental Mindoro on the island of Mindoro and the province of Negros Oriental on the island of Negros in the Visayas. It grows as an epiphyte at elevations up to 300 metres.

Photo: David Titmuss

Aerides quinquevulnera var. ***flava***

Photo: David Banks

Aerides quinquevulnera var. ***purpurata***

Agrostophyllum Blume

Carl Blume established this genus of approximately 100 species in 1825 in *Bijdragen tot de Flora van Nederlandsch Indie*. The generic name refers to the grass-like foliage of some of the species.

The type species for the genus is ***Agrostophyllum javanicum*** Bl.

This genus is found in the Seychelles islands, Thailand, Peninsular Malaysia, Sumatra, Indonesia, Borneo, New Guinea and the islands of the Pacific Ocean. In the Philippines there are at least 13 species.

Agrostophyllum inocephalum

(Schauer) Ames

This species was first described by Schauer as ***Diploconchium inocephalum.*** Professor Oakes Ames transferred the generic name to ***Agrostophyllum*** in 1908 in *Studies in the Family Orchidaceae*, Fascicle II. The generic name may refer to the fibre-like cluster from which the flowers appear.

Growth habit: upright, sympodial. **Pseudobulbs:** flattened, up to 25 cm long by 2.5 cm wide. **Leaves:** two-ranked, linear to oblong, up to 15 cm long by 2.5 cm wide, unequally lobed at the tips. **Inflorescences:** terminal, densely clustered, only one or two flowers open at a time. **Flower colour:** white. **Dorsal sepal:** oblong to lanceolate, up to 5 mm long by 2 mm wide. **Petals:** linear, up to 4 mm long by .75 mm wide. **Lateral sepals:** oblong to lanceolate, up to 5 mm long by 2 mm wide. **Labellum:** three-lobed; side lobes upright; mid-lobe ovate to heart-shaped; overall up to 6 mm long by 4 mm wide.

Habitat and Distribution: ***Agrostophyllum inocephalum*** is endemic to the Philippines and has been recorded from the provinces of Bataan and Benguet on Luzon and the provinces of Agusan del Norte, Agusan del Sur, Lanao del Norte and Lanao del Sur on Mindanao. It grows as an epiphyte at elevations up to 500 metres.

Photo: Jim Cootes

Agrostophyllum inocephalum

Amesiella Schlechter *ex* Garay

In 1926 Dr. Rudolf Schlechter proposed the generic name of ***Amesiella*** but died before his new name could be published. In 1972 Dr. Leslie Garay discovered Schlechter's unpublished name, which he duly published in the *Botanical Museum Leaflets of Harvard University*. The genus was named in honour of Professor Oakes Ames, a botanist who did a lot of work with Philippine orchids, describing many new species. There are three species in this genus.

The type species for this genus is ***Amesiella philippinensis*** (Ames) Garay.

Amesiella monticola

J.E. Cootes and D.P. Banks

David Banks and myself described this species in the October 1998 issue of *Orchids Australia*. The specific name refers to the mountainous habitat where this distinctive species lives.

Growth habit: upright, monopodial, 2.5 cm tall by 16 cm across the leaves. **Leaves:** leathery, linear to obovate, unequally bilobed at their tips, up to 9.5 cm long by 2.4 cm wide. **Inflorescences:** up to three, bearing up to five flowers about 6 cm in diameter. **Flower colour:** pure white; some clones have faint yellow to orange stripes in the labellum; pedicel pale

Photo: David Banks

Amesiella monticola

Photo: David Banks

Amesiella philippinensis

green. **Dorsal sepal:** elliptic, concave, strong ridge down centre, up to 2.5 cm long by 1.8 cm wide. **Petals:** elliptic, concave, up to 2.5 cm long by 2.1 cm wide. **Lateral sepals:** elliptic, strong ridge down centre, 2.9 cm long by 2.1 cm wide. **Labellum:** three-lobed; side lobes semicircular, 7 mm long by 5 mm high; mid-lobe semicircular, appearing folded, 1 cm wide by 7 mm long. **Nectary:** up to 11 cm long (when measured from the tip of the mid-lobe) by 2.5 mm in diameter.

Habitat and Distribution: ***Amesiella monticola*** is endemic to the Philippines and is only known from the mountains of central Luzon. It grows as an epiphyte on the fringes of rainforest at elevations between 1,800 and 2,200 metres.

Notes: ***Amesiella monticola*** is readily separated from ***Amesiella philippinensis*** by the following features:

- The nectary (spur) of ***A. monticola*** is always longer.
- The flowers are usually larger (minimum 6 cm across the lateral sepals) with wider petals and sepals.
- The flowers of ***A. monticola*** are always pure, glistening white and perfumed (more so in the evening). ***A. philippinensis*** generally has off-white to light cream blooms, without a discernible scent.
- The lateral lobes of the labellum in ***A. monticola*** are straight whereas the lateral lobes of ***A. philippinensis*** are rounded (this feature is most noticeable when the labella are flattened).
- The plants are usually larger and more robust in their growth habit. The leaves of ***A. monticola*** are wider, longer and thicker.
- ***A. monticola*** comes from much higher elevations than ***A. philippinensis***, which is generally found at elevations between 400 and 1,400 metres.

Amesiella philippinensis

(Ames) Garay

Professor Oakes Ames originally named this lovely species ***Angraecum philippinensis***. This name was published in the *Philippine Journal of Science* in 1907. The specific name refers to the Philippines, where this plant is found. Elmer D. Merrill collected the type specimens on Mount Halcon, Mindoro, in 1906.

Growth habit: upright, monopodial. The growth habit of the plant is typically vandaceous, but it is a miniature grower seldom getting any taller than 7.5 cm. **Leaves:** very thick and leathery, about 7.5 cm long by 2.5 cm wide. **Inflorescences:** semi-pendulous, carry up five flowers about 6 cm in diameter. **Flower colour:** white with yellow marks in the labellum. **Dorsal sepal:** ovate, up to 2.5 cm long by 1.5 cm wide. **Petals:** ovate, 2.5 cm long by 1.5 cm wide. **Lateral sepals:** ovate, up to 2.5 cm long by 1.5 cm wide. **Labellum:** three-lobed; nectary (spur) up to 7 cm long.

Habitat and Distribution: Since Elmer D. Merrill first discovered ***Amesiella philippinensis*** in 1906, it has subsequently been found in the mountains of Albay, Benguet, Mountain Province and Nueva Vizcaya provinces on Luzon. It grows as an epiphyte at elevations between 400 and 1,400 metres.

Notes: The records of ***Amesiella philippinensis*** from Nueva Vizcaya are incorrect identifications of the recently described ***Amesiella monticola*** Cootes and Banks. I have never seen living specimens of ***Amesiella philippinensis*** from Benguet or the Mountain Province, but I suspect they could be ***Amesiella monticola***.

Appendicula Blume

Carl Blume established this large genus of approximately 150 species in *Bijdragen tot de Flora van Nederlandsch Indie* in 1825. The generic term is Latin and refers to small appendages on the upper surface of the labellum.

There is no designated type species for this genus, as several species were described at the same time. The genus has its centre of distribution in Indonesia and New Guinea and is also found in India, Myanmar (Burma), Thailand, Vietnam, Malaysia, Sulawesi (Celebes) and Borneo. Approximately 23 species occur in the Philippines.

Photo: David Titmuss

Appendicula buxifolia

Photo: David Titmuss

Appendicula micrantha

Appendicula buxifolia Blume

Carl Blume named this species in 1825 in *Bijdragen tot de Flora van Nederlandsch Indie.* The specific name refers to the grass-like foliage of this orchid.

Growth habit: upright, semi-pendulous in large plants, sympodial. The plant grows to over 1 metre in length and branches toward the ends of the canes. **Leaves:** up to 4 cm long by 5 mm wide. **Inflorescences:** semi-pendulous, at the apex of the growths, 3 cm long, bear up to 20 flowers 4 mm in diameter. Only a couple of blooms are open at one time. **Flower colour:** greenish with red stripes on the sepals and petals. **Dorsal sepal** and **petals:** linear-oblong, similar, 2 mm long by 1.5 mm wide. **Lateral sepals:** triangular, up to 2.5 mm long by 1 mm wide. **Labellum:** ovate with the narrow end at the base.

Habitat and Distribution: ***Appendicula buxifolia*** has been found in the provinces of Benguet, Laguna and Sorsogon on the island of Luzon; the island of Samar in the Visayas; and in the provinces of North and South Cotabato, Davao del Norte, Davao Oriental, Davao del Sur, Lanao del Norte, Lanao del Sur, Misamis Occidental and Misamis Oriental on the island of Mindanao. It grows as an epiphyte at altitudes over 500 metres. It is also recorded from Java.

Appendicula micrantha Lindley

Dr. John Lindley named this species in the *Annals and Magazine of Natural History* published in 1845. The specific name refers to the small flowers of this plant.

Growth habit: upright, sympodial, up to 30 cm tall. **Leaves:** oval, about 3 cm long by 1.5 cm wide, two-ranked. **Inflorescences:** short, bear up to six blooms 4 mm in diameter. **Flower colour:** creamy white. **Dorsal sepal** and **petals:** similar, triangular, about 2 mm long by 1.5 mm wide. **Lateral sepals:** elliptic-ovate, 2 mm long by 1.5 mm wide. **Labellum:** oblong, 3 mm long by 2 mm wide.

Habitat and Distribution: ***Appendicula micrantha*** appears to be endemic to the Philippines. It has been recorded from the provinces of Albay, Bataan, Laguna, Pampanga, Quezon and Rizal on Luzon; the islands of Leyte, Negros (province of Negros Oriental) and Panay (the provinces of Antique and Capiz) in the Visayas; the island of Polillo; and the provinces of Agusan del Norte, Agusan del Sur, Bukidnon, Davao del Norte, Davao Oriental, Davao del Sur, Surigao del Norte, Surigao del Sur, Zamboanga del Norte and Zamboanga del Sur on Mindanao. It grows as an epiphyte, occasionally a terrestrial, at elevations between 15 and 700 metres.

Notes: Dr. Valmayor, author of the 1984 opus on Philippine orchid species *Orchidiana Philippiniana*, lists ***Appendicula micrantha*** as occurring in Borneo, Java and Peninsular Malaysia. Recent orchid floras of Java (Jim Comber, 1990) and Malaysia (Gunnar Seidenfaden and Jeffrey Wood, 1992) and *A Checklist of the Orchids of Borneo* by Jeffrey Wood and Phillip Cribb in 1994 do not include any reference to this orchid.

Arachnis Blume

Carl Blume established this genus in 1825 in *Bijdragen tot de Flora van Nederlandsch Indie.* The name is from the Greek for spider, *arachne,* and alludes to the shape of the flowers.

The type species for this genus is ***Arachnis flosaeris*** (Linn.) Rchb. *f.*

The genus is widely spread throughout Asia, being found in the Himalayas, Southeast Asia, the Philippines, New Guinea and across to the Solomon Islands. Two species are described from the islands of the Philippines.

Arachnis longicaulis

(Schlechter) L.O. Williams

Louis O. Williams named this orchid in 1937 in the *Botanical Museum Leaflets of Harvard University.* The specific name refers to the long stem of the plant.

Growth habit: upright, monopodial; large plants semi-pendulous. The plant is similar to ***Arachnis flosaeris*** in all aspects but the stem may be a bit thicker. In the wild this species can attain lengths of several metres. **Leaves:** leathery, oblong, up to 12 cm long by 3 cm wide, tips unequally bilobed. **Inflorescences:** short, up to 10 cm long, carry up to four blooms 5 cm in diameter. **Flower colour:** bright red and yellow. **Dorsal sepal:** elliptic, 3 cm long by 1 cm wide. **Petals:** spathulate, 2.5 cm long by 1 cm wide. **Lateral sepals:** spathulate, 1.8 cm long by 1 cm wide. **Labellum:** small, movable, three-lobed; lateral lobes obtuse; mid-lobe erect and curved inward.

Habitat and Distribution: ***Arachnis longicaulis*** is endemic to the Philippines. It has been found in the provinces of Isabela, Nueva Ecija, Quezon and Rizal on Luzon; the province of Lanao del Sur on Mindanao; and on the southern islands of Tawitawi at elevations around 1,000 metres. It grows as an epiphyte, very high in trees, in localities where there is a lot of air movement and high humidity. It normally receives full sun for some portion of the day.

Photo: Jim Cootes

Arachnis longicaulis

Ascocentrum Schlechter

Dr. Rudolf Schlechter proposed this genus in 1913 in *Fedde, Repertorium Specierum Novarum, Beihefte.* The generic name refers to the bag-like spur of the labellum.

The type species for this genus is ***Ascocentrum miniatum*** (Lindl.) Schltr.

There are approximately eight species found throughout the Himalayas, Southeast Asia, the Philippines and Taiwan. The Philippines has only one species from the genus ***Ascocentrum.***

Ascocentrum aurantiacum subsp. ***philippinense*** E.A. Christenson

The German taxonomist and orchid collector Dr. Rudolf Schlechter named this species in 1913 in volume one of *Fedde, Repertorium Specierum Novarum, Beihefte.* The name was based on the ***Saccolabium aurantiacum*** Schlechter described in 1911. The specific name refers to the orange colour of the flowers. The Philippine plants differ from the original Sulawesi (Celebes) populations by being slightly smaller in all measurements. Dr. Eric Christenson named the new subspecies ***Ascocentrum aurantiacum*** subsp. ***philippinense*** in *Lindleyana* in 1992.

Photo: David Titmuss

Ascocentrum aurantiacum subsp. ***philippinense***

Growth habit: upright, monopodial, reaches a height of 25 cm, of typical vandaceous growth habit. **Leaves:** up to 16 cm long by 1 cm wide, arch gracefully. **Inflorescences:** upright, bear up to 40 blooms about 9 mm in diameter. **Flower colour:** bright orange; side lobes and throat of spur darker orange; anther cap dark red. **Petals** and **sepals:** elliptic, up to 7 mm long by 3 mm wide. **Labellum:** mid-lobe tongue-shaped; lateral lobes up to 2.5 mm high; spur noticeably curved forward.

Habitat and Distribution: ***Ascocentrum aurantiacum*** subsp. ***philippinense*** is endemic to the Philippines where it is widely spread throughout the archipelago. This orchid has been found in the provinces of Nueva Vizcaya and Rizal on Luzon and the provinces of Agusan del Sur, Davao del Norte, Davao del Sur and Lanao del Sur on Mindanao. It grows as an epiphyte at elevations to 1,200 metres. I have observed this subspecies growing in quite deeply shaded locations in its natural habitat.

Ascoglossum Schlechter

Dr. Rudolf Schlechter proposed the generic name for this plant in *Die Orchideen von Deutsch Neu Guinea* in 1914. The name is derived from the Greek words *askos* meaning bag and *glossa,* tongue, and refers to the bag-like labellum of the flower.

There is only one species in this genus, which occurs in the southern Philippines and New Guinea.

Ascoglossum calopterum

(Reichenbach *f.*) Schlechter

Professor Heinrich G. Reichenbach originally named this species ***Saccolabium calopterum*** in 1882 in the *Gardeners' Chronicle.* The specific name is derived from two Greek words, *calo* for beautiful and *pteron,* winged, in reference to the beautifully coloured flowers.

Growth habit: upright, monopodial. **Leaves:** rigid, thick, about 20 cm long, 2.5 cm wide. **Inflorescences:** upright, branch readily, flowers about 1.2 cm across the petals by 2 cm high. There can be 60 to 80 blooms per raceme and on a large, healthy plant there can be several inflorescences. **Flower colour:** varying shades of purple. **Dorsal sepal:** oblanceolate, up to 7 mm long by 3 mm wide. **Petals:** lanceolate, up to 8 mm long by 4 mm wide; all segments reflex. **Lateral sepals:** spathulate, about 12 mm long by 4 mm wide. **Labellum:** small, attached immovably to a spur which is straight, about 1 cm long by 2 mm in diameter.

Habitat and Distribution: ***Ascoglossum calopterum*** grows as an epiphyte on the southern island of Dinagat, north of Mindanao, but would be much more widely distributed on the island of Mindanao. It is also found in New Guinea, at low elevations to 300 metres.

Notes: Dr. Eric A. Christenson believes this species is bird-pollinated. The bright colour of the flowers, the leathery leaves and the plant's exposed habitat, which is chiefly on the trunks of trees, are all factors for bird-pollination.

Photo: David Titmuss

Ascoglossum calopterum

Bulbophyllum Thouars

Aubert du Petit Thouars named this large genus in 1822. The name was originally published in *Orchidees des Iles Australes de l' Afrique.* The generic name is derived from two Greek words, *bolbos* for bulb and *phyllon,* a leaf, referring to the growth habit of many of the species in this wonderful, and at times bizarre, genus.

The type species for this genus is ***Bulbophyllum nutans*** Thouars, which originates in Africa.

There are numerous orchid genera that have large numbers of species such as ***Dendrobium***, ***Eria***, ***Pleurothallis***, ***Encyclia*** and ***Masdevallia,*** but the genus ***Bulbophyllum*** stands alone for several reasons. ***Bulbophyllum*** has the largest number of members of any genus of orchids (some taxonomists and botanists estimate up to 2,000 species), and it is one of the few genera that has representatives in Africa, Asia, Australia and the Americas. They grow in all types of habitat but are most common at elevations above 1,000 metres. They normally grow as epiphytes or lithophytes; I do not know of any terrestrial species. There are approximately 120 species found in the Philippines.

Bulbophyllum alagense Ames

Section: ***Nematorhizis*** Schlechter

Professor Oakes Ames named this small-growing species in the *Philippine Journal of Science* in 1907. The specific name refers to the Alag River on the island of Mindoro. Elmer D. Merrill collected the type specimens in 1906.

Photo: Dr. Mark Clements

Bulbophyllum alagense

Growth habit: upright, sympodial. **Rhizome:** 1 mm in diameter, bears the pseudobulbs at intervals of up to 2 cm. **Pseudobulbs:** tapered, up to 6 mm high by 4 mm in diameter, unifoliate. **Leaves:** ovate, up to 2 cm long by 1 cm wide. **Inflorescences:** upright, up to 2 cm long, bear a single bloom about 1 cm in diameter. **Flower colour:** light yellow to pale orange. **Dorsal sepal:** triangular to lanceolate, up to 8 mm long by 2 mm wide. **Petals:** small, triangular, up to 3 mm long by 1.5 mm wide. **Lateral sepals:** triangular to lanceolate, up to 8 mm long by 2 mm wide. **Labellum:** curved and three-lobed, 2.5 mm long by 1.5 mm wide.

Habitat and Distribution: *Bulbophyllum alagense* has been recorded from the provinces of Benguet, Camarines Sur, Laguna and Rizal on Luzon; the province of Oriental Mindoro on the island of Mindoro; Leyte island in the Visayas; and throughout Mindanao. This lovely little species grows as an epiphyte at elevations to 500 metres.

Bulbophyllum antenniferum

(Lindley) Reichenbach *f.*

Section: ***Hyalosema*** Schlechter

Dr. John Lindley first named this species ***Cirrhopetalum antenniferum*** in 1843. This name was published in the *Botanical Register.* Professor Heinrich G. Reichenbach gave the plant its currently accepted name in 1861 in *Walpers Annales Botanices.* The specific name refers to the antennae on the tips of the petals.

Photo: David Titmuss

Bulbophyllum antenniferum

Growth habit: upright, sympodial. **Rhizome:** up to 5 mm in diameter. **Pseudobulbs:** four-angled, 2 to 5 cm high, unifoliate. **Leaves:** leathery, up to 20 cm long by 5 cm wide. **Inflorescences:** almost as long as the leaf, upright, bear a single bloom. Flowers are large for bulbophyllums, up to 12 cm high; the width is about 2 cm. Flowers are borne singly along the rhizome and can appear from any position. **Flower colour:** greenish brown with numerous little transparent spots (like opaque windows) on the dorsal sepal behind the column. **Dorsal sepal:** ovate to lanceolate the first 4 cm; remaining 6.8 cm linear, concave, 4 mm wide. **Petals:** very small, lanceolate, tipped with a small, hair-like appendage. **Lateral sepals:** joined along the central edge most of their length, about 7.4 cm. **Labellum:** recurved, hairy along edges, small, highly mobile, about 5 mm long by 2.5 mm wide.

Habitat and Distribution: *Bulbophyllum antenniferum* is found at elevations up to 500 metres in Java, Borneo and New Guinea. In the Philippines it has been found in Bataan, Laguna, Quezon and Sorsogon provinces on Luzon; Leyte and Sibuyan islands in the Visayas; and in Lanao del Sur province on Mindanao. It normally grows as an epiphyte, usually on the trunks of trees, but has been observed by the author growing on rocks in dappled light.

Photo: Gary Yong Gee

Bulbophyllum biflorum

Bulbophyllum biflorum

Teijsmann and Binnendijk
Section: ***Biflorae*** Garay, Hamer and Siegerist

Johannes Teijsmann and Simon Binnendijk named this orchid in 1853 in *Natuurkundig Tijdschrift voor Nederlandsch-Indie*. The specific name refers to the inflorescence, which bears two flowers.

Growth habit: upright, sympodial. **Rhizome:** about 2 mm in diameter, carries the pseudobulbs at close intervals. **Pseudobulbs:** distinctly four-angled, up to 3.5 cm high by 2.5 cm across, unifoliate. **Leaves:** up to 12.5 cm long by 2.5 cm wide. **Inflorescences:** semi-pendulous, appear from the base of the pseudobulbs, bear two blooms up to 10 cm long by 1.5 cm wide. **Flower colour:** white or cream with varying degrees of reddish purple spotting; labellum yellowish brown with purple spots. The blossoms have a peculiar odour. **Dorsal sepal:** ovate, extending to a thread-like appendage with a thickened tip, overall length up to 2.5 cm. **Petals:** small, 1 cm long by 5 mm wide. **Lateral sepals:** about 8 cm long by 7 mm wide, joined but sometimes separated, drawn out to slender filaments. **Labellum:** oblong to ovate, fleshy, highly mobile, up to 6 mm long.

Habitat and Distribution: ***Bulbophyllum biflorum*** is found in Thailand, Malaysia, Indonesia and Borneo. In the Philippines it has been recorded from the provinces of Aurora, Quezon and Rizal on Luzon, but it is undoubtedly of wider distribution among the islands. It grows as an epiphyte at elevations to 800 metres, frequently in brightly lit situations.

Notes: The plant of the Philippine species differs from those of other countries. The pseudobulbs are carried close together on the rhizome and are of a different shape. In plants of this species that I have from Thailand and Malaysia, the pseudobulbs are carried at intervals of up to 6 cm on the rhizome, and the pseudobulbs are much longer and narrower. The plant resembles ***Bulbophyllum puguahaanense*** Ames (though much larger), but the flowers are similar to ***Bulbophyllum biflorum*** Teijsm. and Binn. The plant illustrated here is possibly an undescribed species.

Bulbophyllum carunculatum

Garay, Hamer and Siegerist
Section: ***Lepidorhiza*** Schlechter

Photo: David Banks

Bulbophyllum carunculatum

Doctors Leslie Garay, Fritz Hamer and Emly Siegerist named this large-flowered species in *Lindleyana* in 1995. The specific name refers to the surface of the labellum, which appears warty.

Growth habit: upright, sympodial; plant large, up to 45 cm tall; rhizome short. **Pseudobulbs:** ovoid, unifoliate, up to 6 cm tall. **Leaves:** distinctly petiolate, elliptic to obtuse; petiole up to 5 cm long; leaf blade up to 36 cm long, 7 cm wide. **Inflorescences:** upright, up to 45 cm long with several tubular bracts up to 5 cm long; flowers large, appear in succession with up to 12 blooms. **Flower colour:** golden-yellow; labellum dark purple (almost black). **Dorsal sepal:** ovate to oblong, acute, up to 4 cm long, 1.5 cm wide. **Petals:** ovate, acute, up to 2.5 cm long, 1.2 cm wide. **Lateral sepals:** falcate, up to 5 cm long, 1.5 cm wide. **Labellum:** fleshy with wart-like texture, 2.5 cm long, three-lobed; lateral lobes short and toothed. **Column:** erect, about 7 mm long with two forward-pointing stelidia (arms).

Habitat and Distribution: ***Bulbophyllum carunculatum*** was described from a plant originating in the Philippines. Other clones of this species (some with striations on the petals) have recently bloomed in Australia and the United States from recent importations from Sulawesi (Celebes).

Bulbophyllum catenulatum

Kränzlin
Section: ***Leptopus*** Schlechter

Dr. Fritz Kränzlin named this species in 1921 in *Fedde, Repertorium Specierum Novarum, Beihefte.* The specific name refers to the chain-like habit of the pseudobulbs.

Growth habit: upright, sympodial. **Rhizome:** about 1.5 mm in diameter, bears pseudobulbs at intervals of about 1.5 cm. **Pseudobulbs:** squarish, about 8 mm across by 1.5 cm high, unifoliate. **Leaves:** lanceolate, up to 5 cm long by 2 cm wide. **Inflorescences:** upright, up to 5 cm long, very thin, bear a single bloom about 1.2 cm across at the widest point (the lateral sepals). **Flower colour:** lateral sepals red, basally fading to yellow; dorsal sepal yellow with three red stripes, edged with short white hairs; petals yellow spotted with red at the tips and edged with white hairs; labellum red. **Dorsal sepal:** lanceolate, about 1 cm long by 4 mm wide. **Petals:** lanceolate, 7 mm long by 2.5 mm wide. **Lateral sepals:** ovate with acute tips, 1 cm long by 4 mm wide. **Labellum:** fleshy, tongue-shaped, 4 mm long by 1.5 mm wide. **Column:** 2 mm long with two short stelidia.

Photo: David Titmuss

Bulbophyllum catenulatum

Habitat and Distribution: ***Bulbophyllum catenulatum*** is found in the mountains of Nueva Ecija and Nueva Vizcaya in central Luzon at elevations between 1,000 and 1,800 metres. It grows as an epiphyte, usually on the branches and trunks of trees, but has also been observed growing on rocks.

Bulbophyllum cephalophorum

Garay, Hamer and Siegerist
Section: ***Desmosanthes*** Blume

Doctors Leslie Garay, Fritz Hamer and Emly Siegerist described this species in *Lindleyana* in 1996. The specific name refers to the spherical inflorescence.

Growth habit: upright, sympodial. **Rhizome:** diameter about 3 mm, bears pseudobulbs at intervals of about 5 cm. **Pseudobulbs:** ovate to elliptic, about 2.5 cm long, 1.5 cm in diameter, unifoliate. **Leaves:** very leathery, ovate to elliptic, up to 7.5 cm long, 3 cm wide. **Inflorescences:** upright, slender, bear up to 15 blooms about 1.2 cm in diameter. **Flower colour:** sepals and petals dull orange; labellum brown. **Dorsal sepal:** linear to lanceolate, 1 cm long, 3 mm wide. **Petals:** lanceolate, up to 7 mm long, 2.5 mm wide. **Lateral sepals:** ovate to lanceolate, up to 1.1 cm long, 3 mm wide. **Labellum:** fleshy, small, up to 4 mm long, 2 mm wide.

Habitat and Distribution: ***Bulbophyllum cephalophorum*** grows as an epiphyte at elevations around 1,200 metres in the mountains of Nueva Ecija and Nueva Vizcaya in central Luzon.

Photo: David Titmuss

Bulbophyllum cephalophorum

Photo: David Banks

Bulbophyllum sp. aff. *cheiri*

Bulbophyllum sp. aff. ***cheiri*** Lindley

Section: ***Sestochilus*** (Breda) Bentham and Hooker *f.*

I had originally identified this species as ***Bulbophyllum cheiri*** Lindley because the flower perfectly matched the photograph in Vermeulen 1991 plate 17B (see References). It was only much later that closer study of the blooms revealed enough differences to convince me that this is a different species. It differs from ***B. cheiri*** Lindley in the colour of the flowers, in being sweetly perfumed, and in the shape of the labellum and the column foot.

Growth habit: upright, sympodial. **Rhizome:** about 5 mm in diameter, without remnants of sheaths at the internodes; roots from each internode. **Pseudobulbs:** small, up to 10 cm apart, unifoliate. **Leaves:** leathery, elliptic, up to 14 cm long by 5 cm wide, petiole up to 2.5 cm long. **Inflorescences:** up to 6.5 cm high, single-flowered, appearing from the pseudobulb. **Flower colour:** base colour of sepals and petals off-white, spotted with maroon on the inside; outside of outer segments brownish; labellum deep red, whitish underneath. Blooms are sweetly scented. **Dorsal sepal:** triangular, up to 4.3 cm long by 9 mm wide. **Petals:** triangular, up to 3.9 cm long by 5 mm wide. **Lateral sepals:** falcate, up to 4 cm long by 1.2 cm wide. **Labellum:** bent, ovate with a linear tip, 2 cm long by 1.3 mm wide (all measurements are when flower is flattened).

Habitat and Distribution: ***Bulbophyllum*** sp. aff. ***cheiri*** is known only from Quezon province, where it grows on rocks, and also trees, at elevations of about 300 metres in light shade.

Notes: The flowers of this species only open widely first thing in the morning and are usually closed by the middle of the day.

Bulbophyllum cleistogamum

Ridley

Section: ***Intervallatae*** Ridley

This species was named by Henry Ridley in the *Journal of the Linnean Society* published in 1896. The specific name refers to a self-pollinating (cleistogamous) clone from which the original description was made. Thankfully clones that self-pollinate are not common.

Growth habit: upright, sympodial. **Rhizome:** 4 mm in diameter, pseudobulbs up to 5 cm apart. **Pseudobulbs:** up to 6 cm high, 1.5 cm in diameter, unifoliate. **Leaves:** leathery, up to 17.5 cm long by 5 cm wide.

Photo: David Banks

Bulbophyllum cleistogamum

Inflorescences: upright, wiry, longer than the leaves. **Flower colour:** white striped with dark purple; labellum black; flowers 4.5 cm wide. **Dorsal** and **lateral sepals:** subulate, about 2 cm long by 5 mm wide. **Petals:** minute, only 2 mm long by 3 mm wide. **Labellum:** large, roughly surfaced, shiny, 2 cm long by 6 mm wide.

Habitat and Distribution: ***Bulbophyllum cleistogamum*** grows in shady locations on both rocks and trees at elevations of about 500 metres in peninsular Malaysia and Borneo. In the Philippines this species has only been recorded from Quezon province on Luzon, but I expect it is much more widely distributed throughout eastern Luzon.

Bulbophyllum cootesii

M.A. Clements

Section: ***Lepidorhiza*** Schlechter

Dr. Mark Clements named this species in the August 1999 edition of the *Australian Orchid Review.* The plant was named in honour of the author of this book.

Growth habit: epiphytic, upright. **Rhizome:** short, 4 mm in diameter. **Pseudobulbs:** ovate, 3 cm long by 2.5 cm wide, wrinkled, bear a single leaf. **Leaves:** leathery, linear, 20 cm long by 3.3 cm wide including petiole; petiole 3 cm long. **Inflorescence:** at right angles to pseudobulb, 9 cm long with three short (1.5 cm) sheaths bearing several flowers that open sequentially. **Flower colour:** like honey, veined with pale brown; tips of the segments bright yellow; labellum basally white, front portion reddish brown; flower 7 cm high by 5 cm wide across the petals; all segments shiny, as if varnished. **Dorsal sepal:** lanceolate, 4 cm long by 1.5 cm wide, ridge down centre. **Petals:** lanceolate, 2.6 cm long by 1.2 cm wide. **Lateral sepals:** falcate, 4.2 cm long by 8 mm wide, deeply ridged as if folded. Joined at base to form a very short, almost negligible, mentum; if flattened this segment would be approximately 1.4 cm wide. **Labellum:** fleshy, 1.5 cm long, 5 mm wide; mid-lobe oblanceolate; two short arms at the back of the lip. **Pedicel and ovary:** 2.5 cm long by 3 to 4 mm in diameter.

Habitat and Distribution: ***Bulbophyllum cootesii*** is endemic to the Philippines. It grows as an epiphyte at elevations up to 500 metres on the island of Mindanao.

Bulbophyllum cupreum Lindley

Section: ***Careyana*** Pfitzer

Dr. John Lindley named this species in the *Botanical Register* in 1838. The specific name refers to the coppery-coloured flowers.

Growth habit: upright, sympodial. **Rhizome:** 1.5 mm in diameter, with persistent sheaths. **Pseudobulbs:** ovate, up to 2.5 cm high by 1.5 cm in diameter; appear at intervals of up to 4.5 cm on the rhizome; covered

Photo: David Titmuss

Bulbophyllum cootesii

Photo: David Banks

Bulbophyllum cupreum

with a papery sheath, unifoliate. **Leaves:** leathery, oblong to elliptic, 5 to 9.2 cm long by 1.4 to 1.8 cm wide. **Inflorescences:** upright, four equally spaced bracts, 14 cm long with last 4 cm bending down, up to 20 blooms 6 mm across by 1 cm long. **Flower colour:** orange. **Dorsal sepal:** ovate, up to 6 mm long by 4 mm wide, concave. **Petals:** lanceolate, up to 3 mm long by 1.5 mm wide. **Lateral sepals:** lanceolate, up to 8 mm long by 3.5 mm wide, concave. **Labellum:** fleshy, tongue-shaped, deeply grooved, 6 mm long by 2.2 mm wide; two sharp tooth-like projections on the basal portion.

Habitat and Distribution: ***Bulbophyllum cupreum*** has been found in Thailand. Section *Careyana* has not previously been recorded from the Philippines. The illustrated plant came from Aurora province on the east coast of Luzon. It was collected by the author from the trunk of a fallen tree at an altitude of about 500 metres. No other specimens were found in the area.

Notes: Hugh Cuming collected the original specimens from which this species was named. He stated that they were found in the city of Manila on Luzon. The locality data of Cuming has been questioned by Dr. Seidenfaden and Jaap Vermeulen (Seidenfaden and Wood 1992, pages 488 and 491) who suggest that the Philippines is far outside the range of this section.

Alcohol-preserved flowers are held at the Centre for Plant Biodiversity Research (CPBR), Canberra, Australia, under the number MAC 9321. A dried, flowering specimen is also held at the CPBR under the number MAC 9321.

Bulbophyllum dearei Reichenbach *f.*

Section: ***Sestochilus*** (Breda) Bentham and Hooker *f.*

Professor Heinrich G. Reichenbach named this magnificent orchid in *Flora* 71 in 1888. The specific name honours Colonel Deare, the first person to flower this species in England.

Growth habit: upright, sympodial. **Rhizome:** up to 5 mm in diameter, pseudobulbs at intervals of approximately 3 cm. **Pseudobulbs:** up to 5 cm high by 2 cm in diameter, unifoliate. **Leaves:** leathery, up to 20 cm long by 5 cm wide. **Inflorescences:** upright, single-flowered. **Flower colour:** orange-yellow marked with purple; labellum reddish purple. **Dorsal sepal:** ovate, up to 4 cm long by 2 cm wide. **Petals:** triangular, up to 3 cm long by 1 cm wide, reflex badly but do not detract from the beauty of this species. **Lateral sepals:** triangular, ovate, up to 3 cm long by 1.8 cm wide. **Labellum:** three-lobed, highly mobile; mid-lobe ovate, up to 1 cm long by 8 mm wide; lateral lobes point backward, about 6 mm long.

Habitat and Distribution: ***Bulbophyllum dearei*** has been found in the provinces of Davao del Sur and Lanao del Sur on Mindanao and on the island of Palawan. It has also been recorded from Borneo and Peninsular Malaysia. It grows at elevations from 700 to 1,200 metres, usually as an epiphyte, but also as a lithophyte.

Photo: David Banks

Bulbophyllum dearei

Bulbophyllum elassoglossum

Siegerist
Section: **Leptopus** Schlechter

This lovely species was first described by Dr. Emly Siegerist in the *American Orchid Society Bulletin*, September 1991 (pages 865 to 866). The specific name refers to the small labellum.

Growth habit: upright, sympodial; plants up to 11 cm tall. **Rhizome:** slender, creeping, produces roots at the base of pseudobulbs that are up to 4.5 cm apart. **Pseudobulbs:** ovoid, more or less tetragonal, up to 3 cm high by 1.5 cm wide at base, unifoliate. **Leaves:** elliptic to narrow-elliptic, up to 8.5 cm long by 2 cm wide. **Inflorescences:** upright, single, hair-like, one-flowered with a small dorsal sheath up to 11 mm long; flower rather showy, large, with spreading segments. **Flower colour:** dorsal and lateral sepals white, heavily blotched with purple toward the base; petals and lip deep maroon. **Dorsal sepal:** elliptic, concave, acute, up to 2.4 cm long by 1 cm wide. **Petals:** elliptic, rounded to obtuse, parallel with one another; lateral margins somewhat infolded and tips recurved, up to 1 cm long by 5 mm wide. **Lateral sepals:** oblong to elliptic, acute, up to 2.3 cm long by 7 mm wide. **Labellum:** fleshy, tongue-shaped, up to 5 mm long by 2 mm wide. **Column:** short with erect, forward pointing stelidia.

Habitat and Distribution: ***Bulbophyllum elassoglossum*** originates from the Philippines, but the exact locality is unknown. This species was described from plants grown in the United States.

Photo: David Banks

Bulbophyllum elassoglossum

Bulbophyllum emiliorum

Ames and Quisumbing
Section: **Sestochilus** (Breda) Bentham and Hooker *f.*

Professor Oakes Ames and Dr. Eduardo Quisumbing named this species in the *Philippine Journal of Science* in 1931. The specific name honours Emilio Quisumbing (son of the plant's co-author), collector of the type specimens.

Growth habit: upright, sympodial. **Rhizome:** up to 6 mm in diameter. Each internode, about 1 cm apart, has persistent bristles about 1 cm long. **Pseudobulbs:** up to 2 cm high by 8 mm in diameter, unifoliate. **Leaves:** leathery, up to 15 cm long by 5 cm wide. **Inflorescences:** upright, about 4 cm long, carry a single flower 3.5 cm in diameter; blooms sweetly perfumed, non-resupinate (blooms have labellum uppermost). **Flower colour:** base colour creamy white heavily spotted with maroon; labellum yellow. **Dorsal sepal** and **petals:** oblong-lanceolate, up to 2.5 cm long by 6 mm wide. **Lateral sepals:** lanceolate, up to 2.5 cm long by 1 cm wide, joined near the tips. **Labellum:** small, about 8 mm long by 3 mm wide.

Habitat and Distribution: ***Bulbophyllum emiliorum*** is found throughout the lowlands of the Philippines in the provinces of Laguna, Quezon and Rizal on Luzon; the island of Samar in the Visayas; and the provinces of Agusan del Norte and Agusan del Sur on Mindanao. This plant has recently been recorded from Papua New Guinea. It grows as an epiphyte, normally on tree trunks at elevations between 100 and 400 metres.

Photo: Elias Javier

Bulbophyllum emiliorum

Photo: David Banks

Bulbophyllum facetum

Bulbophyllum facetum

Garay, Hamer and Siegerist
Section: ***Sestochilus*** (Breda) Bentham and Hooker *f.*

Doctors Leslie Garay, Fritz Hamer and Emly Siegerist described this magnificent species in *Lindleyana* in 1996. The specific name refers to the elegantly coloured flowers.

Growth habit: upright, sympodial. **Rhizome:** up to 4 mm in diameter with each internode about 1 cm apart; pseudobulbs spaced at intervals of up to 5 cm on the rhizome. **Pseudobulbs:** ovoid, up to 3.5 cm high by 2.5 cm in diameter, unifoliate. Persistent bristles, up to 3 cm long, are at the base of each pseudobulb. **Leaves:** oblong to elliptic, leathery, up to 20 cm long by 5 cm wide; petiole up to 5 cm long. **Inflorescences:** upright, about 6 cm long, appear from the base of the pseudobulbs, carry a single flower 7 cm in diameter; tubular bract at the base of the pedicel. **Flower colour:** base colour creamy white heavily spotted and tessellated with reddish brown. **Dorsal sepal:** ovate to elliptic, acute, up to 4 cm long by 1.5 cm wide. **Petals:** obovate to lanceolate, up to 3.2 cm long by 1.2 cm wide. **Lateral sepals:** obliquely ovate to oblong, up to 3.3 cm long by 2 cm wide. **Labellum:** heart-shaped, curved, about 1 cm long by 8 mm wide.

Habitat and Distribution: ***Bulbophyllum facetum*** is found in a limited area in the mountains of Nueva Ecija on Luzon at elevations of about 1,200 metres. It grows on the trunks of trees in fairly shady positions.

Notes: When cultivated, the flowers open normally in the morning only to be fully closed by midday. This happens for the life of the bloom, about four days.

Bulbophyllum sp. aff. *fenixii* Ames

Section: ***Leptopus*** Schlechter

I tentatively identified this plant as ***Bulbophyllum leptocaulon***, which was named by Kränzlin, until an examination of the microfiche of the type specimen convinced me that ***Bulbophyllum leptocaulon*** was the wrong name. Further study of the microfiches revealed ***Bulbophyllum fenixii*** has close affinities with this species, if in fact they are not the same taxa. This species differs from ***Bulbophyllum fenixii*** in the wider distance separating the pseudobulbs, the colour of the blooms and the shape of the petals.

Growth habit: upright, sympodial. **Rhizome:** thin, carries the pseudobulbs at about 2 cm intervals. **Pseudobulbs:** four-angled, tapered, up to 1 cm tall by 5 mm across. **Leaves:** ovate, up to 5 cm long by 8 mm wide. **Inflorescences:** upright, up to 4 cm long, bear a single flower; bloom about 1 cm across at the widest point, the lateral sepals. **Flower colour:** base colour white to pale pink with three to five dark red stripes; labellum maroon. **Dorsal sepal:** ovate, concave, 8 mm long by 3.5 mm wide, edged with tiny hairs. **Petals:** narrowly lanceolate, 5 mm long by 1.5 mm wide, edged with small white hairs, curve in toward the column. **Lateral sepals:** ovate, 8 mm long by 4 mm wide. **Labellum:** thick and fleshy, tongue shaped, recurved, 3 mm long by 1.5 mm wide.

Photo: David Banks

Bulbophyllum sp. aff. ***fenixii***

Habitat and Distribution: ***Bulbophyllum*** sp. aff. ***fenixii*** appears to be endemic to the Philippines where it is found at elevations of approximately 1,200 metres in the provinces of Nueva Ecija and Nueva Vizcaya on the island of Luzon. It grows as an epiphyte.

Bulbophyllum halconense Ames

Section: ***Nematorhizis*** Ridley

Professor Oakes Ames named this small-growing species in the *Philippine Journal of Science* in 1907. The specific name refers to Mount Halcon on the island of Mindoro.

Growth habit: upright, sympodial. **Pseudobulbs:** conical, 5 mm high by 5 mm in diameter; borne on a wiry rhizome at intervals of up to 5 cm. **Leaves:** leathery, elliptic to oblong, up to 2.5 cm long by 6 mm wide. **Inflorescences:** upright, appear along the rhizome, equal to or slightly longer than the leaf, very wiry, bear a single bloom. **Flower colour:** base colour translucent white, all the segments striped lengthwise with a deep purple that appears almost black. **Dorsal sepal:** narrowly lanceolate, 1.3 cm long by 3 mm wide. **Petals:** ovate, 6 mm long by 3 mm wide. **Lateral sepals:** narrowly lanceolate, 1.4 cm long by 3 mm wide. **Labellum:** fleshy, heart-shaped, 3.5 mm long by 2 mm wide.

Habitat and Distribution: ***Bulbophyllum halconense*** appears to be endemic to Mount Halcon in the north of the province of Oriental Mindoro on the island of Mindoro. It grows as an epiphyte on the branches and trunks of trees at elevations between 1,500 and 2,500 metres.

Photo: David Banks

Bulbophyllum halconense

Bulbophyllum lasioglossum

Rolfe *ex* Ames

Section: ***Hirtula*** Ridley

Robert Rolfe first described this species, but Professor Oakes Ames provided a more detailed description in *Studies in the Family Orchidaceae* Fascicle I in 1905. The specific name refers to the hairy labellum. H.N. Whitford collected the type specimens in the province of Bataan on Luzon in May 1904.

Photo: Dr. Mark Clements

Bulbophyllum lasioglossum

Growth habit: upright, sympodial. **Pseudobulbs:** tapered, up to 1 cm tall by 1 cm wide, unifoliate. **Leaves:** ovate, leathery, up to 7 cm long by 2.5 cm wide; petiole 1 cm long. **Inflorescences:** upright, up to 30 cm long, bear up to 50 blooms about 1 cm wide. **Flower colour:** dorsal sepal, lateral sepal and petals white to greenish; labellum almost black with whitish hairs. **Dorsal sepal:** lanceolate, up to 5 mm long by 2.5 mm wide, edged with short hairs. **Petals:** oblong to lanceolate, up to 2.5 mm long by 1.5 mm wide, edged with hairs. **Lateral sepals:** oblong, up to 5 mm long by 2.5 mm wide. **Labellum:** lanceolate, up to 5 mm long by 2.5 mm wide, edges covered with short hairs.

Habitat and Distribution: ***Bulbophyllum lasioglossum*** is found in Bataan and Laguna on Luzon. It grows as an epiphyte at an elevation of about 1,000 metres.

Bulbophyllum laxiflorum

Bulbophyllum lepantense

Bulbophyllum laxiflorum

(Blume) Lindley
Section: ***Corymbosia*** (Bl.) Pfeiffer

Carl Blume first named this species ***Diphyes laxiflora*** in 1825. Dr. John Lindley gave this plant its currently accepted name in 1830 in *Genera and Species of Orchidaceous Plants.* The specific name refers to the loosely flowered inflorescence.

Growth habit: upright, sympodial. **Rhizome:** about 2 mm in diameter, carries the pseudobulbs at intervals of about 5 cm. **Pseudobulbs:** up to 4 cm high by 1 cm wide, unifoliate. **Leaves:** leathery, about 7 cm long by 2 cm wide. **Inflorescences:** upright, appear anywhere along the rhizome or the base of the pseudobulbs, longer or equal to the length of the leaves; up to 20 blooms 1.5 cm in diameter. **Flower colour:** pale cream to yellow; labellum yellow; blooms sweetly scented. **All the segments:** subulate, about 1 cm long by 2 mm wide. **Labellum:** obtuse, about 2 mm long.

Habitat and Distribution: *Bulbophyllum laxiflorum* is widely spread throughout Thailand, Myanmar (Burma), Laos, Malaysia, Indonesia, Borneo and Sulawesi (Celebes). In the Philippines this species has been found in the provinces of Cagayan, Laguna, Nueva Vizcaya and Zambales on Luzon; the island of Leyte in the Visayas; and the provinces of Agusan del Sur and Surigao del Sur on Mindanao. It grows as an epiphyte at elevations between 120 and 1,000 metres.

Bulbophyllum lepantense Ames

Section: ***Aphanobulbon*** Schlechter

Professor Oakes Ames named this species in the *Philippine Journal of Science* in 1912. The specific name refers to Lepanto, which was once a sub-province of Benguet province on Luzon, where the first specimens were collected.

Growth habit: upright, sympodial. **Pseudobulbs:** very small, about 3.5 mm in diameter, unifoliate. **Leaves:** fleshy, lanceolate, up to 5.5 cm long by 1.6 cm wide. **Inflorescences:** up to 10 cm long, bear up to six flowers. **Flower colour:** orange. **Dorsal sepal:** strap-shaped, acute, concave, up to 5 mm long by 3 mm wide. **Petals:** rhombic-spathulate, obtuse, margin slightly hairy, up to 3.5 mm long by 1.5 mm wide. **Lateral sepals:** triangular-lanceolate, up to 5 mm long by 3 mm wide. **Labellum:** tongue-shaped, fleshy, up to 3 mm long.

Habitat and Distribution: *Bulbophyllum lepantense* is endemic to the Philippines and has been recorded from the provinces of Benguet, Ifugao and Mountain Province on Luzon. It grows as an epiphyte on the mossy trunks and branches of trees at elevations between 2,200 and 2,450 metres.

Photo: David Banks

Bulbophyllum lobbii

Bulbophyllum lobbii Lindley

Section: ***Sestochilus*** (Breda) Bentham and Hooker *f.*

Dr. John Lindley named this species in the *Botanical Register* in 1847. The plant was named for Thomas Lobb, who collected the first plants, in Borneo, for Veitch and Sons of Britain.

Growth habit: upright, sympodial. **Rhizome:** quite stout, up to 7 mm in diameter, pseudobulbs at 5 cm intervals. **Pseudobulbs:** up to 3 cm high by 2.5 cm in diameter, unifoliate. **Leaves:** up to 20 cm long by 5 cm wide. **Inflorescences:** upright, appear from the base of the pseudobulbs, up to 15 cm tall by 7 cm wide, sweetly perfumed. **Flower colour:** orange-yellow. **Dorsal sepal:** triangular, up to 5 cm long by 1.5 cm wide. **Petals:** lanceolate, up to 3.5 cm long by 8 mm wide; reflex badly in some clones. **Lateral sepals:** triangular, up to 4 cm long by 1.5 cm wide. **Labellum:** highly mobile, heart-shaped, about 1.2 cm wide by 1.5 cm long.

Habitat and Distribution: *Bulbophyllum lobbii* is closely related to ***Bulbophyllum dearei*** but is more widely spread, being found in Thailand, Peninsular Malaysia, Indonesia and Borneo. In the Philippines it has been found in the provinces of Benguet, Mountain Province and Ifugao on Luzon and the island of Palawan. This species grows as an epiphyte at elevations between 1,000 and 1,500 metres.

Bulbophyllum maquilingense

Ames and Quisumbing

Section: ***Leptopus*** Schlechter

Professor Oakes Ames and Dr. Eduardo Quisumbing named this species in the *Philippine Journal of Science* in 1932. They named it after Mount Makiling in Laguna province on Luzon, where the type specimens were collected.

Growth habit: upright, sympodial. **Pseudobulbs:** conical, about 1.5 cm tall by 1 cm in diameter, unifoliate. **Leaves:** leathery, up to 5 cm long by 1.5 cm wide. **Inflorescences:** upright, single-flowered, appear from the pseudobulbs, which can produce many blooms over several years; blooms about 1.5 cm long by 1 cm wide. **Flower colour:** pink heavily mottled with maroon; labellum maroon and white. Flowers are quite interesting in that only the lateral sepals open fully. **Dorsal sepal:** elliptic, 9 mm long by 5 mm wide, hooded. **Petals:** elliptic, 6.5 mm long by 3 mm wide, remain inside the dorsal sepal. **Lateral sepals:** obliquely oblong, 8 mm long by 3 mm wide, point down. **Labellum:** highly mobile, narrow, fleshy, 6 mm long by 1.5 mm wide.

Habitat and Distribution: *Bulbophyllum maquilingense* is endemic to the mountains of Laguna and Quezon provinces on Luzon where it grows as an epiphyte, usually at the base of small trees and shrubs, at elevations to 1,000 metres.

Photo: David Titmuss

Bulbophyllum maquilingense

Photo: David Titmuss

Bulbophyllum maxillare

Bulbophyllum maxillare

(Lindley) Reichenbach *f.*
Section: ***Ephippium*** Schlechter

Dr. John Lindley first named this species in 1843 as ***Cirrhopetalum maxillare***. Professor Heinrich G. Reichenbach transferred the generic name to ***Bulbophyllum*** in 1861 in W.G. Walpers, *Annales botanices systematicae.* The generic name refers to the resemblance of the flowers to some species of the South American genus ***Maxillaria***.

Growth habit: upright, sympodial. **Rhizome:** up to 3 mm in diameter, carrying the pseudobulbs at intervals of about 5 to 7 cm. **Pseudobulbs:** tapered, about 3 cm high by 1.5 cm in diameter, unifoliate. **Leaves:** ovate to elliptic, leathery, up to 15 cm long by 2.5 cm wide. **Inflorescences:** upright, almost as long as the leaf, bear a single bloom that resembles a ***Masdevallia***; flower up to 7 cm high by 2.5 cm across at widest point. **Flower colour:** base colour yellowish; dorsal and lateral sepals heavily marked with dark purplish red; labellum reddish brown. **Dorsal sepal:** ovate, acuminate, 2.5 cm long by 6 mm wide, edged with white hairs. **Petals:** ovate, about 6 mm long by 2 mm wide, tips pointed, almost enclosing the column. **Lateral sepals:** falcate, 4 cm long by 1 cm wide; tails drawn out into a rounded tip. **Labellum:** ovate with a linear tip, fleshy, up to 1 cm long by 2 mm wide.

Habitat and Distribution: *Bulbophyllum maxillare* has been found in Borneo, where it is uncommon. In the Philippines it has been recorded from the province of Cavite on Luzon, the island of Mindoro and the islands of Bohol and Leyte in the Visayas. It grows as an epiphyte at elevations of up to 300 metres, usually on the trunks of trees.

Bulbophyllum mearnsii Ames

Section: ***Lepidorhiza*** Schlechter

Professor Oakes Ames named this species in 1913 in the *Philippine Journal of Science*. The plant was named after its discoverer, Major E.A. Mearns.

Growth habit: upright, sympodial. **Pseudobulbs:** cylindrical, tapered, up to 3 cm high by 2.3 cm in diameter, unifoliate. **Leaves:** leathery, up to 25 cm long by 5.5 cm wide; petiole grooved, up to 5 cm long. **Inflorescences:** up to 26 cm long, bear several funnel-shaped bracts 1 cm long, up to four flowers that open in succession. **Flower colour:** golden yellow with small red spots; labellum brownish red. **Dorsal sepal:** lanceolate, up to 3 cm long by 1 cm wide. **Petals:** triangular, reflex toward the column, 1.5 cm long by 7 mm wide. **Lateral sepals:** conspicuously ridged, triangular, up to 4 cm long by 1 cm wide. **Labellum:** fleshy, grooved lengthwise, covered with small warts, basal third smooth.

Photo: Anton Sieder

Bulbophyllum mearnsii

Habitat and Distribution: ***Bulbophyllum mearnsii*** is endemic to the Philippines and has been recorded from the province of Sorsogon on Luzon; Catanduanes island; Bohol island in the Visayas; and the provinces of Agusan del Norte, Agusan del Sur, Misamis Occidental, Zamboanga del Norte and Zamboanga del Sur on Mindanao. It grows as an epiphyte at elevations of around 900 metres.

Bulbophyllum negrosianum Ames

Section: ***Hirtula*** Ridley

Professor Oakes Ames named this species in 1912 in *Leaflets of Philippine Botany*. The specific name refers to Negros island where this plant was first collected.

Growth habit: upright, sympodial. **Pseudobulbs:** ovoid, 2.5 cm long by 1.2 cm in diameter, unifoliate. **Leaves:** leathery, elliptic, up to 16 cm long by 4 cm wide. **Inflorescences:** up to 42 cm long, carrying up to 45 blooms. **Flower colour:** sepals and petals reddish brown; labellum reddish brown, fading to yellow ochre at edges. **Dorsal sepal:** recurved, elliptic, up to 6 mm long by 2.2 mm wide. **Petals:** ovate, up to 3 mm long by 0.7 mm wide, hairy-edged. **Lateral sepals:** recurved, ovate, up to 6 mm long by 3 mm wide. **Labellum:** three-lobed; mid-lobe elliptic to obovate, up to 4 mm long by 1.5 mm wide and hairy around the edges; side lobes erect, semicircular, margins hairy.

Habitat and Distribution: ***Bulbophyllum negrosianum*** is endemic to the Philippines where it is found in the province of Negros Oriental on the island of Negros in the Visayas. It grows as an epiphyte at medium elevations, around 1,400 metres.

Notes: An interesting feature is that the flowers begin opening at the very tip of the rachis first, then open back toward the plant.

Photo: Anton Sieder

Bulbophyllum negrosianum

Photo: David Titmuss

Bulbophyllum nymphopolitanum

Bulbophyllum nymphopolitanum Kränzlin

Section: ***Lepidorhiza*** Schlechter

Dr. Fritz Kränzlin named this very beautiful species in *Annalen des K.K. Naturhistorischen Hofmuseums, Wien* in 1916. The specific name refers to Nymphenburg where the Munich Botanical Gardens are located.

Growth habit: upright, sympodial. **Pseudobulbs:** oval in shape, 3 cm high by 2 cm in diameter, unifoliate. **Leaves:** leathery, up to 15 cm long by 3 cm wide. **Inflorescences:** semi-erect, appear from the base of the pseudobulbs, bear up to four blooms, only one or two open at a time; flowers large, up to 8.5 cm high by 4 cm wide; blooms have an unpleasant odour, but only if you put your nose too close. **Flower colour:** brownish red, appear varnished; labellum almost black. **Dorsal sepal:** subulate, about 4 cm long by 8 mm wide. **Petals:** subulate, up to 2 cm long, reflex toward the pedicel. **Lateral sepals:** subulate, up to 4 cm long. **Labellum:** fleshy, about 2 cm long by 5 mm wide.

Habitat and Distribution: ***Bulbophyllum nymphopolitanum*** is endemic to the Philippines where it is found in the southern Luzon province of Albay at elevations to 1,000 metres as an epiphyte and occasionally as a lithophyte.

Notes: Professor Oakes Ames considered this species synonymous with ***Bulbophyllum levanae***, which he also described, but I am of the opinion the two species are readily separable due to the vast difference in the size of the flowers and their totally different colour.

Bulbophyllum odoratum

(Blume) Lindley
Section: ***Aphanobulbon*** Schlechter

Carl Blume first named this species ***Diphyes odorata*** in 1825. Dr. John Lindley transferred the generic name to the more widely accepted ***Bulbophyllum*** in 1830 in his *Genera and Species of Orchidaceous Plants.* The specific name refers to the scent of the blooms.

Growth habit: upright, sympodial. **Rhizome:** up to 8 mm in diameter, small pseudobulbs at intervals of 5 to 8 cm. **Pseudobulbs:** almost nonexistent—slight thickenings on the rhizome, unifoliate. **Leaves:** leathery, up to 25 cm long by 6 cm wide. **Inflorescences:** upright, up to 30 cm long, up to 120 perfumed flowers about 1.5 cm in diameter. **Flower colour:** creamy white. **Dorsal sepal:** triangular with tips drawn out into tails; total length about 7 mm. **Petals:** ovate, up to 2 mm long. **Lateral sepals:** triangular with tips drawn out into tails; total length about 1 cm. **Labellum:** thick, elliptic to ovate in shape, about 2 mm long.

Habitat and Distribution: ***Bulbophyllum odoratum*** is a widely distributed species found in Peninsular Malaysia, throughout Indonesia, Sulawesi and the Moluccas. In the Philippines it has been found in the provinces of Ifugao and Rizal on Luzon; the island of Samar in the Visayas; and in the Mindanao provinces of Agusan del Norte, Agusan del Sur, Davao Del Norte, Davao Oriental, Davao del Norte, Surigao del Norte and Surigao del Sur. It grows as an epiphyte at elevations to 1,000 metres.

Photo: David Banks

Bulbophyllum odoratum

Bulbophyllum orthoglossum

Wendland and Kränzlin
Section: ***Lepidorhiza*** Schlechter

Hermann Wendland and Dr. Fritz Kränzlin named this species in the *Gardeners' Chronicle* in 1896. The specific name refers to the erect lateral lobes of the labellum.

Growth habit: upright, sympodial. **Rhizome:** very short. **Pseudobulbs:** up to 5 cm long by 2.5 cm in diameter, unifoliate with persistent bracts. **Leaves:** leathery, oblong, up to 20 cm long by 5 cm wide. **Inflorescence:** upright, up to 25 cm long, bear the blooms in succession. **Flower colour:** yellowish green with brown striping; labellum red. **Dorsal sepal:** lanceolate to acuminate, 4 cm long by 1.4 cm wide. **Petals:** oblong to ovate, reflexing, 1.8 cm long by 8 mm wide. **Lateral sepals:** acuminate, 5 cm long by 1 cm wide, make a half turn from the column foot. **Labellum:** three-lobed, 2 cm long by 6 mm wide; lateral lobes small, bent forward, toothed at margin, covered with small swellings; mid-lobe fleshy with a tongue-shaped apex.

Habitat and Distribution: ***Bulbophyllum orthoglossum*** is a widely distributed species that has been found in Sulawesi (Celebes) and the Moluccas. In the Philippines it has only been recorded from the islet of Sarangani, southwest of the entrance to Davao Gulf, Mindanao. It grows as an epiphyte at elevations to 300 metres.

Bulbophyllum papulosum Garay

Section: ***Lepidorhiza*** Schlechter

Doctor Leslie A. Garay named this lovely species in 1999 in *Harvard Papers in Botany.* The specific name refers to the wart-like protuberances on the labellum.

Growth habit: upright, sympodial. **Rhizome:** about 4 mm in diameter. **Pseudobulbs:** close together, about 2.5 cm in diameter, somewhat flattened, unifoliate. **Leaves:** quite leathery, up to 15 cm long by 4 cm wide; petiole 5 cm in length. **Inflorescences:** about 5 cm long, semi-erect, appearing from the base of the pseudobulbs; blooms quite large, up to 9 cm high by 2.5 cm wide, and appear successively so flowering season lasts several weeks. **Flower colour:** greenish yellow, very shiny; labellum brownish red. **Dorsal sepal:** lanceolate, up to 5 cm long by 1 cm wide. **Petals:** ovate, up to 2 cm long by 6 mm wide; reflex toward the back of the flower. **Lateral sepals:** triangular, 5 cm long by 1.2 cm wide at the base. **Labellum:** three-lobed, 1.5 cm long by 5 mm wide,

Photo: Charles Nishihira

Bulbophyllum orthoglossum

Photo: David Banks

Bulbophyllum papulosum

fleshy with a warty surface; two upright arms about 3 mm long, near the column.

Habitat and Distribution: ***Bulbophyllum papulosum*** is endemic to the Philippines but the exact locality of this beautiful species is not known. It grows as an epiphyte at elevations up to 500 metres.

Bulbophyllum pardalotum

Garay, Hamer and Siegerist
Section: ***Leptopus*** Schlechter

Doctors Leslie Garay, Fritz Hamer and Emly Siegerist named this attractive species in *Lindleyana* in 1995. The specific name means spotted like a leopard, referring to the floral segments.

Growth habit: upright, sympodial. **Rhizome:** thin, 1 to 2 mm in diameter, purplish when young. The plant is a bit of a rambler having a distance of 3 to 4 cm between the pseudobulbs. **Pseudobulbs:** four-angled, 1.2 cm high by 1 cm wide. **Leaves:** ovate, leathery, up to 6 cm long by 1.5 cm wide. **Inflorescences:** upright, longer than the leaves, carry a single showy flower about 2 cm across the widest point—the lateral segments. **Flower colour:** bright orange spotted with varying amounts of reddish brown; labellum bright red with a yellow tip. **Dorsal sepal:** lanceolate, 1.4 cm long by 6 mm wide. **Petals:** linear-lanceolate, 8 mm long by 2.5 mm wide. **Lateral sepals:** linear-lanceolate, 1.5 cm long by 5 mm wide. **Labellum:** 6 mm long by 2.5 mm wide, tongue-shaped, highly mobile, fleshy, shiny.

Habitat and Distribution: ***Bulbophyllum pardalotum*** is endemic to the Philippines and is found in the mountains of Nueva Ecija and Nueva Vizcaya in central Luzon at elevations of about 1,200 metres where it grows as an epiphyte. It normally grows in well-lit positions.

Notes: The flowers open widely for the first few hours of the day only to be fully closed by midday. This happens for the life of the blooms—about 10 days.

Photo: David Titmuss

Bulbophyllum pardalotum

Photo: David Titmuss

Bulbophyllum piestoglossum

Bulbophyllum piestoglossum Vermeulen

Section: ***Sestochilus*** (Breda) Bentham and Hooker *f.*

Jaap Vermeulen named this large flowered species in 1994 in ***Blumea***. The specific name refers to the narrow labellum, which is true in herbarium specimens. The labellum of living clones is quite broad.

Growth habit: upright, sympodial. **Rhizome:** 4.5 mm in diameter, covered with sheaths which disappear with age. **Roots:** appear from the internodes of the rhizome. **Pseudobulbs:** tapered, up to 3.2 cm long by 1.5 cm diameter, spaced 5 to 6 cm intervals apart on the rhizome, unifoliate. **Leaves:** oblong to elliptic, up to 14 cm long by 5.5 cm wide; petiole 2 to 2.3 cm long. **Inflorescences:** upright, up to 7 cm long, appear from the pseudobulbs with three basal sheaths; single flower up to 5 cm in diameter. **Flower colour:** base colour orange-yellow with reddish brown stripes and spots; lateral sepals greenish brown with some white spots; labellum deep yellow, brown on the underside. **Dorsal sepal:** lanceolate, 2.8 cm long by 1.3 cm wide. **Petals:** linear, pointed at the tips, 2.4 cm long by 8 mm wide. **Lateral sepals:** ovate, keeled, 3 cm long by 8 mm wide. **Labellum:** fleshy, tongue-shaped, highly mobile, curved, 1.5 cm long by 7 mm wide.

Habitat and Distribution: *Bulbophyllum piestoglossum* is endemic to the Philippines. It has been found in Laguna and Quezon provinces on Luzon. It grows as an epiphyte, usually on tree trunks at elevations between 600 and 1,300 metres.

Bulbophyllum profusum Ames

Section: ***Oxysepalum*** (Wight) Reichenbach *f.*

Professor Oakes Ames named this small-growing species in 1912 in the *Philippine Journal of Science.* The specific name relates to the many flowers produced by this species.

Growth habit: semi-pendulous, sympodial. **Pseudobulbs:** very small, mere thickenings on the rhizome, at about 2 cm intervals. **Leaves:** leathery, ovate, up to 3.5 cm long by 1.5 cm wide. **Inflorescences:** many, single-flowered, appear from along the rhizome. **Flower colour:** white; dorsal and lateral sepals tipped with yellow; labellum cream. **Dorsal sepal:** triangular-linear, tips drawn out, 5 mm long by 1.5 mm wide. **Petals:** elliptic to ovate, up to 2 mm long by 1.5 mm wide. **Lateral sepals:** triangular-linear, tips drawn out, 5 mm long by 1.5 mm wide. **Labellum:** lanceolate, fleshy, 1.5 mm long, smooth.

Habitat and Distribution: *Bulbophyllum profusum* has been recorded from Guam. In the Philippines it has been found in the provinces of Cagayan, Laguna, Nueva Vizcaya, Quezon, Rizal and Sorsogon on Luzon; the islands of Palawan and Polillo; Leyte, Samar and Sibuyan islands in the Visayas; the provinces of Agusan del Norte and Surigao del Sur on Mindanao; and the island of Basilan. It grows as an epiphyte, often in shady localities, at elevations to 500 metres.

Bulbophyllum santosii Ames

Section: ***Sestochilus*** (Breda) Bentham and Hooker *f.*

Professor Oakes Ames named this plant in 1915 in his *Studies in the Family Orchidaceae,* Fascicle V. It was named after F. Santos, who collected the type specimens in December 1913 on Mount Iriga in the province of Camarines Sur, Luzon.

Growth habit: upright, sympodial. **Rhizome:** about 5 mm in diameter; pseudobulbs appear at 5 cm intervals. **Pseudobulbs:** tapered, up to 4 cm tall by 1.5 cm in diameter, unifoliate. **Leaves:** linear, about 12 cm long by 1.5 cm wide. **Inflorescences:** upright, shorter than the leaves, carry up to four flowers approximately 3 cm in diameter. **Flower colour:** milky white, labellum creamy white with pinkish markings. **Dorsal sepal:**

Photo: David Banks

Bulbophyllum profusum

Photo: Jim Cootes

Bulbophyllum santosii

lanceolate, 1.5 cm long by 1 cm wide. **Petals:** lanceolate, 1.2 cm long by 8 mm wide. **Lateral sepals:** ovate, up to 1.5 cm long by 1.2 cm wide. **Labellum:** very fleshy with a rough surface, 1 cm long by 6 mm wide.

Habitat and Distribution: ***Bulbophyllum santosii*** is endemic to the Philippines and has been found in the provinces of Aurora, Camarines Sur, Laguna and Quezon on Luzon; the island of Leyte in the Visayas; and in the province of Bukidnon in Mindanao. It grows as an epiphyte at elevations to 800 metres.

Photo: David Titmuss

Bulbophyllum saurocephalum

Bulbophyllum saurocephalum

Reichenbach *f.*

Section: ***Globiceps*** Schlechter

Professor Heinrich G. Reichenbach named this plant in 1886 in the *Gardeners' Chronicle*. The name refers to the flowers' fanciful resemblance to a lizard's head.

Growth habit: upright, sympodial. **Rhizome:** diameter up to 2.5 mm; pseudobulbs at intervals of up to 5 cm. **Pseudobulbs:** up to 5 cm tall, 2 cm in diameter, topped by a single leaf—rarely two—up to 15 cm long, 1.5 cm wide. **Inflorescences:** upright; remarkable in that they are swollen in a similar manner to some African members of this genus; up to 40 blooms that open sequentially, per inflorescence. **Flower colour:** inflorescence dull red; small flowers pale yellow striped with brown. **Dorsal sepal:** oblong, up to 5 mm long, 3 mm wide. **Petals:** linear, 3 mm long, 1 mm wide. **Lateral sepals:** joined, 5 mm long, 3 mm wide. **Labellum:** small, heart-shaped, 3 mm long, 2 mm wide.

Habitat and Distribution: ***Bulbophyllum saurocephalum*** is endemic to the Philippines where it grows as an epiphyte in the mountains of Nueva Ecija, Nueva Vizcaya and Quezon provinces on Luzon at elevations of about 1,000 metres.

Photo: David Titmuss

Bulbophyllum savaiense ssp. *subcubicum*

Bulbophyllum savaiense subsp. *subcubicum* (J.J. Smith) J.J. Vermeulen

Section: ***Pelma*** (Finet) Schlechter

Dr. Rudolph Schlechter originally described this species in 1911. Jaap Vermeulen described the subspecies in 1993 in *Orchid Monographs* Volume 7. The sub-specific name refers to the shape of the labellum.

Growth habit: pendulous, up to 40 cm long, sympodial. **Rhizome:** small, covered by a papery sheath, small oval pseudobulbs at intervals of 1 to 1.5 cm. **Pseudobulbs:** up to 1 cm long by 2 mm in diameter, unifoliate. **Leaves:** lanceolate, up to 6 cm long by 1 cm wide. **Inflorescences:** pendulous, carry up to six blooms about 3 mm in diameter. **Flower colour:** cream to greenish; labellum bright orange. **Dorsal sepal:** ovate, 2 mm long by 1.5 mm wide. **Petals:** elliptic, up to 1.5 mm long by 1 mm wide. **Lateral sepals:** ovate, up to 2.5 mm long by 1.75 mm wide. **Labellum:** small with a distinct cavity at the front.

Habitat and Distribution: ***Bulbophyllum savaiense*** subsp. ***subcubicum*** is widely distributed throughout Indonesia and New Guinea. In the Philippines it has been recorded from the island of Leyte. It grows as an epiphyte at elevations to 2,000 metres.

Notes: Flowers of this subspecies are the smallest of any Philippine ***Bulbophyllum*** with which I am familiar.

Calanthe R. Brown

Robert Brown established this genus in the *Botanical Register* in 1821. The generic name refers to the beautiful flowers. The type species for this lovely genus of terrestrial orchids is ***Calanthe triplicata*** (Willem.) Ames.

Calanthes are found throughout South Africa, Madagascar, all of tropical Asia, Australia, the Pacific Islands, Tahiti and one member in Central America. There are approximately 150 species in the genus of which 14 have been recorded from the Philippines.

Calanthe hennisii Loher

Augustus Loher named this species in 1909 in the *Gardeners' Chronicle.* I have not been able to find out who was honoured in the naming of this plant.

Growth habit: upright, sympodial. **Pseudobulbs:** tapered, up to 10 cm high by 3.5 cm in diameter. **Leaves:** two to four at top of pseudobulb, ovate to lanceolate, up to 30 cm long by 8 cm wide, of a thin texture. **Inflorescences:** upright, up to 40 cm tall, covered with short white hairs; bear up to eight blooms 5 cm in diameter. **Flower colour:** white with a yellow blotch near the column. **Dorsal sepal:** ovate, up to 2 cm long by 5 mm wide; back of segment hairy. **Petals:** obovate, up to 2.2 cm long by 6 mm wide. **Lateral sepals:** ovate, up to 2.3 cm long by 5 mm wide; back of segment hairy. **Labellum:** 2.5 cm long by 3.4 cm wide, three-lobed; lateral lobes squarish; mid-lobe bilobed; hairy spur curves toward front of labellum.

Habitat and Distribution: ***Calanthe hennisii*** is endemic to the Philippines and has been found in Quezon and Rizal provinces on Luzon and North and South

Photo: David Titmuss

Calanthe hennisii

Cotabato provinces on Mindanao. It grows as an epiphyte at the base of trees at elevations of between 500 and 1,000 metres.

Calanthe rubens Ridley

Henry Ridley named this species in 1890 in the *Gardeners' Chronicle*. The specific name refers to the reddish marking on the labellum.

Photo: David Banks

Calanthe rubens

Growth habit: upright, sympodial. **Pseudobulbs:** up to 15 cm tall by 5 cm square, tapered. The top third of the pseudobulb constricts noticeably and will break off easily, but this will grow if placed in sphagnum moss. **Leaves:** ovate to lanceolate, up to 30 cm long by 5 cm wide, usually deciduous when flowering. **Inflorescences:** upright, covered with short white hairs, up to about 35 cm tall, carry up to 12 blooms. **Flower colour:** cream or pale pink; labellum has maroon markings; blooms about 3 cm across the petals. **Dorsal sepal** and **petals:** oblong-ovate. **Lateral sepals:** lanceolate, reflex toward the pedicel. **Labellum:** large, three-lobed; spur long, slender, curved.

Habitat and Distribution: ***Calanthe rubens*** is found in Abra, Benguet and Mountain Province on Luzon and on Palawan island. It grows as a terrestrial at elevations to 1,800 metres. It also occurs in Thailand and Peninsular Malaysia.

Ceratocentron Senghas

Dr. Karlheinz Senghas established this genus, of only one species, in 1989 in *Die Orchidee*. The generic name refers to the horn-like protuberance just below the labellum.

Ceratocentron are only found in a small, restricted area in the mountains of the central Philippines.

Ceratocentron fesselii Senghas

Dr. Karlheinz Senghas named this species in 1989 in *Die Orchidee*. This delightful miniature species was named in honour of Dr. Hans Fessel, a taxonomist and botanist from Germany.

Growth habit: upright, monopodial. The plant is of vandaceous growth habit but is very small, only reaching 2.5 cm high and about 4 cm across the leaves. **Leaves:** very leathery, 2.5 cm long by 1 cm wide. **Inflorescences:** upright; flowers very large in comparison to the plant, blooms up to 1 cm in diameter, up to four per inflorescence, usually at least two inflorescences. **Flower colour:** very bright red or orange; labellum white with pinkish purple markings. **Dorsal sepal:** lanceolate, up to 7 mm long by 4 mm wide. **Petals:** ovate, 5 mm long by 4 mm wide. **Lateral sepals:** ovate, up to 5 mm long by 4 mm wide. **Labellum:** small, firmly attached to the column, little horn just below the entrance to the nectary, spur about 6 mm long.

Photo: David Banks

Ceratocentron fesselii

Habitat and Distribution: ***Ceratocentron fesselii*** is of limited distribution in the mountains of Nueva Ecija on Luzon where it grows as an epiphyte at elevations of around 1,000 metres.

Notes: Dr. Eric A. Christenson believes this species is bird-pollinated. The bright colour of the flowers, the leathery leaves and the plant's exposed habitat, which is chiefly on the trunks of trees, are all factors for bird-pollination.

Ceratostylis Blume

Carl Blume established this genus in 1825 and the name was published in *Bijdragen tot de Flora van Nederlandsch Indie.* The generic name refers to the fleshy column.

There is no designated type species, as several species were mentioned when the description for this genus was made.

There are approximately 60 species in this genus, which are found in India, Southeast Asia, the Philippines and the islands of the Pacific Ocean. The Philippines has approximately 13 species, but I believe more await description.

Photo: David Banks

Ceratostylis incognita

Ceratostylis incognita

J.T. Atwood and J. Beckner

John T. Atwood and John Beckner described this species in 1999 in *Selbyana* 19(2): 265-266. The specific name means unknown and describes the species' previously unknown status.

Growth habit: semi-pendulous, sympodial; branched stems up to 30 cm long, covered with light brown bracts. **Leaves:** terete, up to 25 cm long by 4 mm in diameter. **Inflorescences:** bear solitary flowers on a short stem, up to 1.5 cm in diameter. **Flower colour:** white petals and sepals, some clones have red edging on the basal half of the petals; labellum bright yellow. **Dorsal sepal:** elliptic, up to 1 cm long by 6 mm wide; exterior of dorsal and lateral sepals covered in tiny black hairs. **Petals:** narrowly lanceolate, up to 7 mm long by 3 mm wide. **Lateral sepals:** broadly lanceolate, up to 9 mm long by 5 mm wide. **Labellum:** triangular, 4 mm long by 3 mm wide.

Habitat and Distribution: ***Ceratostylis incognita*** is endemic to the Philippines. It is found in the provinces of Nueva Ecija and Nueva Vizcaya on Luzon growing at elevations between 750 and 1,500 metres where it lives as an epiphyte, usually in very brightly lit situations.

Ceratostylis retisquama

Reichenbach *f.*

Professor Heinrich G. Reichenbach named this pretty species in 1857 in *Bonplandia.* The specific name refers to the scale-like bracts on the rhizome.

Growth habit: semi-pendulous, sympodial; branched stems up to 40 cm long, covered with overlapping, brown papery bracts. **Leaves:** leathery, up to 12 cm long by 1 cm wide. **Inflorescences:** bear solitary blooms up to 4 cm in diameter that appear from the base of the plant. **Flower colour:** bright reddish orange; labellum white. **Dorsal sepal:** lanceolate, up to 2 cm long by 6 mm wide. **Petals:** oblanceolate, pointed, up to 2 cm long by 5 mm wide. **Lateral sepals:** oblong-lanceolate, pointed, up to 2 cm long by 5 mm wide. **Labellum:** tiny, pointed, 3 mm long by 1.5 mm wide.

Habitat and Distribution: ***Ceratostylis retisquama*** is endemic to the Philippines where it is found in the provinces of Bataan, Camarines Norte, Camarines Sur, Ilocos Norte, Quezon, Rizal and Zambales on Luzon and the provinces of Agusan del Norte and Lanao del Sur on Mindanao. It grows as an epiphyte at elevations to 500 metres.

Notes: ***Ceratostylis retisquama*** is the largest-flowered member for this genus that I know of and the most widely cultivated.

Ceratostylis retisquama

Ceratostylis senilis Reichenbach *f.*

Professor Heinrich G. Reichenbach described this plant in 1878 in *Otia Botanica Hamburgensia.* The specific name refers to the white hairs at the back of the dorsal and lateral sepals.

Growth habit: upright, sympodial. **Pseudobulbs:** up to 2 cm long, covered by persistent bracts, unifoliate. **Leaves:** terete, fleshy, grooved, up to 11 cm long by 3.5 mm in diameter. **Inflorescences:** up to 1.5 cm long, bear a single flower about 7 mm high by 5 mm wide. **Flower colour:** white. **Dorsal** and **lateral sepals:** oblong to lanceolate, up to 5 mm long by 2 mm wide. **Petals:** linear to lanceolate, up to 4 mm long by 1.5 mm wide; inner portion edged with pink. **Labellum:** oblanceolate, up to 3 mm long by 1.5 mm wide; backs of the dorsal and lateral sepals covered with short white hairs.

Ceratostylis senilis

Habitat and Distribution: ***Ceratostylis senilis*** is endemic to the Philippines and has been found in the provinces of Bataan, Camarines Sur, Laguna, Nueva Vizcaya, Pangasinan, Quezon, Rizal and Sorsogon on Luzon and in Surigao del Norte on Mindanao. It grows as an epiphyte in bright light at elevations of about 1,000 metres.

Photo: David Titmuss

Ceratostylis subulata

Ceratostylis subulata Blume

Carl Blume named this species in 1825 in *Bijdragen tot de Flora van Nederlandsch Indie.* The specific name refers to the piercer-shaped leaves.

Growth habit: upright, sympodial; stems do not branch but are covered with dark brown bracts. **Leaves:** terete, up to 25 cm long, 3 mm wide, upper one-fifth of leaves channelled; blooms located where channelling begins. **Inflorescences:** short, hairy; flowers small, up to 5 mm in diameter. **Flower colour:** dull white to yellow; labellum bright yellow. **Dorsal** and **lateral sepals:** ovate, 3 mm long, 2 mm wide. **Petals:** lanceolate, 3 mm long, 1.5 mm wide. **Labellum:** slipper-shaped, 3 mm long, 2 mm wide, tip fleshy.

Habitat and Distribution: *Ceratostylis subulata* is probably the most widely distributed species for this genus, being found in India, Thailand, Peninsular Malaysia, Sumatra, Java, Borneo, New Guinea, the Solomon Islands and Vanuatu. In the Philippines it has been found in the provinces of Bataan, Benguet, Ifugao, Nueva Vizcaya, Pampanga and Rizal on Luzon; the provinces of Occidental and Oriental Mindoro on the island of Mindoro; Panay in the Visayas; and in the provinces of Agusan del Norte, Agusan del Sur, Bukidnon, Davao del Norte, Davao Oriental, Davao del Sur, Lanao del Norte, Lanao del Sur, Zamboanga del Norte and Zamboanga del Sur on Mindanao. It grows as an epiphyte at elevations up to 2,100 metres.

Chelonistele Pfitzer

Ernst Pfitzer established this genus in 1907 in A. Engler (ed.) *Das Pflanzenreich Regni Vegetabilis Conspectus* IV 50 II B7. The generic name alludes to the winged column.

The type species for this genus is ***Chelonistele sulphurea*** (Bl.) Pfitzer.

Plants for this genus are found throughout Peninsular Malaysia, Sumatra, Java, Borneo and the southern Philippines. There are 11 species, 10 of which are endemic to Borneo and the other being widespread.

Chelonistele sulphurea

(Blume) Pfitzer

Carl Blume first named this species ***Chelonanthera sulphurea*** in 1825. Ernst Pfitzer transferred the generic name in 1907 in A. Engler (ed.) *Das Pflanzenreich Regni Vegetabilis Conspectus* IV 50 II B7. The specific name refers to the sulphur yellow markings on the labellum.

Growth habit: upright, sympodial. **Pseudobulbs:** oval, up to 3.5 cm high by 1.5 cm in diameter. **Leaves:** elliptic to lanceolate, 15.5 cm long by 4 cm wide, leathery, petiole up to 8 cm long. **Inflorescences:** appear with new growth, gracefully arch; bear up to 12 flowers 1.4 cm in diameter. **Flower colour:** flesh-coloured; labellum cream with orange to yellow markings. **Dorsal sepal:** concave, ovate to lanceolate, up to 1 cm long by 4 mm wide. **Petals:** linear, curve toward the pedicel, up to 1 cm long by 1.2 mm wide. **Lateral sepals:** concave, ovate to lanceolate, up to 1 cm long by 5 mm wide. **Labellum:** three-lobed, up to 1.5 cm long by 1 cm wide.

Habitat and Distribution: ***Chelonistele sulphurea*** has been recorded from Peninsular Malaysia, Sumatra, Java and Borneo. In the Philippines this species has been recorded from throughout Mindanao and the island of Palawan. It grows as an epiphyte at elevations between 600 and 2,700 metres.

Notes: Dr. Valmayor did not include this genus in *Orchidiana Philippiniana.*

Photo: David Banks

Chelonistele sulphurea

Cirrhopetalum Lindley

Dr. John Lindley proposed this genus in 1830 in his *Genera and Species of Orchidaceous Plants.* The generic name possibly refers to the thread-like extensions of the petals or may allude to the lateral sepals, which in some species are curled.

The type species for the genus is ***Cirrhopetalum umbellatum*** (Forst. *f.*) Hooker and Arn.

There are about 75 species from the genus ***Cirrhopetalum***, which are found in Africa, Madagascar, India, Sri Lanka, Myanmar (Burma), Thailand, Malaysia, Indonesia, Borneo, New Guinea, northern Australia and the Pacific Islands. In the Philippines there are approximately 16 species, nine of which are endemic.

Cirrhopetalum brevibrachiatum Schlechter

Section: ***Cirrhopetalum*** Lindley

Dr. Rudolf Schlechter named this lovely orchid in 1911 in *Fedde, Repertorium Specierum Novarum, Beihefte.* The specific name refers to the short arms at the tip of the column.

Growth habit: upright, sympodial; likes to ramble, pseudobulbs on a small rhizome up to 3 cm apart. **Pseudobulbs:** round, about 1 cm in diameter. **Leaves:** up to 12 cm long by 2 cm wide. **Inflorescences:** upright, to 15 cm in length, carry up to 12 blooms, 4 cm long by 6 mm wide, on a typically ***Cirrhopetalum*** umbel. **Flower colour:** base colour creamy yellow, liberally spotted with pale red; labellum bright red on the apical half, bright yellow near the column. **Dorsal sepal:** concave, 5 mm long by 3.5 mm wide, edged with orange hairs. **Petals:** lanceolate, 4 mm long by 3 mm wide, edged with small hairs. **Lateral sepals:** acuminate, 4 cm in length by 3 mm in diameter, joined for entire length. **Labellum:** recurved, small, highly mobile, 3 mm long by 1 mm wide.

Photo: David Banks

Cirrhopetalum brevibrachiatum

Photo: David Banks

Cirrhopetalum cumingii

Photo: David Titmuss

Cirrhopetalum loherianum

Habitat and Distribution: ***Cirrhopetalum brevibrachiatum*** was originally named from plants found in Sulawesi (Celebes). Specimens from the Philippines that I recently flowered are very close to this species. The photograph is of a plant that originated in the mountains of Nueva Ecija and Nueva Vizcaya in central Luzon. It grows as an epiphyte on the trunks of trees at altitudes between 500 and 1,400 metres.

Cirrhopetalum cumingii Lindley

Section: ***Cirrhopetalum*** Lindley

Dr. John Lindley named this orchid in 1838 in the *Botanical Register*, in honour of Hugh Cuming, who collected the type specimens.

Growth habit: upright, sympodial. **Rhizome:** up to 3 mm in diameter, carries pseudobulbs at intervals of about 3 cm. **Pseudobulbs:** up to 2 cm high by 1.5 cm in diameter, unifoliate. **Leaves:** leathery, up to 10 cm long by 2 cm wide. **Inflorescences:** upright, appear from base of pseudobulb, up to 15 cm long; flowers about 2.5 cm long by 4 mm wide, on an umbel, up to 12 blooms common. **Flower colour:** reddish purple. **Dorsal sepal:** concave, edged with little hairs. **Petals:** lanceolate, edged with hairs. **Lateral sepals:** narrowly elliptic, joined for their length, about 2 cm. **Labellum:** small, highly mobile, 3 mm long by 1.3 mm wide.

Habitat and Distribution: ***Cirrhopetalum cumingii*** has been recorded from Sabah (Borneo). In the Philippines it has been recorded from Cagayan, Cavite, Laguna, Quezon and Rizal on Luzon; Palawan island; and Negros and Samar islands in the Visayas. It is found at about 500 metres, growing as an epiphyte.

Cirrhopetalum loherianum

Kränzlin

Section: ***Cirrhopetalum*** Lindley

Dr. Fritz Kränzlin named this species in *Annalen des K. K. Naturhistorischen Hofmuseums, Wien* in 1916. The species was named after Mr. Augustus Loher, collector of the type specimens.

Growth habit: upright, sympodial. **Rhizome:** about 3 mm in diameter, carry pseudobulbs at 2 cm intervals. **Pseudobulbs:** up to 3 cm high by 2 cm in diameter, squarish, unifoliate. **Leaves:** leathery, up to 8 cm long by 2 cm wide. **Inflorescences:** upright, longer than the leaves, reddish; flowers of typical ***Cirrhopetalum*** shape, about 2.5 cm long, up to 13 blooms on the umbel. **Flower colour:** yellow with a reddish blush; labellum yellow. **Dorsal sepal:** ovate, concave, edged with small purplish hairs. **Petals:** triangular, edged with hairs. **Lateral sepals:** linear, joined for most of their length. **Labellum:** highly mobile, fleshy, 3.5 mm long by 1.5 mm wide.

Habitat and Distribution: ***Cirrhopetalum loherianum*** has been found in Camarines Sur, Laguna and Quezon on Luzon; Leyte island in the Visayas; and Surigao del Norte and Surigao del Sur on Mindanao. It grows as an epiphyte at around 1,000 metres.

Cirrhopetalum makoyanum

Reichenbach *f.*
Section: ***Cirrhopetalum*** Lindley

Professor Heinrich G. Reichenbach named this beautiful species in 1897, in the *Gardeners' Chronicle*. He named it after a Mr. Makoy who provided him with the specimen from which the descriptions were made.

Growth habit: upright, sympodial. **Rhizome:** about 3 mm in diameter, carry pseudobulbs at 3 cm intervals. **Pseudobulbs:** oval, up to 3 cm long by 1.5 cm in diameter, unifoliate. **Leaves:** leathery, up to 10 cm long by 3 cm wide. **Inflorescences:** upright, 21.5 cm long, longer than the leaves, carry up to 12 blooms about 4 cm long. **Flower colour:** yellow with small red dots; labellum dull yellow. **Dorsal sepal:** ovate, concave, 4 mm long by 2.5 mm wide; edged with small, dark red hairs. **Petals:** ovate, 4 mm long by 2.5 mm wide; edged with short, dark red hairs. **Lateral sepals:** joined for entire 4 cm length, about 2 mm in diameter, taper gradually toward the tip. **Labellum:** fleshy, highly mobile, 4 mm long by 1.6 mm wide.

Habitat and Distribution: *Cirrhopetalum makoyanum* is widely spread in the Philippines and has been found in the provinces of Ifugao, Rizal and Zambales on Busuanga island, Oriental Mindoro on Mindoro; Palawan and Polillo islands; and in Agusan del Norte and Agusan del Sur provinces on Mindanao. It also occurs in Peninsular Malaysia and Borneo. It grows as an epiphyte at elevations up to 300 metres.

Cirrhopetalum nutans Lindley

Section: ***Cirrhopetalum*** Lindley

Dr. John Lindley named this species in 1839 in the *Botanical Register*. The specific name means nodding or hanging and refers to the habit of the flowers.

Photo: David Banks

Cirrhopetalum makoyanum

Photo: David Titmuss

Cirrhopetalum nutans

Growth habit: upright, sympodial. **Rhizome:** about 1.5 mm in diameter, carry the pseudobulbs up to 6 cm apart. **Pseudobulbs:** ovate, 1 cm in diameter. **Leaves:** leathery, up to 5 cm long by 2 cm wide, dull purplish on underside. **Inflorescences:** upright, to 20 cm long, carry up to 15 flowers on an umbel; blooms about 3 cm long by 6 mm wide. **Flower colour:** cream; labellum bright yellow. **Dorsal sepal:** lanceolate, slightly concave. **Petals:** ovate, toothed for their basal third. **Lateral sepals:** linear, joined for most of their length. **Labellum:** fleshy, small, 3 mm long by 1 mm wide.

Habitat and Distribution: *Cirrhopetalum nutans* is endemic to the Philippines where it is found in the provinces of Albay, Quezon and Sorsogon on Luzon and the island of Leyte in the Visayas. This species grows epiphytically, usually at elevations to 300 metres.

Cirrhopetalum trisetum

(Ames) Garay, Hamer and Siegerist
Section: ***Cirrhopetalum*** Lindley

Professor Oakes Ames first named this species ***Bulbophyllum trisetum***. Doctors Garay, Hamer and Siegerist transferred it to ***Cirrhopetalum*** in 1994 in the *Nordic Journal of Botany*. The specific name refers to the prominent hairs on the tip of the dorsal sepal and petals.

Photo: David Titmuss

Cirrhopetalum trisetum

Growth habit: Upright, sympodial. **Rhizome:** about 3 mm in diameter; pseudobulbs up to 15 cm apart. **Pseudobulbs:** four angled or pear-shaped, up to 3.5 cm high by 2 cm in diameter, unifoliate. **Leaves:** up to 15 cm long by 4 cm wide. **Inflorescences:** upright, streaked with dark purple, usually longer than leaves, can carry up to 10 very attractive blooms; flowers up to 4 cm long by 1 cm across petals. **Flower colour:** creamy yellow, varying degrees of reddish purple spotting on dorsal sepal and petals; dorsal sepal and petals spotted purple; labellum purple and green. **Dorsal sepal:** circular, concave, tipped by small hair-like appendage with a thickened tip. **Petals:** small, triangular, edged with short hairs, tips curl backward. **Lateral sepals:** linear, joined for the majority of their length, up to 3 cm long by 5 mm wide. **Labellum:** small, highly mobile, 4 mm long by 1.5 mm wide.

Habitat and Distribution: ***Cirrhopetalum trisetum*** is endemic to the Philippines and has been found in Benguet, Nueva Ecija and Nueva Vizcaya on Luzon growing as an epiphyte at elevations to 1,200 metres.

Notes: In its natural habitat this species can become very large, covering the whole trunk of its host tree.

Cirrhopetalum weberi

(Ames) Senghas
Section: ***Cirrhopetalum*** Lindley

Photo: David Banks

Cirrhopetalum weberi

Professor Oakes Ames named this lovely species in 1912 in *The Philippine Journal of Science* as ***Bulbophyllum weberi***. Dr. Karlheinz Senghas transferred the generic name to ***Cirrhopetalum*** in *Schlechter's Orchideen* in 1983. The specific name honours C.M. Weber, the collector of the type specimens.

Growth habit: upright, sympodial. **Rhizome:** stout, up to 3 mm in diameter. **Pseudobulbs:** squarish, up to 2.5 cm high, 2 cm across, unifoliate. **Leaves:** leathery, ovate, up to 10 cm long, 5 cm wide; petiole up to 1.5 cm long. **Inflorescences:** upright, bright green, bear about 10 blooms up to 2.5 cm long, on an umbel. **Flower colour:** dorsal sepal base colour yellow, heavily spotted with dark brown; petals similar to dorsal sepal; lateral sepals base colour creamy yellow, lightly spotted with reddish brown; labellum cream with dark red spotting. **Dorsal sepal:** ovate, concave, 4 mm long, 3.5 mm wide, tipped with a 2 mm long thread. **Petals:** triangular, 4 mm long, 2 mm wide, edged with short hairs on apical half, tipped with a 1.5 mm long thread. **Lateral sepals:** lanceolate, joined along their length on upper edges, 2 cm long, 5 mm wide (flattened). **Labellum:** fleshy, sharply bent in the middle, about 6 mm long, 2 mm wide.

Habitat and Distribution: ***Cirrhopetalum weberi*** is endemic to the Philippines. It has been recorded from Laguna, Mountain Province, Nueva Ecija, Nueva Vizcaya and Rizal on Luzon and Agusan del Norte, Agusan del Sur, Bukidnon, Davao del Norte, Davao Oriental and Davao del Sur on Mindanao. It grows as an epiphyte, usually on tree trunks, at elevations to 1,500 metres.

Notes: It is quite remarkable that this species, widely distributed on Luzon and Mindanao, has not been recorded from the Visayan islands. These lie between Luzon and Mindanao, but only Leyte island has been explored botanically and this was early in the 1900s.

Cleisostoma Blume

Carl Blume established this genus in 1825 in *Bijdragen tot de Flora van Nederlandsch Indie.* The generic name refers to the callus at the entrance of the spur.

The type species for this genus of small-flowered monopodial orchids is ***Cleisostoma sagittatum*** Bl.

Cleisostoma species are found in Australia, India, throughout continental Southeast Asia, Indonesia, New Guinea and the islands of the Pacific. This is the second largest genus of monopodial orchids, with over 100 species. There are at least seven species found in the Philippines.

Cleisostoma subulatum Blume

Carl Blume named this species in 1825 in *Bijdragen tot de Flora van Nederlandsch Indie.* The specific name refers to the tips of the leaves, which are subulate: narrow, pointed and more or less flattened.

Growth habit: semi-pendulous, monopodial. **Stems:** up to 6 mm in diameter by up to 45 cm long. **Leaves:** leathery, linear, up to 20 cm long by 1 cm wide. **Inflorescences:** pendulous, occasionally branch, bear up to 40 flowers about 1 cm in diameter. **Flower colour:** cream, heavily striped with brown; labellum bright pink. **Dorsal sepal:** ovate, up to 5 mm long by 2.5 mm wide. **Petals:** similar to dorsal sepal but a little shorter. **Lateral sepals:** as per dorsal sepal. **Labellum:** three-lobed, spurred; mid-lobe arrow-shaped.

Habitat and Distribution: ***Cleisostoma subulatum*** is a widely distributed species and has been found in India, Malaysia, Indonesia, Thailand, Vietnam, Borneo and Sulawesi (Celebes). In the Philippines it has been recorded from the islands of Mindoro and Palawan and the provinces of Davao del Norte, Davao Oriental and Davao del Sur on Mindanao. It grows as an epiphyte in a wide range of habitats—from full shade to almost full sun—at elevations to 300 metres.

Photo: Jim Cootes

Cleisostoma subulatum

Cleisostoma weberi (Ames) Garay

Professor Oakes Ames originally named this species ***Sarcanthus***. Dr. Leslie Garay transferred the generic name to ***Cleisostoma*** in 1972 in the *Botanical Museum Leaflets of Harvard University.* The specific name honours C.M. Weber, collector of the type specimen, which was found in the province of Davao del Sur on Mindanao in September 1911.

Growth habit: upright, monopodial. **Leaves:** terete, 6.5 cm long by 3 mm in diameter. **Inflorescence:** semi-pendulous, branching. **Flower colour:** petals and sepals pale brown with darker stripes; labellum

Photo: Gary Yong Gee

Cleisostoma weberi

white and pale purple. **Dorsal sepal:** concave, elliptic, 3 mm long by 1.25 mm wide. **Petals:** spathulate, 2.5 mm long by 1 mm wide. **Lateral sepals:** ovate to oblong, 3.5 mm long by 1.5 mm wide. **Labellum:** 4 mm long by 2 mm wide, three-lobed; lateral lobes upright; mid-lobe pointed, longer than lateral lobes.

Habitat and Distribution: ***Cleisostoma weberi*** is endemic to the Philippines and has been found in the province of Bataan on Luzon; the island of Palawan; and the provinces of North and South Cotabato, Davao del Norte and Davao del Sur on Mindanao. It grows as an epiphyte at elevations to 300 metres.

Coelogyne Lindley

Dr. John Lindley established this genus in 1822 in *Collectanea Botanica*. The generic name refers to the hollow cavity of the stigma.

The type species for this genus is ***Coelogyne cristata*** Lindley.

This genus of showy orchids is widely distributed throughout Asia. Members are found in China, the Himalayas, continental Southeast Asia, Indonesia, Borneo, New Guinea, the Philippines and the islands of the Pacific. Current estimates are of approximately 150 species. ***Coelogyne*** generally grow as epiphytes, although a number of lithophytic and terrestrial species do occur. Twenty-one species are recorded from the Philippines.

Coelogyne asperata Lindley

Section: ***Verrucosae*** Pfitzer

Dr. John Lindley named this orchid in the *Journal of the Horticultural Society* in 1849. The specific name possibly refers to the roughened surface of the labellum.

Growth habit: upright, sympodial. **Pseudobulbs:** tapered, up to 20 cm long and up to 5 cm wide. **Leaves:** oblanceolate, up to 60 cm by 10 cm wide. **Inflorescences:** arching, appear with the new growth, bear up to 15 blooms 6 cm in diameter. **Flower colour:** creamy white; labellum marked with brown and orange. **Dorsal sepal:** oblong-lanceolate, 3 cm long by 1 cm wide. **Petals:** linear-lanceolate, about 3 cm long by 7 mm in width. **Lateral sepals:** oblong-lanceolate, about 3 cm long by 1 cm wide with a central ridge. **Labellum:** three-lobed, up to 3 cm long by 2 cm wide (when flattened); lateral lobes rounded; mid-lobe oblong to ovate. Two ridges run along the length of the labellum with wart-like growths on their surface.

Habitat and Distribution: ***Coelogyne asperata*** is widely distributed, being found in Malaysia, Indonesia, Borneo and New Guinea. In the Philippines it has been found on the islands of Basilan in the Sulu Archipelago; the islands of Leyte, Negros Oriental and Samar in the Visayas; and the provinces of Agusan del Norte, Agusan del Sur, Bukidnon, Davao del Norte, Davao del Sur, Surigao del Norte and Surigao del Sur on Mindanao. It grows as an epiphyte and lithophyte, often close to creeks, at elevations to 500 metres.

Photo: David Banks

Coelogyne asperata

Photo: David Titmuss

Coelogyne bilamellata (brown)

Photo: David Titmuss

Coelogyne bilamellata (white)

Coelogyne bilamellata Lindley

Section: ***Longifoliae*** Pfitzer

Dr. John Lindley named this orchid in 1854 in *Folia Orchidacea*. The specific name refers to the two lamellae (flattened ridges) that are in the labellum.

Growth habit: upright, sympodial. **Pseudobulbs:** up to 12 cm tall, 3 cm in diameter, pear-shaped. **Leaves:** two, ovate, up to 20 cm long, 5 cm wide. **Inflorescences:** pendulous, can reach over 60 cm in length; flowers up to 3 cm across petals, flowers continuously. One clone in my collection recently finished flowering after producing 54 blooms over a two-year period. **Flower colour:** range from pure white to caramel brown. Each bloom has a bract behind it, but this falls off early and does not interfere with the flower opening fully. **Dorsal** and **lateral sepals:** oblong, acute, up to 1.5 cm long, 5 mm wide. **Petals:** linear, up to 1.5 cm long, 1.5 mm wide. **Labellum:** three-lobed, 1.6 cm long, 7.5 mm (when flattened); lateral lobes triangular; mid-lobe oblong to ovate with two ridges running lengthwise.

Habitat and Distribution: *Coelogyne bilamellata* occurs in Borneo and the Philippines, where it has been found in the provinces of Aurora, Laguna and Quezon on Luzon; Leyte island in the Visayas; and in the provinces of Agusan del Norte, Agusan del Sur, Bukidnon, North and South Cotabato, Surigao del Norte, Surigao del Sur, Zamboanga del Norte and Zamboanga del Sur on Mindanao at elevations to 1,000 metres. It grows as either a lithophyte or epiphyte.

Notes: Once established this species seems to be continually in flower.

Photo: David Banks

Coelogyne chloroptera

Coelogyne chloroptera

Reichenbach *f.*

Section: ***Lentiginosae*** Pfitzer

Professor Heinrich G. Reichenbach named this plant in the *Gardeners' Chronicle* in 1883. The specific name refers to green wings, but I am not certain which feature Reichenbach had in mind when he named this attractive species.

Growth habit: upright, sympodial. **Pseudobulbs:** pear-shaped, about 6 cm tall, bifoliate. **Leaves:** leathery, about 20 cm long, 6 cm wide. **Inflorescences:** upright, appear with the new growth, up to 25 cm long; carry up to 12 flowers up to 4 cm in diameter. **Flower colour:** apple-green; labellum white and brown. **Dorsal** and **lateral sepals:** lanceolate, up to 2 cm long, 7 mm wide. **Dorsal sepal:** hooded over the column. **Petals:** linear, up to 2 cm long by 2 mm wide, reflex toward the pedicel. **Labellum:** three-lobed, 1.9 cm long, 1.2 cm wide (when flattened); lateral lobes rounded; mid-lobe semi-circular. Three ridges run lengthwise along the labellum.

Coelogyne marmorata

Coelogyne merrillii

Habitat and Distribution: ***Coelogyne chloroptera*** is endemic to the mountains of Apayao, Bataan, Benguet, Cagayan, Kalinga, Mountain Province, Nueva Ecija, Nueva Vizcaya, Pampanga, Pangasinan and Quezon on Luzon and Negros Oriental in the Visayas. I have seen this species growing in the wild both as an epiphyte and as a lithophyte at elevations around 1,200 metres.

Notes: The record of this species from Negros island may be an incorrect identification of ***Coelogyne confusa*** Ames 1915.

Coelogyne marmorata

Reichenbach *f.*

Section: ***Lentiginosae*** Pfitzer

Professor Heinrich G. Reichenbach named this plant in *Linnaea* in 1877. The specific name refers to the marbled lateral lobes of the labellum.

Growth habit: upright, sympodial. **Rhizome:** short, 7 mm in diameter. **Pseudobulbs:** pear-shaped, 7 cm high, 3 cm in diameter. **Leaves:** leathery, lanceolate, 18 cm long by 3.5 cm wide. **Inflorescences:** appear with the new growth, upright, bear up to five blooms 4.5 cm in diameter. **Flower colour:** sepals and petals apple-green; labellum white with chocolate brown markings; column green edged with orange. **Dorsal sepal:** ovate, 3.7 cm long, 1.6 cm wide. **Petals:** linear, 3.1 cm long, 2 mm wide. **Lateral sepals:** oblong, acute, distinct ridge down centre, 3.4 cm long, 1.2 cm wide. **Labellum:** three-lobed, 3.5 cm long, 2.3 cm wide (flattened); lateral lobes rounded; mid-lobe round with a notch at the tip, five wavy ridges run longitudinally along the length.

Habitat and Distribution: ***Coelogyne marmorata*** is endemic to the Philippines and has been found in Benguet and Pangasinan on Luzon and on Camiguin island, north of Mindanao. It grows as an epiphyte, occasionally a lithophyte, at elevations over 1,000 metres.

Coelogyne merrillii Ames

Section: ***Lentiginosae*** Pfitzer

Professor Oakes Ames first described this species in 1911 after Dr. Elmer Merrill, who collected the first specimens. The name was published in the *Philippine Journal of Science*. He named the plant

Growth habit: upright, sympodial. **Pseudobulbs:** pear-shaped, up to 5 cm high, bifoliate. **Leaves:** quite leathery, up to 25 cm long by 7 cm wide. **Inflorescences:** upright, appear with the new growth, attain a length of 25 cm; blooms up to 5 cm across the petals. Each inflorescence can have up to four flowers. **Flower colour:** white; labellum has yellow markings. **Dorsal sepal:** lanceolate, up to 5.1 cm long by 1.3 cm wide. **Petals:** linear, up to 5 cm long by 1.5 mm wide. **Lateral sepals:** oblong, acute, up to 5 cm long by 1 cm wide. **Labellum:** three-lobed, 4.5 cm long by 2.8 cm wide (when flattened); lateral lobes rounded, mid-lobe heart-shaped. Three ridges run the length of the labellum; central ridge is shorter.

Habitat and Distribution: ***Coelogyne merrillii*** is endemic to the Philippines and has been found in Benguet and Rizal on Luzon and Negros Occidental and Negros Oriental on Negros island in the Visayas. It grows at elevations between 300 and 2,300 metres, usually as a lithophyte.

Coelogyne quinquelamellata

Coelogyne quinquelamellata

Ames

Section: ***Longifoliae*** Pfitzer

Professor Oakes Ames named this unusual species in 1920 in *Studies in the Family Orchidaceae*. The specific name refers to the five lamellae (flattened ridges) that are in the labellum.

Growth habit: upright, sympodial. **Pseudobulbs:** close together on stout rhizome, up to 9 cm tall, 2 cm in diameter, pear-shaped, slightly four-angled, persistent bracts that are fibrous. **Leaves:** two; leathery, oblong to elliptic, up to 30 cm long, 4.5 cm wide. **Inflorescences:** semi-pendulous, flattened, appear from the apex of the pseudobulb, can reach over 50 cm in length; flowers up to 2 cm across the petals. **Flower colour:** sepals and petals reddish brown; labellum cream. Each bloom has a bract behind it, but this falls off early and does not interfere with the flower opening fully. **Dorsal sepal:** ovate to lanceolate, acute, concave, 1.2 cm long, 7 mm wide. **Petals:** linear, up to 1.1 cm long, 1 mm wide. **Lateral sepals:** lanceolate, acute, 1.1 cm long, 5 mm wide. **Labellum:** simple, transversely elliptic, 9 mm long, 1.4 cm wide; five ridges run along the back half of the labellum.

Habitat and Distribution: ***Coelogyne quinquelamellata*** is endemic to the Philippines where it has been found in the provinces of Surigao del Norte and Surigao del Sur on Mindanao at elevations to 150 metres. It grows as an epiphyte in lowland rain forest.

Coelogyne rochussenii de Vriese

Section: ***Tomentosae*** Pfitzer

Willem de Vriese named this spectacular species in 1854 in *Illustrationes des Orchidees des Indes Neerlandaises*. The specific name honours J.J. Rochussen, Minister of the Colonies of the Dutch East Indies during the 1850s.

Coelogyne rochussenii

Growth habit: upright, sympodial. **Pseudobulbs:** cylindrical, up to 15 cm long by 2 cm in diameter, spaced along a 5 mm diameter rhizome at intervals of up to 5 cm. **Leaves:** two, broadly lanceolate, up to 20 cm long by 9 cm wide. **Inflorescences:** pendulous, appear with the new growth, can bear up to 30 fragrant blooms about 4 cm in diameter. **Flower colour:** greenish yellow to lemon-yellow. **Dorsal** and **lateral sepals:** narrowly lanceolate, about 2.5 cm long by 5 mm wide. **Petals:** of similar size and shape but narrower than the dorsal and lateral sepals. **Labellum:** three-lobed, about 3 cm long by 1.5 cm wide (when flattened); mid-lobe drawn out to a sharp point.

Habitat and Distribution: ***Coelogyne rochussenii*** is widely distributed throughout Thailand, Peninsular Malaysia, Indonesia and Borneo. In the Philippines it has been recorded from the province of Pampanga on Luzon and the provinces of Lanao del Norte and Lanao del Sur on Mindanao, where it is found at elevations to 500 metres. This species is quite rare in the Philippines.

Cymbidium Swartz

Olof Swartz established the genus ***Cymbidium*** in 1799. The name was published in *Dianome Epidendri Generis Linn* and refers to the boat-like labellum.

The type species for this popular genus is ***Cymbidium aloifolium*** (L.) Sw.

There are approximately 44 species in the genus and they are found in India, China, Japan, Taiwan, Indochina (Laos, Cambodia, Vietnam), Thailand, Myanmar (Burma), Peninsular Malaysia, Indonesia, New Guinea, the Philippines and Australia. The Philippines has six species for this genus, which grow as epiphytes, lithophytes and terrestrials.

Photo: Andres Golamco Jr.

Cymbidium aliciae

Photo: Elias Javier

Cymbidium atropurpureum

Cymbidium aliciae Quisumbing

Section: ***Maxillarianthe*** Schlechter

Dr. Eduardo Quisumbing named this plant in1940 in *The Philippine Journal of Science.* The specific name honours Alice Day, collector of the type specimens.

Growth habit: upright, sympodial. **Pseudobulbs:** small, covered with sheathing leaf bases. **Leaves:** linear, up to 60 cm long by 2 cm wide. **Inflorescences:** upright, bearing up to eight blooms about 5 cm in diameter. **Flower colour:** pale yellow to greenish; labellum white marked with pale purple blotches. **Dorsal sepal:** narrowly linear-lanceolate, up to 3.5 cm long by 6 mm wide. **Petals:** linear-lanceolate, up to 3 cm long by 5 mm wide. **Lateral sepals:** spreading, slightly falcate, up to 3.5 cm long by 6 mm wide. **Labellum:** three-lobed, up to 2.4 cm long by 1.6 cm wide; lateral lobes erect, rounded; mid-lobe oblong to ovate, reflexing under the lateral lobes.

Habitat and Distribution: *Cymbidium aliciae* is endemic to the Philippines and has been found in Nueva Vizcaya and Rizal on Luzon and Negros Occidental on Negros island in the Visayas. It grows as a terrestrial at elevations between 300 and 2,750 metres in shaded locations.

Cymbidium atropurpureum

(Lindley) Rolfe

Section: ***Cymbidium*** P. Hunt

Dr. John Lindley first described this dark-flowered species in 1854 as ***Cymbidium pendulum*** var. ***atropurpureum***. Robert Rolfe decided in 1903 that this plant was worthy of specific status and named the plant ***Cymbidium atropurpureum*** in *The Orchid Review.* The specific name refers to the colour of the blooms (dark purple).

Growth habit: upright, sympodial. **Pseudobulbs:** not very pronounced. **Leaves:** very leathery, up to 60 cm long by 2 cm wide. **Inflorescences:** pendulous, up to one metre long with up to 30 flowers; blooms about 3 cm in diameter. **Flower colour:** very dark purple; labellum white with a few purple spots. **Dorsal** and **lateral sepals:** linear-oblong, up to 1.5 cm long by 6 mm wide. **Petals:** of similar shape and size to the dorsal and lateral sepals. **Labellum:** three-lobed, up to 2.5 cm long by 1.5 cm wide; lateral lobes erect, shorter than column; apices obtuse; mid-lobe broadly ovate.

Habitat and Distribution: ***Cymbidium atropurpureum*** is found in Thailand, Peninsular Malaysia, Sumatra, Java and Borneo. In the Philippines it has been found in the provinces of Benguet, Laguna, Mountain Province and Rizal on Luzon; the island of Leyte in the Visayas; the provinces of Bukidnon, Davao del Norte, Davao del Sur, Surigao del Norte and Surigao del Sur on Mindanao; and Tawitawi in the Sulu Archipelago. It usually grows as an epiphyte, occasionally as a lithophyte, at elevations to 2,200 metres. The majority of the plants I have seen in the wild were growing in full sun very high in trees.

Cymbidium dayanum Reichenbach *f.*

Section: ***Himantophyllum*** Schlechter

Professor Heinrich G. Reichenbach named this plant in 1869; the name was published in the *Gardeners' Chronicle*. It was named after J. Day who collected the first plants in India.

Growth habit: upright, sympodial. **Pseudobulbs:** small, not very pronounced. **Leaves:** up to 60 cm long by 1.5 cm wide. **Inflorescences:** pendulous, carry up to 12 flowers approximately 3 cm in diameter. **Flower colour:** white with a red line down the centre of each segment; labellum dark red with red and yellow striping at the back. **Dorsal** and **lateral sepals:** elliptic-lanceolate, up to 2.5 cm long by 5 mm wide. **Petals:** smaller than the sepals but of similar shape, up to 1.5 cm long by 4 mm wide. **Labellum:** three-lobed, up to 2.2 cm long by 1.5 mm wide (when flattened); lateral lobes erect, as long as the column; mid-lobe ovate, recurved.

Habitat and Distribution: ***Cymbidium dayanum*** is widely distributed, being found in India, China, Taiwan, Japan, Thailand, Indochina, Peninsular Malaysia, Sumatra and Sabah. In the Philippines it has been found in the provinces of Benguet, Mountain Province and Nueva Vizcaya on Luzon. It usually grows as a terrestrial but is also found in hollow logs on trees at elevations between 300 and 1,800 metres.

Photo: David Titmuss

Cymbidium dayanum

Cymbidium finlaysonianum

Lindley

Section: ***Cymbidium*** P. Hunt

Dr. John Lindley named this plant in 1833 and the name was published in the *Genera and Species of Orchidaceous Plants*. He named it after a Mr. Finlayson who collected the first plants in Vietnam.

Growth habit: upright, sympodial. **Pseudobulbs:** not very distinct. **Leaves:** very leathery and stiff, up to 1 metre in length by 5 cm wide. **Inflorescences:** pendulous, up to 1.4 metres in length; up to 40 flowers up to 4 cm in diameter. **Flower colour:** dull yellow; labellum white with purple and yellow marks. **Dorsal** and **lateral sepals:** linear-oblong, up to 2 cm long by 5 mm wide. **Petals:** oblong-elliptic, a little shorter than the sepals. **Labellum:** three-lobed, up to 2.8 cm long by 1.8 cm wide (when flattened); lateral lobes erect, longer than the column; mid-lobe broadly elliptic.

Habitat and Distribution: ***Cymbidium finlaysonianum*** is widely distributed, being found in Indochina, Thailand, Peninsular Malaysia and Indonesia. In the

Photo: Jim Cootes

Cymbidium finlaysonianum

Photo: Elias Javier

Cymbidium pubescens

Philippines it has been recorded from the provinces of Albay, Bataan, Benguet, Quezon and Rizal on Luzon; the island of Polillo off the coast of Luzon; Bohol in the Visayas; the island of Dinagat north of Mindanao; the provinces of Surigao del Norte, Surigao del Sur, Zamboanga del Norte and Zamboanga del Sur on Mindanao; and the island of Tawitawi in the Sulu Archipelago. It grows as an epiphyte and occasionally as a lithophyte. It is found at elevations from sea level to 300 metres, usually growing in very well-lit situations.

Cymbidium pubescens Lindley

Section: ***Cymbidium*** P. Hunt

Dr. John Lindley named this species in 1840 in the *Botanical Register*. The specific name alludes to the downy nature of the labellum.

Growth habit: upright, sympodial. **Pseudobulbs:** not very pronounced. **Leaves:** leathery, up to 50 cm long, 2 cm wide. **Inflorescences:** pendulous; carry up to 20 blooms about 4 cm in diameter. **Flower colour:** red with yellowish margins; labellum creamy yellow with red spotting. **Dorsal** and **lateral sepals:** oblanceolate, up to 2 cm long, 5 mm wide. **Petals:** lanceolate, up to 2 cm long, 4 mm wide. **Labellum:** three-lobed, up to 1.8 cm long, 1.5 cm wide (flattened); lateral lobes equal in length to column, apices pointed; mid-lobe elliptic, with tiny hairs on both surfaces.

Habitat and Distribution: *Cymbidium pubescens* is widely distributed, being found in Peninsular Malaysia, Sumatra, Java, Borneo and Sulawesi (Celebes). In the Philippines this plant has been recorded from Agusan del Norte and Agusan del Sur on Mindanao. It grows as an epiphyte at elevations to 300 metres.

Cystorchis Blume

Carl Blume established this genus of terrestrial orchids in 1858 in *Florae Javae*. The generic name refers to the inflated base of the labellum.

The type species for this genus is ***Cystorchis variegata*** Bl.

There are approximately eight species in the genus and they have been found in China, Peninsular Malaysia, Indonesia, Borneo and New Guinea. The Philippines has three species.

Cystorchis luzonensis Ames

Professor Oakes Ames named this species in 1915 in Fascicle V of his *Studies in the Family Orchidaceae*. The specific name refers to the island of Luzon.

Growth habit: upright, sympodial. **Pseudobulbs:** none; the plant grows from a rhizome to a height of about 25 cm. **Leaves:** ovate-lanceolate, up to 10, up to 2.5 cm long by 1.5 cm wide, purplish in colour. **Inflorescences:** up to 15 cm long, white, hairy, bear up to six blooms. **Flower colour:** yellowish white. **Dorsal sepal:** lanceolate, some hairs on the outside, erect, 6 mm long by 2 mm wide, adhering to petals. **Petals:** linear, about 1 mm wide. **Lateral sepals:** oblong, 7 mm long by 2.5 mm wide, some hairs on the outside. **Labellum:** erect, short conical spur protrudes between lateral sepals.

Habitat and Distribution: *Cystorchis luzonensis* is endemic to the Philippines and has only been recorded from the province of Laguna on Luzon and the province of Oriental Mindoro on the island of Mindoro. It grows as a terrestrial in shaded damp localities at elevations between 800 and 1,600 metres.

Photo: Dr. Henrik Pedersen

Cystorchis luzonensis

Dendrobium Swartz

Olof Swartz named the genus ***Dendrobium*** in 1799. The name was published in *Nova Acta Societatis Scientiarum Upsaliensis*. The generic name refers to the epiphytic habit of the majority of the species. Current estimates as to the number of species in the genus vary from between 800 and 1,100 members.

The type species for this large, commonly grown genus is ***Dendrobium moniliforme*** (L.) Sw.

Dendrobiums are one of the most widely distributed orchid genera. They are found as far north as Korea and Japan, throughout China, west to India and Sri Lanka, down through Peninsular Malaysia, Indonesia, Borneo, the Philippines, Guam, New Guinea (which has the most species of any country) to Australia and as far east as Tahiti in the Pacific Ocean. There are at least 90 species found in the Philippines.

Dendrobium aloifolium

(Blume) Reichenbach *f.*
Section: ***Aporum*** (Blume) Lindley

Carl Blume first described this species as ***Macrostomium aloefolium*** in 1825. Professor Heinrich G. Reichenbach transferred the generic name in 1861 in *Walpers Annales Botanices*. The specific name refers to the aloe-like foliage of this plant.

Growth habit: semi-pendulous, sympodial. **Stems:** flattened, up to 35 cm long. **Leaves:** triangular, up to 2.5 cm long, 6 mm wide. **Inflorescences:** bear single

Photo: David Titmuss

Dendrobium aloifolium

Photo: David Titmuss

Dendrobium amethystoglossum

Photo: Jim Cootes

Dendrobium anosmum

flowers about 3 mm in diameter. **Flower colour:** creamy white. **Dorsal sepal:** oblong, up to 2 mm long, 1.5 mm wide. **Petals:** linear, up to 2 mm long, 1 mm wide; apart from labellum, all segments reflex badly. **Lateral sepals:** triangular, up to 3 mm long, 2 mm wide. **Labellum:** straight, up to 3 mm long, 1.5 mm wide.

Habitat and Distribution: ***Dendrobium aloifolium*** is widespread, being found in Thailand, Peninsular Malaysia, Indonesia and Borneo. In the Philippines I have seen this plant growing in Nueva Vizcaya on Luzon, where it grows in quite bright light. Dr. Valmayor, author of *Orchidiana Philippiniana*, does not list any localities for this species in the Philippines. It grows as an epiphyte at elevations up to 300 metres.

Dendrobium amethystoglossum Reichenbach *f.*

Section: ***Calcarifera*** J.J. Smith

Professor Heinrich G. Reichenbach named this beautiful species in the *Gardeners' Chronicle* in 1872. The specific name refers to the colour of the labellum.

Growth habit: upright, sympodial. **Pseudobulbs:** up to 1 metre tall and 2 cm in diameter. **Leaves:** up to 10 cm long by up to 2 cm wide, striped with purple when young. **Inflorescences:** semi-pendulous, appear from the leafless pseudobulbs, carry up to 20 blooms about 3 cm in diameter. **Flower colour:** milky white; labellum amethyst-purple. **Dorsal sepal** and **petals:** lanceolate, up to 1.2 cm long by 4 mm wide. **Lateral sepals:** lanceolate, join to form a spur about 1.5 cm long. **Labellum:** lanceolate, about 2 cm long by 4 mm wide.

Habitat and Distribution: ***Dendrobium amethystoglossum*** is endemic to the mountains of Benguet, the Mountain Province, Nueva Ecija, Nueva Vizcaya and Pangasinan in northern Luzon. During a recent trip to southern Mindoro (September 1999) I saw plants of ***Dendrobium amethystoglossum*** being sold by the Mangyans, the original natives of Mindoro. This is a long extension range for this attractive species. It grows as an epiphyte, usually in quite bright light at elevations over 1,000 metres.

Dendrobium anosmum Lindley

Section: ***Dendrobium*** Swartz

Synonym: ***Dendrobium superbum*** Rchb. *f.*

Dr. John Lindley named this species in the *Botanical Register* in 1845. The specific name means without scent, which is a bit of a misnomer as this plant has a strong perfume.

Growth habit: pendulous, sympodial. **Pseudobulbs:** up to 1.5 metres long by 1 cm in diameter. **Leaves:** deciduous, last only 12 months, up to 7.5 cm long by 1.5 cm wide. **Inflorescences:** short, appearing from the leafless nodes, bear up to three flowers to 7.5 cm across petals. **Flower colour:** normally pale purple with two darker purple blotches on labellum. **Dorsal** and **lateral sepals:** lanceolate, up to 3.5 cm long by 1.5 cm wide. **Petals:** ovate, up to 3.5 cm long by 2 cm wide. **Labellum:** circular (if flattened), up to 5 cm long by 3.5 cm wide with a pointed tip, hairy within.

There are two well-known varieties of this strikingly beautiful species:

1) ***Dendrobium anosmum*** var. ***dearei*** is the albino variety.

2) ***Dendrobium anosmum*** var. ***huttonii*** has white petals and sepals and purple markings in the labellum.

Both varieties are similar to the type species in size. Each flowering node can produce several blooms, and a plant that is a metre long with 20 flowering nodes is a spectacular sight.

Habitat and Distribution: ***Dendrobium anosmum*** is widely distributed and is found in Thailand, the Malay Archipelago, Indonesia, Borneo and New Guinea. In the Philippines it has been found in Abra, Benguet, Mountain Province and Nueva Vizcaya on Luzon and Davao del Sur in Mindanao. Variety ***dearei*** only occurs in the Philippines and has been found in Rizal and Mountain Province on Luzon. The variety ***huttonii*** is found in the Malay Archipelago, Papua New Guinea and the Philippines, where it is very rare and has only been recorded from Rizal on Luzon. It grows as an epiphyte at elevations to 1,000 metres.

Dendrobium auriculatum

Ames and Quisumbing
Section: ***Dolichocentrum*** Schlechter

Professor Oakes Ames and Dr. Eduardo Quisumbing named this species in 1932 in the *Philippine Journal of Science*. The specific name refers to ear-like appendages, which in this species are at the base of leaves.

Growth habit: upright, sympodial. **Stems:** up to 80 cm tall and 2 mm in diameter, usually erect and never branching. **Leaves:** up to 12 cm long by 6 mm wide, very pointed. **Inflorescences:** single flowered; blooms very showy, about 3.5 cm in diameter. **Flower colour:** white with a short green spur. **Dorsal sepal:** ovate, up to 2.2 cm long by 1.1 cm wide. **Petals:** oblong, up to 2 cm long by 8 mm wide. **Lateral sepals:** oblong-lanceolate, joined at their base to form a spur about 1 cm long; overall length 3.3 cm by 1.1 cm wide. **Labellum:** simple, 3.4 cm long by 1.7 cm wide (at the

Photo: Elias Javier

Dendrobium anosmum var. ***dearei***

Photo: Kelvin Manubay

Dendrobium anosmum var. ***huttonii***

Photo: David Banks

Dendrobium auriculatum

Photo: Hans Fessel

Dendrobium balzerianum

widest point when flattened), lower third circular; the front portion broadly heart-shaped when flattened.

Habitat and Distribution: ***Dendrobium auriculatum*** is endemic to the Philippines where it is found in the provinces of Bulacan, Mountain Province, Nueva Ecija and Nueva Vizcaya on Luzon and in the province of Davao del Sur on Mindanao at elevations of about 900 metres. It grows as either an epiphyte or a lithophyte.

Notes: This species is often confused with ***Dendrobium papilio*** Loher.

Dendrobium balzerianum

Fessel and Lückel
Section: ***Fugacia*** J.J. Smith

Hans Fessel and Emil Lückel named this species in the July-August 1997 issue of *Die Orchidee.* It is named in honour of Peter Balzer who has added much to our knowledge of the orchids of the island of Leyte.

Growth habit: semi-pendulous, sympodial. **Pseudobulbs:** clavate (narrow at the base, gradually thickening upward), square in section, up to 16 cm long by 1 cm across, bifoliate. **Leaves:** lanceolate, up to 10 cm long by 2.2 cm wide. **Inflorescences:** single-flowered. **Flower colour:** sepals and petals white; labellum white with pink markings, orange blotch at the base. **Dorsal sepal:** ovate, tip pointed, 1.5 cm long by 4.5 mm wide. **Petals:** subspathulate, tips pointed, 1 cm long by 3 mm wide. **Lateral sepals:** broadly ovate, tips pointed, 1.2 cm long by 7 mm wide. **Labellum:** three-lobed, all lobes circular; lateral lobes of similar size; mid-lobe slightly larger, 1 cm long by 1 cm wide.

Habitat and Distribution: ***Dendrobium balzerianum*** is found on Leyte island in the Visayas where it grows as an epiphyte at elevations of about 500 metres.

Notes: The illustration of this species has the labellum forced down to show its features.

Dendrobium bicaudatum

Reinwardt *ex* Lindley
Section: ***Spatulata*** Lindley

Reinwardt first introduced this species to orchid growers by an illustration. Dr. John Lindley formally named this species in 1859 in the *Journal of the Linnean Society.* The specific name refers to two tails, or possibly the petals.

Photo: Gary Yong Gee

Dendrobium bicaudatum

Photo: Hans Fessel

Dendrobium blumei

Growth habit: upright, sympodial. **Pseudobulbs:** terete, up to 30 cm long by 9 mm in diameter. **Leaves:** leathery, ovate, up to 10 cm long by 2.3 cm wide, at the top of the pseudobulb. **Inflorescences:** usually upright, bear up to eight blooms. **Flower colour:** dorsal and lateral sepals yellowish green; petals brownish; labellum white, striped with purple. **Dorsal sepal:** triangular to ovate, 1.6 cm long by 5.5 mm wide. **Petals:** linear, 2.8 cm long by 3 mm wide. **Lateral sepals:** triangular to falcate, 1.8 cm long by 8 mm wide. **Labellum:** three-lobed; side lobes triangular; mid-lobe oval, central portion constricting to a point.

Habitat and Distribution: ***Dendrobium bicaudatum*** is widely distributed and has been found in northern Sulawesi (Celebes) and the Moluccas. In the Philippines it has been found in Negros Oriental on Negros island in the Visayas and in Bukidnon on Mindanao. It grows as an epiphyte at elevations up to 300 metres.

Dendrobium blumei Lindley

Section: ***Crumenata*** Pfitz.

Dr. John Lindley named this species after Carl Blume, a botanist of the early to mid-1800s, in his *Genera and Species of Orchidaceous Plants*, published in 1830.

Growth habit: upright, sympodial. **Pseudobulbs:** up to 30 cm tall; narrow at the base, swollen for the next two nodes; somewhat flattened, then slender for the remainder of the pseudobulb. **Leaves:** leathery, alternate, lanceolate, up to 5 cm long, 1 cm wide. **Inflorescences:** single-flowered, grow from near top of stem. **Flower colour:** white; segments tipped with purple; labellum white striped with purple, pale yellow blotch near mid-lobe. **Dorsal sepal:** ovate, 7 mm long, 4 mm wide. **Petals:** linear, slightly notched around margin, 7 mm long, 3 mm wide. **Lateral sepals:** triangular, 1.2 cm long, 5 mm wide. **Labellum:** three-lobed; side lobes circular; mid-lobe fringed, 1.2 cm long by 6 mm wide.

Habitat and Distribution: ***Dendrobium blumei*** is widely distributed. It has been recorded from Borneo, Java, Peninsular Malaysia, Sumatra and Thailand. In the Philippines it has been found on Samar island in the Visayas; the provinces of Agusan del Norte, Agusan del Sur, Bukidnon, Lanao del Norte, Lanao del Sur, Surigao del Norte, Surigao del Sur, Zamboanga del Norte and Zamboanga del Sur on Mindanao; and the island of Basilan. It grows as an epiphyte, sometimes in swamps, at elevations to 1,000 metres.

Photo: David Banks

Dendrobium bullenianum

Dendrobium bullenianum

Reichenbach *f.*

Section: ***Pedilonum*** (Blume) Lindley

Professor Heinrich G. Reichenbach named this species in 1852 in *Botanische Zeitung*, after Mr. Bullen, the first person in Britain to flower the plant.

Growth habit: pendulous, sympodial. **Pseudobulbs:** up to 1 metre long, 1 cm wide. **Leaves:** deciduous, up to 10 cm long by 1.5 cm wide. **Inflorescences:** pendulous, short, appear at the leafless nodes, can bear up to 25 blooms 1.5 cm in diameter. **Flower colour:** bright orange to yellow; petals and labellum striped with red. **Dorsal sepal:** lanceolate, up to 8 mm long by 3.5 mm wide. **Petals:** lanceolate, up to 8 mm long by 3 mm wide. **Lateral sepals:** ovate, joined at their base to form a short spur about 1 cm long; overall length 1.7 cm by 5 mm wide. **Labellum:** elliptic, about 1.6 cm long by 5 mm wide.

Habitat and Distribution: ***Dendrobium bullenianum*** is endemic to the Philippines where it has been found in Camarines Sur, Isabela and Quezon on Luzon; Leyte island in the Visayas; and Bucas Grande island near Mindanao. It grows as an epiphyte at elevations to 300 metres.

Dendrobium carinatum

(Linnaeus) Willdenow

Section: ***Crumenata*** Pfitz.

This species has had a long taxonomic life. It was first named ***Bontiana luzonica*** by Petiver in *Gazophylacium* in the early 1700s, then Carl Linnaeus named it ***Epidendrum carinatum*** in his classic publication *Species Plantarum*. In 1805 Karl Willdenow made the generic transfer to ***Dendrobium***. The specific name refers to the keels (carinae) in the labellum.

Growth habit: upright, sympodial. **Pseudobulbs:** up to 10 cm long, three to four nodes, total length of plant over 40 cm. **Leaves:** terete, up to 10 cm long by 2 mm wide. **Inflorescences:** single-flowered, at the junction of the leaves and stems. **Flower colour:** cream to yellow with reddish striations; labellum similar. **Dorsal sepal:** triangular, 1.2 cm long by 6 mm wide. **Petals:** oblong to lanceolate, 1.1 cm long by 3 mm wide. **Lateral sepals:** triangular, 1.4 cm long by 1.2 cm wide, curve slightly forward to form a spur. **Labellum:** three-lobed, 1.7 cm long by 1.2 cm wide (when flattened); mid-lobe circular; lateral lobes enclose the column; two keels extend back toward the column.

Habitat and Distribution: ***Dendrobium carinatum*** is endemic to the Philippines and has been found in the provinces of Mountain Province, Nueva Vizcaya, Rizal and Zambales on Luzon and Oriental Mindoro on Mindoro island. It grows as an epiphyte to 1,000 metres.

Dendrobium ceraula Reichenbach *f.*

Section: ***Calcarifera*** J.J. Smith

Professor Heinrich G. Reichenbach named this species in *Linnaea* in 1876. The specific name refers to the circle formed by the spur, as *ceraula* means trumpeter or horn-blower.

Growth habit: pendulous, sympodial. **Pseudobulbs:** up to 30 cm long by 7 mm in diameter; each node tapers and size increases as the plant gets longer. **Leaves:** deciduous, 7 cm long by 2 cm wide. **Flower colour:** extremely variable, ranging from pink to pale blue to creamy white; labellum has maroon stripes; flowers up to 3.5 cm in diameter, each flowering node bears up to four blooms. **Dorsal sepal:** narrowly oblong, 2.3 cm long by 7 mm wide, tapers to a point. **Petals:** lanceolate, up to 2.7 cm long by 9 mm wide. **Lateral sepals:** linear-lanceolate, 2.7 cm long by 8 mm wide, joined at the base to form a curved spur about 1.5 cm long. **Labellum:** narrowly oblong, about 3.3 cm long by 1 cm wide.

Photo: David Banks

Dendrobium carinatum

Photo: David Banks

Dendrobium ceraula

Photo: David Banks

Dendrobium ceraula (white)

Habitat and Distribution: ***Dendrobium ceraula*** is endemic to the Philippines in the southern province of Albay on Luzon where it grows as an epiphyte at elevations above 1,300 metres.

Notes: ***Dendrobium ceraula*** is an earlier name for the species better known as ***Dendrobium gonzalesii*** Quis.

Photo: David Banks

Dendrobium cerinum

Dendrobium cerinum Reichenbach *f.*

Section: ***Calcarifera*** J.J. Smith

Professor Heinrich G. Reichenbach named this charming species in the *Gardeners' Chronicle* in 1879. The specific name refers to the waxy appearance of the flowers.

Growth habit: upright to semi-pendulous, sympodial. **Pseudobulbs:** up to 40 cm long by 1 cm wide. **Leaves:** lanceolate, to 10 cm long by 3 cm wide, tinged with purple when young, **Inflorescences:** pendulous, appear from the leafless pseudobulbs, carry up to eight blooms about 3.5 cm in diameter. **Flower colour:** shiny yellow with olive, pale purple and green markings in the spur; some clones have red markings in the labellum, others are plain yellow. **Dorsal sepal:** lanceolate, up to 1.8 cm long by 8 mm wide. **Petals:** oblong-elliptic, up to 1.8 cm long by 1 cm wide. **Lateral sepals:** lanceolate, join to form a spur about 2 cm long; overall length 4 cm by 9 mm wide. **Labellum:** simple, up to 3 cm long by 1.6 cm wide, edge is scalloped.

Photo: David Banks

Dendrobium cerinum fma. *album*

Photo: David Titmuss

Dendrobium chameleon

Dendrobium cerinum fma. *album* Hort.

This variety has not been formally described. The plant is the same as the species, but the flowers lack colour in the labellum.

Habitat and Distribution: ***Dendrobium cerinum*** and the variety appear to be endemic to the Philippines where they have been recorded from the province of Rizal on Luzon. It grows as an epiphyte at altitudes of approximately 800 metres.

Notes: Several recent authors, G. Seidenfaden and J.J. Wood (1992), Jim Comber (1990) and G. Seidenfaden (1985), all list ***Dendrobium cerinum*** as a synonym of ***Dendrobium sanguinolentum*** Lindl. I feel that this is incorrect because the plants of both species are quite distinct. ***Dendrobium sanguinolentum*** is a very large growing plant—at least a metre long in mature specimens—where ***Dendrobium cerinum*** barely reaches 40 cm in height. The colour of the flowers is also different. ***Dendrobium sanguinolentum*** has blooms that are salmon pink with purple tips to all the floral segments, but there are clones with no purple tips. The flowers of ***Dendrobium cerinum*** are usually shiny yellow. I admit that the flowers have similarities; at worst ***Dendrobium cerinum*** should be considered a variety of the earlier named ***Dendrobium sanguinolentum,*** not a synonym.

Dendrobium chameleon Ames

Section: ***Calcarifera*** J.J. Smith

Professor Oakes Ames named this orchid in 1908. The name was published in *Studies in the Family Orchidaceae*, Fascicle II. The specific name is from the Greek word *chamaileon*, which is a lizard that can change its colour, in reference to the flowers that are pale yellow to white with reddish purple stripes. R.S. Williams collected the type specimens in Benguet province on the island of Luzon in September 1904.

Growth habit: pendulous, sympodial. **Pseudobulbs:** much branched and tapering, up to 60 cm long and 7 mm wide. **Leaves:** up to 7 cm long by 2 cm wide, deciduous. **Inflorescences:** short, appear from the leafless nodes; each node produces several blooms about 3 cm in diameter. **Flower colour:** pale yellow to white, striped with reddish purple. There is also an albino form of this species that is rare and seldom seen. **Dorsal sepal:** lanceolate, up to 1.5 cm long by 8 mm wide. **Petals:** oblong, up to 2 cm long by 5 mm wide, arch toward the column. **Lateral sepals:** lanceolate, join to form a spur about 3 cm long. **Labellum:** elliptic-oblong, up to 2 cm long by 4 mm wide.

Habitat and Distribution: ***Dendrobium chameleon*** occurs in the mountains of southern Taiwan and in Benguet, Laguna, Mountain Province, Pampanga and Sorsogon on Luzon. It grows as an epiphyte in fairly shaded locations at elevations over 1,000 metres.

Dendrobium chameleon var. *ovatilabium* Ames and Quisumbing

Professor Oakes Ames and Dr. Eduardo Quisumbing named this variety in *The Philippine Journal of Science* in 1931. The specific name refers to the ovate labellum.

Growth habit: pendulous, sympodial. **Pseudobulbs:** much branched and tapering, up to 60 cm long, 7 mm in diameter. **Leaves:** up to 7 cm long, 2 cm wide, deciduous. **Inflorescences:** short, appear from the leafless nodes, each node produces several blooms about 3 cm in diameter. **Flower colour:** pale yellow to white, without striping. **Dorsal sepal:** lanceolate, up to 1.5 cm long, 8 mm wide. **Petals:** oblong, up to 2 cm long, 5 mm wide, arch toward the column. **Lateral sepals:** lanceolate, join to form a spur about 3 cm long. **Labellum:** ovate, up to 2 cm long, 4 mm wide.

Habitat and Distribution: ***Dendrobium chameleon*** var. ***ovatilabium*** occurs in the mountains of Benguet and Pampanga on Luzon. It grows as an epiphyte in fairly shaded locations at elevations over 1,000 metres.

Dendrobium conanthum

Schlechter
Section: ***Spatulata*** Lindley

Dr. Rudolf Schlechter first described this species in 1912 in *Fedde, Repertorium Specierum Novarum, Beihefte.* The specific name means standing tightly, possibly referring to the upright petals.

Growth habit: upright, sympodial. **Pseudobulbs:** up to 2 metres tall by 1 cm in diameter. **Leaves:** leathery, elliptic, up to 7.5 cm long by 4.5 cm wide. **Inflorescences:** upright, up to 50 cm long, bear up to 50 blooms. **Flower colour:** dorsal and lateral sepals cream with darker veining; petals brownish; labellum pale brown with purple veining. **Dorsal sepal:** oblong-lanceolate, 2.3 cm long by 9 mm wide. **Petals:** linear, upright, twisted, up to 3.3 cm long by 5 mm wide. **Lateral sepals:** triangular, up to 2.3 cm long by 1.1 cm wide. **Labellum:** three-lobed, 2.7 cm long by 2 cm wide (when flattened); side lobes semi-ovate; mid-lobe square.

Habitat and Distribution: ***Dendrobium conanthum*** is a widely distributed species recorded from Papua New Guinea, the Solomon Islands, Vanuatu, the Bismarck Archipelago and the Admiralty Islands. In the Philippines it has been found on the islands of Palawan, Busuanga and Calamianes. It grows as an epiphyte at elevations to 300 metres, usually in very bright light.

Dendrobium crumenatum Swartz

Section: ***Crumenata*** Pfitz.

Olof Swartz named this plant in 1799 in *Schrader's Journal fur die Botanik.* The specific name refers to the purse-shaped spur of the flower.

Growth habit: upright, sympodial. **Pseudobulbs:** lower portion swollen for three or four nodes; middle portion carries the leaves; top portion carries the flowers; overall up to 1 metre in length. **Leaves:** very leathery, about 7 cm long by 2 cm wide. **Flower**

Photo: David Titmuss

Dendrobium chameleon var. ***ovatilabium***

Photo: Danny Tiu

Dendrobium conanthum

colour: pure white with bright yellow markings in the labellum; blooms about 2.5 cm in diameter, sweetly perfumed. **Dorsal sepal:** oblong lanceolate, up to 1.5 cm long by 7 mm wide. **Petals:** oblanceolate, 1.5 cm long by 8 mm wide. **Lateral sepals:** lanceolate, join to form a short spur. **Labellum:** three-lobed, scalloped edges.

Photo: David Banks

Dendrobium crumenatum

Photo: David Banks

Dendrobium cultratum

Habitat and Distribution: ***Dendrobium crumenatum*** is widely spread throughout Asia, being found in India, Sri Lanka, the Andaman Islands, Myanmar (Burma), Thailand, Indochina, Peninsular Malaysia, Indonesia, Borneo, Sulawesi (Celebes) and Taiwan. In the Philippines this species is probably the most widely spread of any orchid. I have visited almost half of the provinces in the Philippine archipelago and have always seen ***Dendrobium crumenatum*** growing somewhere. Frequently this orchid colonises coconut plantations and is a commonly cultivated garden plant. In northeast Mindoro this is the most commonly found orchid in the forest. It grows in almost every locality from full sun to deep shade. The plant often produces small plants along old inflorescences.

Notes: This orchid flowers nine days after a sudden drop in temperature, usually after a thunder storm. Blooms only last one day.

Dendrobium cultratum Schlechter

Section: ***Oxystophyllum*** (Blume) Miquel

Dr. Rudolf Schlechter named this small-growing plant in 1911 in *Feddes Repertorium Speciarum Novarum Regnis Vegetabilis.* The specific name means shaped like a knife blade and possibly refers to the leaves.

Growth habit: upright, sympodial, up to 6 cm tall. **Leaves:** two-ranked, fleshy, sharply pointed, up to 2.5 cm long by 4 mm wide. **Inflorescences:** short; flowers appear from near the top of the plant from dried bracts. **Flower colour:** dark red to claret. **Dorsal sepal:** lanceolate, up to 3 mm long by 1.5 mm wide. **Petals:** oblong, up to 3 mm long by 1 mm wide. **Lateral sepals:** triangular, join to form a small spur about 4 mm long. **Labellum:** linear-oblong, fringed margin.

Habitat and Distribution: ***Dendrobium cultratum*** is widely spread and has been found in Myanmar (Burma), Peninsular Malaysia, Borneo, Sulawesi (Celebes) and Papua New Guinea. In the Philippines it has been found in Apayao, Bataan, Ifugao, Kalinga, Nueva Ecija, Pangasinan and Rizal on Luzon; on the islands of Negros (Occidental Negros and Oriental Negros provinces) and Panay in the Visayas; Culion and Polillo islands; Basilan and Sibutu islands in the Sulu Archipelago; and Agusan del Norte, Agusan del Sur, Bukidnon, Surigao del Norte, Surigao del Sur, Zamboanga del Norte and Zamboanga del Sur on Mindanao. It grows as an epiphyte at elevations to 1,000 metres, usually in fairly bright light.

Photo: David Banks

Dendrobium dearei

Photo: David Banks

Dendrobium distichum

Dendrobium dearei Reichenbach *f.*

Section: ***Formosae*** (Bentham and Hooker *f.*) Hooker *f.*

Professor Heinrich G. Reichenbach named this species in the *Gardeners' Chronicle* in 1882. He named it after Colonel Deare, who collected the first plants.

Growth habit: upright, sympodial. **Pseudobulbs:** up to 90 cm tall by 1 cm in diameter, always upright. **Leaves:** leathery, up to 7.5 cm long by 1.5 cm wide. **Inflorescences:** semi-pendulous; bear up to 10 long-lasting, shapely blooms about 5 cm in diameter. Some clones flower several times a year. **Flower colour:** pure white; apple-green markings in labellum. **Dorsal** and **lateral sepals:** lanceolate, taper to a point, up to 3 cm long by 1 cm wide. **Petals:** oblong-ovate, about 2 cm long by 1.7 cm wide. **Labellum:** three-lobed; lateral lobes frame the column.

Habitat and Distribution: ***Dendrobium dearei*** is endemic to the Philippines where it is found along the eastern coast of the entire archipelago growing as an epiphyte at elevations to 500 metres.

Notes: This species is almost extinct in parts of its native habitat due to over-collection and the activities of illegal loggers.

Dendrobium distichum

(Presl) Reichenbach *f.*

Section: ***Aporum*** (Blume) Lindley

Karel Presl originally described this species as ***Schizmoceras disticha*** in 1827. Professor Heinrich G. Reichenbach gave the plant its currently accepted name in *Linnaea*, published in 1876. The specific name refers to the leaves, which are arranged in two opposite rows (distichous).

Growth habit: semi-pendulous, sympodial. **Stems:** flattened, up to 40 cm long by 3 cm wide across the leaves. **Leaves:** triangular, clasp the stem. In profile the foliage has the appearance of saw teeth. **Inflorescences:** terminal; flowers, about 7 mm in diameter, appear from dried bracts. **Flower colour:** greenish yellow with red striping in the dorsal and lateral segments. **Dorsal sepal:** oblong-lanceolate, up to 3 mm long by 1.5 mm wide. **Petals:** oblong, about 3 mm long by 1 mm wide. **Lateral sepals:** ovate, up to 3 mm long by 2 mm wide. **Labellum:** creamy white, noticeable callus.

Habitat and Distribution: ***Dendrobium distichum*** is endemic to the Philippines where it is found in the provinces of Albay, Bataan, Benguet, Camarines Norte, Camarines Sur, Ifugao, Laguna, Quezon, Rizal, Sorsogon and Zambales on Luzon; the island of Polillo; the provinces of Surigao del Norte and Surigao del Sur on Mindanao; and in the Sulu Archipelago. It grows as an epiphyte at elevations from sea level (in mangrove swamps) to about 900 metres.

Notes: I feel this species is closely related to ***Dendrobium indivisum*** (Bl.) Miq. and may prove to be synonymous on further investigation.

Photo: David Titmuss

Dendrobium distichum fma. *album*

Photo: David Banks

Dendrobium equitans

Photo: David Titmuss

Dendrobium fairchildiae

Dendrobium distichum fma. *album* Hort.

Section: ***Aporum*** (Blume) Lindley

This form has not been formally described. The plant is the same as the species, but the flowers lack the coloured stripes in the sepals and petals.

Habitat and Distribution: *Dendrobium distichum* fma. ***album*** is a rare plant and its habitat and distribution are unknown.

Dendrobium equitans Kränzlin

Section: ***Crumenata*** Pfitz.

Dr. Fritz Kränzlin named this plant in 1910 in A. Engler (ed.) *Das Pflanzenreich. Regni vegetabilis conspectus* IV 50 II B 21. The specific name refers to the habit of the foliage of this species.

Growth habit: upright, sympodial. **Stems:** very slender for two to three nodes, which thicken for a single node to make a small **pseudobulb** and then narrow again for a length of up to 25 cm. **Leaves:** leathery, two-ranked, pointed, about 5 cm long by 5 mm wide, flattened. **Inflorescences:** short; the single flowers appear on the leafless portion of the stems, at or near the top of the pseudobulb. **Flower colour:** white; yellow blotch in the base of the labellum. **Dorsal sepal** and **petals:** oblong-lanceolate, 1.5 cm long by 4 mm wide. **Lateral sepals:** oblong-lanceolate, 2 cm long by 5 mm wide, join at base to form a spur 8 mm long. **Labellum:** three-lobed, 1.6 cm long by 5 mm wide; mid-lobe deeply fringed.

Habitat and Distribution: *Dendrobium equitans* is found in the Batanes islands north of Luzon, and also in Taiwan. It grows as an epiphyte or lithophyte at elevations of about 600 metres.

Dendrobium fairchildiae

Ames and Quisumbing

Section: ***Calcarifera*** J.J. Smith

Professor Oakes Ames and Dr. Eduardo Quisumbing named this plant in the *Philippine Journal of Science* in 1932, after a Mrs. Fairchild, who found the first plants of this magnificent species.

Growth habit: upright to semi-pendulous, sympodial. **Pseudobulbs:** up to 1.8 metres long by 1 cm in diameter. **Leaves:** up to 15 cm long by 2 cm wide, deciduous after one season. **Inflorescences:** semi-pendulous, appear from the leafless nodes, bear up to nine blooms 4 cm in diameter. **Flower colour:** white; tip of each segment pale pink to purplish. **Dorsal sepal:** oblong to elliptic, 2.2 cm long by 1.2 cm wide. **Petals:** oblong-elliptic, up to 2.2 cm long by 1.1 cm wide. **Lateral sepals:** triangular, join to form a spur about 3 cm long that is bent at about 80 degrees for the final 1 cm; overall length 4.7 cm by 1.5 cm wide. **Labellum:** obovate, 3.8 cm long by 2 cm wide; lateral lobes upright.

Dendrobium fairchildiae fma. *album* Hort.

This form has not been formally described. Plant as per species, but the flowers are pure white with some yellow in the base of the labellum.

Habitat and Distribution: ***Dendrobium fairchildiae*** is endemic to the mountains of Mountain Province, Nueva Ecija and Rizal provinces on Luzon and Catanduanes island at elevations over 1,200 metres. It grows on rocks or trees in shady positions. The albino form of ***Dendrobium fairchildiae*** is rare and of limited distribution.

Dendrobium gerlandianum

Kränzlin
Section: ***Crumenata*** Pfitz.

Dr. Fritz Kränzlin named this plant in 1909 in *Feddes Repertorium Speciarum Novarum Regnis Vegetabilis.* The specific name is from the Latin word *gerlandia* for Erlangen, a town in Bavaria, Germany, where the plant first flowered in the botanical garden.

Growth habit: upright, sympodial. **Stems:** up to 20 cm tall, first couple of internodes swollen to about 5 mm in diameter to form a **pseudobulb**. The upper portion of the plant is only about 1.5 mm in diameter. **Leaves:** up to 6.5 cm long by 2 mm wide, slightly flattened. **Inflorescences:** short, bear solitary blooms about 2 cm in diameter. **Flower colour:** pale yellowish green; labellum has purple markings on the lateral lobes with a flattened yellow callus in its centre. **Dorsal sepal:** ovate, up to 1 cm long by 5 mm wide. **Petals:** lanceolate, about 8 mm long by 4 mm wide. **Lateral sepals:** triangular, join at the base to form a small spur about 1 cm long; overall length 1.8 cm by 7 mm wide. **Labellum:** simple, 1.8 cm long by 5 mm wide, pointed at the front, wavy edges.

Habitat and Distribution: ***Dendrobium gerlandianum*** was described from plants originating in the Philippines. Its exact locality is unknown.

Photo: David Banks

Dendrobium fairchildiae fma. ***album***

Photo: David Titmuss

Dendrobium gerlandianum

Notes: I have recently (June 1999) been given plants, gathered in Bukidnon, Mindanao, that may prove to be this species.

Dendrobium goldschmidtianum Kränzlin

Section: **Pedilonum** (Blume) Lindley

Dr. Fritz Kränzlin named this species in 1909 in *Feddes Repertorium Speciarum Novarum Regnis Vegetabilis*. The name honours Mr. Goldschmidt, the first person to flower this beautiful species in Europe. ***Dendrobium goldschmidtianum*** is an earlier name of the better known ***Dendrobium miyakei***.

Photo: David Banks

Dendrobium goldschmidtianum

Growth habit: semi-pendulous, sympodial. **Pseudobulbs:** up to 1 metre long, 1 cm in diameter. **Leaves:** up to 9 cm long, 2 cm wide. **Inflorescences:** semi-pendulous, bear up to 20 blooms about 2 cm in diameter; flowers appear from the leafless pseudobulbs. A well-grown plant is extremely attractive when flowering. **Flower colour:** bright purple striped with dark red. **Dorsal sepal:** lanceolate, up to 1.2 cm long, 5 mm wide. **Petals:** lanceolate, up to 1.1 cm long, 5 mm wide. **Lateral sepals:** ovate, join at the base to form short spur about 1 cm long; overall length 1.6 cm, 7 mm wide. **Labellum:** obovate, up to 1.5 cm long, 8 mm wide.

Habitat and Distribution: ***Dendrobium goldschmidtianum*** is found in the mountains of southern Taiwan and Batan island in the northern Philippines. It grows as an epiphyte at elevations over 1,000 metres.

Dendrobium guerreroi

Ames and Quisumbing
Section: ***Calcarifera*** J.J. Smith

Professor Oakes Ames and Dr. Eduardo Quisumbing named this rarely seen species in 1932 in the *Philippine Journal of Science*, in honour of Dr. Leon Ma. Guerrero, a specialist in the medicinal plants of the Philippines.

Growth habit: upright, sympodial. **Pseudobulbs:** up to 50 cm long by 1 cm wide. **Leaves:** up to 10 cm long, 2.5 cm wide, deciduous. **Inflorescences:** pendulous, up to 7 cm long, carry up to 15 blooms about 2 cm in diameter. **Flower colour:** cream to yellowish. **Dorsal sepal:** oblong-lanceolate, 1.2 cm long by 5 mm wide. **Petals:** lanceolate, about 1.2 cm long by 5 mm wide. **Lateral sepals:** oblong-lanceolate, join to form a spur about 2 cm long. **Labellum:** obovate, 3 cm long by 6 mm wide, darker yellow markings in its base.

Photo: Jim Cootes

Dendrobium guerreroi

Habitat and Distribution: ***Dendrobium guerreroi*** is endemic to the Philippines where it has been found on the island of Dinagat and in the province of Oriental Mindoro on the island of Mindoro. It grows as an epiphyte but is never common.

Dendrobium heterocarpum

Lindley
Section: ***Dendrobium*** Swartz
Synonyms: ***Dendrobium aureum*** Lindl.

Dr. John Lindley named this plant in 1830 in his *Genera and Species of Orchidaceous Plants*. The specific name refers to various kinds of seedpods that were on the original specimen.

Growth habit: upright, but some clones become semi-pendulous on older plants; sympodial. **Pseudobulbs:** up to 50 cm tall by 1 cm in diameter. **Leaves:** deciduous, up to 10 cm long by 2.5 cm wide. **Inflorescences:** short, appear from leafless nodes, up to three blooms up to 5 cm wide. **Flower colour:** creamy yellow; labellum darker yellow in the basal portion. **Dorsal sepal:** oblong, up to 2.5 cm long by 8 mm wide. **Petals:** oblong-ovate, up to 2.5 cm long by 1.2 cm wide. **Lateral sepals:** oblong-lanceolate, up to 2.5 cm long by 1 cm wide. **Labellum:** simple, 3 cm long by 2 cm wide (flattened); lateral lobes enclose the column.

Habitat and Distribution: ***Dendrobium heterocarpum*** is widespread, being found in India, Sri Lanka, Myanmar, Thailand, Peninsular Malaysia, Vietnam, Laos, Indonesia, Borneo and Sulawesi. In the Philippines it has been found in Albay, Benguet, Ilocos,

Photo: David Titmuss

Dendrobium heterocarpum

Photo: Peter O'Byrne

Dendrobium hymenanthum

Mountain Province, Nueva Ecija, Nueva Vizcaya, Rizal and Sorsogon on Luzon; Oriental Mindoro on Mindoro island; and Davao del Sur on Mindanao. It grows as an epiphyte at elevations over 1,000 metres.

Notes: There are two forms of this variable species found growing naturally in the Philippines. One grows upright to a height of about half a metre and has a hairy labellum that has a lot of brown markings on it. The other grows semi-pendulously and reaches lengths of up to two metres. The labellum of this species has no hairs and is of a different shape. The two are certainly worth investigating further, as I am of the opinion that there are two distinct species involved. I have not seen any plants that could be considered intermediate.

Dendrobium hymenanthum

Reichenbach *f.*

Section: ***Bolbidium*** Lindley

Professor Heinrich G. Reichenbach named this species in 1855 in *Bonplandia*. The specific name refers to flowers having a membrane. I am not certain which part of the bloom is meant.

Growth habit: upright, sympodial. **Pseudobulbs:** three nodes; lowest node very narrow, upper nodes square; overall up to 6 cm long by 4 mm wide, bifoliate. **Leaves:** leathery, narrowly ovate, up to 3.5 cm long by 1.2 cm wide. **Inflorescences:** single-flowered, grow from bracts between the leaves. **Flower colour:** pure white; bright green spot on labellum; sweetly perfumed. **Dorsal sepal:** triangular, 8.5 mm long by 3 mm wide. **Petals:** linear, 8.5 mm long by 1.5 mm wide. **Lateral sepals:** triangular, 1.6 cm long by 8 mm wide. **Labellum:** wedge-shaped (when flattened), 1.7 cm long by 1.2 cm wide; three ridges run from the base of the labellum to a small callus near the centre. **Column:** 2 cm long by 4 mm wide.

Habitat and Distribution: ***Dendrobium hymenanthum*** has been recorded from Thailand, Malaysia and Myanmar (Burma). In the Philippines it has been found in Bataan, Quezon and Rizal provinces on Luzon; Polillo island; and Bukidnon province on Mindanao. It grows as an epiphyte, sometimes on mangrove trees, at elevations to 300 metres.

Dendrobium ionopus Reichenbach *f.*

Section: ***Calcarifera*** J.J. Smith

Professor Heinrich G. Reichenbach named this species in the *Gardeners' Chronicle* in 1882. The specific name refers to the violet coloured stripes on the leaves and their sheaths.

Growth habit: semi-pendulous; sympodial. **Pseudobulbs:** grow to about 60 cm in length, 1 cm in diameter. **Leaves:** up to 7 cm long by 1.5 cm wide, young leaves have purple striping. **Inflorescences:** pendulous, appear from the leafless pseudobulbs; carry up to 20 blooms 2 cm in diameter. Each pseudobulb can carry several inflorescences. **Flower colour:** variable, ranging from apple green to bronze. **Dorsal sepal:**

Photo: David Titmuss

Dendrobium ionopus

lanceolate, up to 1.3 cm long by 5 mm wide. **Petals:** lanceolate, 1.2 cm long by 4 mm wide. **Lateral sepals:** ovate, join to form a spur 2.5 cm long. **Labellum:** square, about 3 cm long by 6 mm wide.

Habitat and Distribution: ***Dendrobium ionopus*** is an epiphyte that lives in the mountains of Bataan, Benguet, Nueva Ecija, Nueva Vizcaya and Rizal provinces on Luzon at altitudes of about 1,200 metres.

Notes: ***Dendrobium ionopus*** is an earlier name for the species better known as ***Dendrobium epidendropsis*** Krzl.

Dendrobium junceum Lindley

Section: ***Crumenata*** Pfitz.

Dr. John Lindley named this species in the *Botanical Register* in 1842. The specific name refers to the rush-like foliage.

Growth habit: upright, sympodial. **Pseudobulbs:** swollen, at the base for four nodes (approximately 10 cm), up to 1.5 cm in diameter. Upper portion of plant can grow to 60 cm long, 2 mm in diameter. **Leaves:** terete, about 10 cm long, 2 mm wide. **Inflorescences:** short, appear toward the end of pseudobulbs; flowers up to 2.5 cm in diameter. Only a couple of blooms will appear at any one time, but they are extremely showy. **Flower colour:** green with maroon veining in labellum. **Dorsal sepal:** lanceolate, up to 1.5 cm long, 5 mm wide. **Petals:** lanceolate, up to 1.2 cm long, 4 mm wide. **Lateral sepals:** ovate, join at the base to form a short spur about 1.5 cm long; overall length 2.5 cm, 8 mm wide. **Labellum:** three-lobed, 2.7 cm long, 1.4 cm wide (flattened); lateral lobes upright. There are also ridges running toward the base of the column.

Habitat and Distribution: ***Dendrobium junceum*** occurs in Borneo and in the Philippines where it has been found in the provinces of Quezon and Rizal on Luzon; Bucas Grande island northeast of Mindanao; and Surigao del Norte province on Mindanao. It grows as an epiphyte at elevations to 500 metres.

Dendrobium merrillii Ames

Section: ***Aporum*** (Blume) Lindley

Professor Oakes Ames named this species in 1908 in Fascicle 2 of his *Studies in the Family Orchidaceae.* The specific name honours Dr. Elmer D. Merrill, a botanist with the Philippine Bureau of Agriculture.

Growth habit: semi-pendulous, sympodial. **Pseudobulbs:** nonexistent, just flattened stems up to 20 cm

Photo: David Banks

Dendrobium junceum

Photo: David Banks

Dendrobium merrillii

long. **Leaves:** alternate, flattened, up to 2 cm long by 1.5 cm wide. **Inflorescences:** single-flowered, appear from tips of the stems. **Flower colour:** milky white; labellum has pale purple blotches on the side lobes. **Dorsal sepal:** oblong, 3 mm long by 1.5 mm wide. **Petals:** oblong to linear, up to 3 mm long by 1.25 mm wide. **Lateral sepals:** triangular to oblong, 3 mm long by 1.5 mm wide. **Labellum:** broadly oblong, three-lobed, 4 mm long by 1.5 mm wide; mid-lobe rounded at tip; side lobes very small.

Habitat and Distribution: ***Dendrobium merrillii*** is endemic to the Philippines and has been recorded from Ifugao, Pangasinan, Quezon and Rizal on Luzon; Mindoro island; and Samar island in the Visayas. It grows as an epiphyte at elevations to 350 metres.

Notes: This species could be confused with the very widely distributed ***Dendrobium aloifolium,*** but they are readily separable. The mid-lobe of the ***Dendrobium aloifolium*** labellum is two-lobed, whereas the mid-lobe ***Dendrobium merrillii*** labellum is not lobed. The side lobes of ***Dendrobium merrillii*** are purple, but ***Dendrobium aloifolium*** has no colour markings.

Dendrobium milaniae

Fessel and Lückel
Section: ***Platycaulon*** Schltr.

This species was described by Hans Fessel and Emil Lückel in the May-June 1996 issue of *Die Orchidee*, the well-known German orchid magazine. They named the plant in honour of Dr. Paciencia P. Milan of the Visayan State College of Agriculture, Leyte.

Growth habit: semi-pendulous, sympodial. **Pseudobulbs:** about 10 cm long by 2 cm wide by 3 mm thick. **Leaves:** three to six, on the top third of the pseudobulb, lanceolate, green, no purple markings, about 6 cm long by 2.5 cm wide. **Inflorescences:** pendulous, bear up to four blooms about 1.5 cm in diameter. **Flower colour:** white; labellum white with yellow spot in the centre surrounded by purple markings. **Dorsal sepal:** linear, acute, 1 cm long by 5 mm wide. **Petals:** linear, tip rounded, 1.1 cm long by 3.5 mm wide. **Lateral sepals:** triangular, 9 mm long. **Labellum:** about 1.2 cm long, wavy margin; there are three short ridges in the middle; central part round.

Habitat and Distribution: ***Dendrobium milaniae*** is found growing as an epiphyte hanging over water, at elevations of about 600 metres on the island of Leyte in the Visayas.

Photo: Hans Fessel

Dendrobium milaniae

Photo: David Banks

Dendrobium mindanaense

Dendrobium mindanaense Ames

Section: ***Strongyle*** Lindley

Professor Oakes Ames named this species in the *Philippine Journal of Science* in 1913. The specific name refers to Mindanao in the southern Philippines.

Growth habit: upright to semi-pendulous, sympodial. **Pseudobulbs:** non-existent; plant up to 50 cm tall. **Leaves:** terete, curving, purplish tinge, 5 cm long by 1.5 mm wide. **Inflorescences:** single-flowered, grow from the leaf axils. **Flower colour:** white with pink veins; labellum white with a yellow spot. **Dorsal sepal:** ovate, 6 mm long by 3 mm wide. **Petals:** linear to ovate, 6 mm long by 3 mm wide. **Lateral sepals:** triangular, 1.4 cm long by 7 mm wide, join at the base to form a small mentum. **Labellum:** triangular, 1.5 cm long by 8 mm wide; five ridges run the length.

Habitat and Distribution: ***Dendrobium mindanaense*** is endemic to the Philippines and has been recorded from the provinces of Quezon and Rizal on Luzon; the islands of Leyte and Negros (Negros Occidental province) in the Visayas; Culion and Palawan islands; and the provinces of Agusan del Norte, North and South Cotabato, Davao del Norte, Davao del Sur, Surigao del Sur, Zamboanga del Norte and Zamboanga del Sur on Mindanao. It grows at elevations up to 300 metres and has been found in the mangrove swamps of Mindanao.

Photo: David Banks

Dendrobium modestum

Dendrobium modestum

Reichenbach *f.*
Section: ***Crumenata*** Pfitz.

Professor Heinrich G. Reichenbach named this species in 1855 in *Bonplandia.* The specific name refers to the modest-sized flowers of this orchid.

Growth habit: upright, sympodial. **Pseudobulbs:** up to 70 cm tall; lower three nodes swollen up to 9 mm in diameter; upper portion terete; small plants grow from upper nodes. **Leaves:** terete, up to 8 cm long by 2 mm in diameter. **Inflorescences:** single-flowered, grow from the upper portion of pseudobulbs; blooms sweetly perfumed. **Flower colour:** white; labellum has yellow ridges and purple lines on the side lobes. **Dorsal sepal:** triangular-lanceolate, up to 1.2 cm long by 5 mm wide. **Petals:** narrowly elliptic, up to 1 cm long by 3 mm wide. **Lateral sepals:** triangular, up to 1.4 cm long by 1 cm wide. **Labellum:** three-lobed, up to 1.8 cm long by 1.2 cm wide (when flattened); mid-lobe squarish, edges ruffled; side lobes triangular. Three ridges run the length of the labellum.

Habitat and Distribution: *Dendrobium modestum* is endemic to the Philippines and has been found in Nueva Vizcaya, Quezon and Rizal on Luzon, and on Coron island, north of Palawan. It grows as an epiphyte, often in brightly lit situations, at elevations up to 1,200 metres.

Notes: The plant described here differs in several aspects from the original description of ***Dendrobium modestum***. The original description states that the flowers are odourless; my plant has very sweetly perfumed blooms. The size of the floral segments on my plant are much larger than those in the original diagnosis. The plant itself is also much larger. My plant originated in the mountains of central Luzon. Possibly, this species is not described.

Photo: Andres Golamco Jr.

Dendrobium nemorale

Dendrobium nemorale L.O. Williams

Section: ***Conostalix*** Kränzlin

Louis O. Williams named this lovely species in 1937 in *Botanical Museum Leaflets of Harvard University.* The specific name means that the plant grows in shady places, though the only plants I have seen of this species were growing in well-lit positions.

Growth habit: upright, sympodial. **Pseudobulbs:** grow to about 50 cm tall. **Leaves:** up to 5 cm long by 1 cm wide; sheaths clasping the pseudobulbs have short brown hairs. **Inflorescences:** short, appear from along the upper half of the pseudobulb, bear up to four flowers about 2 cm in diameter. **Flower colour:** creamy yellow, beautifully tessellated with brown; labellum yellow. **Dorsal sepal:** lanceolate, up to 1 cm long by 4 mm wide. **Petals:** broadly oblanceolate, up to 1 cm long by 4 mm wide. All these segments reflex for approximately half their length. **Lateral sepals:** oblong, 1 cm long by 4 mm wide, join to form a small spur. **Labellum:** three-lobed, 1.2 cm long by 6 mm wide (when flattened); lateral lobes triangular, prominent; mid-lobe is oval in shape when flattened.

Habitat and Distribution: *Dendrobium nemorale* is endemic to the Philippines and has been found in the provinces of Aurora, Quezon and Rizal on Luzon where it grows as an epiphyte.

Dendrobium orbilobulatum

Fessel and Lückel
Section: ***Fugacia*** J.J. Smith

Hans Fessel and Emil Lückel described this species in the September-October 1996 issue of *Die Orchidee*, the well-known German orchid magazine. The specific name refers to the lateral lobes of the labellum.

Growth habit: semi-pendulous, sympodial. **Stems:** clavate (narrow at the base, then gradually thickening upward), up to 12 cm long by 2 cm wide at the thickest point. **Leaves:** two, broadly ovate, 10 cm long by 4 cm wide. **Inflorescences:** single-flowered. **Flower colour:** white; labellum white with an orange blush on the lateral lobes, darker orange in basal portion. **Dorsal sepal:** ovate, 1.5 cm long by 9 mm wide. **Petals:** narrowly ovate, 1.3 cm long by 4 mm wide. **Lateral sepals:** broadly triangular, 1.5 cm long by 1 cm wide. **Labellum:** three-lobed, 1.1 cm long by 1.6 cm wide (when flattened) with three short ridges.

Habitat and Distribution: ***Dendrobium orbilobulatum*** is found growing as an epiphyte, hanging over water, at elevations of about 600 metres on the island of Leyte in the Visayas.

Dendrobium orbilobulatum

Dendrobium papilio

Dendrobium papilio Loher

Section: ***Dolichocentrum*** Schlechter

Augustus Loher named this plant and the name was published in the *Gardeners' Chronicle* in 1897. The specific name refers to the flower's resemblance to a butterfly.

Growth habit: upright, sympodial. **Stems:** many-branched, up to 45 cm tall. **Leaves:** up to 7.5 cm long by 5 mm wide; sheaths of the leaves that cling to the pseudobulb are purplish. **Inflorescences:** short, bear a single bloom about 5 cm in diameter that is faintly perfumed. **Flower colour:** pure white; labellum white with pale purple and yellow markings. **Dorsal sepal:** oblong-lanceolate, up to 2.9 cm long by 1 cm wide. **Petals:** ovate-lanceolate, up to 2.7 cm long by 1.1 cm wide. **Lateral sepals:** oblong, up to 2.7 cm long by 1.1 cm wide, join at the base to form a spur about 1.5 cm long. **Labellum:** simple, basal half linear, 1.9 cm long by 3 mm wide; front half broadly ovate, 2 cm long by 1.3 cm wide.

Habitat and Distribution: ***Dendrobium papilio*** is endemic to the mountains of Albay, Benguet, Mountain Province and Nueva Vizcaya provinces on Luzon; the provinces of Negros Oriental on Negros island in the Visayas and Davao del Sur and Surigao del Sur on Mindanao. It grows in the wild as both a lithophyte and an epiphyte at elevations over 1,200 metres.

Notes: The plant of ***Dendrobium papilio*** frequently branches along the stems, a feature that distinguishes it from the closely related ***Dendrobium auriculatum***.

Dendrobium pentapterum

Schlechter
Section: ***Oxyglossum*** Schlechter

Dr. Rudolf Schlechter named this charming species in 1905 in *Nachträge zur Flora der Deutschen Schutzgebeite in der Südsee.* The specific name refers to the five-winged ovary.

Growth habit: upright, sympodial. **Pseudobulbs:** fusiform to ovoid, bear up to four leaves. **Leaves:** linear to elliptic, up to 10 cm long by 1.2 cm wide. **Inflorescences:** terminal, usually bear only one flower but occasionally up to three blooms. **Flower colour:** sepals and petals greenish white to creamy yellow; labellum green with orange apex. **Dorsal sepal:** ovate to elliptic, up to 1.6 cm long by 7 mm wide. **Petals:** subspathulate, up to 1.5 cm long by 5.5 mm wide. **Lateral sepals:** obliquely ovate, up to 3 cm long by 1 cm wide. **Labellum:** linear, tip pointed, up to 2.5 cm long by 4.5 mm wide.

Photo: David Titmuss

Dendrobium pentapterum

Photo: David Titmuss

Dendrobium philippinense

Habitat and Distribution: ***Dendrobium pentapterum*** has been recorded from northeast New Guinea and has been found on Mount Apo in Davao del Sur province on Mindanao. It has a wide elevation range, between 500 and 2,080 metres, and usually grows on the mossy branches of trees.

Dendrobium philippinense Ames

Section: ***Crumenata*** Pfitz.

Professor Ames named this plant in the *Philippine Journal of Science* in 1913. The specific name refers to the Philippines, which is this orchid's native land.

Growth habit: upright, sympodial. **Stems:** terete. **Pseudobulbs:** lower nodes swollen to about 6 mm in diameter for about 10 cm. **Leaves:** terete, diameter up to 2 mm. **Inflorescences:** short, bear small, solitary flowers up to 6 mm across petals. Blooms are produced several times a year and the flowers last for about four days. **Flower colour:** yellow to cream; labellum has a large greenish callus. **Dorsal sepal:** triangular, up to 4 mm long, 2.5 mm wide. **Petals:** small, lanceolate up to 4 mm long, 1.5 mm wide. **Lateral sepals:** triangular, up to 6 mm long, 3 mm wide. **Labellum:** simple, obovate, 9 mm long, 5.5 mm wide.

Habitat and Distribution: ***Dendrobium philippinense*** is found in Guam and the Mariana Islands. In the Philippines it is found in Bataan, Bulacan, Laguna, Quezon, Rizal and Zambales on Luzon; Oriental Mindoro on Mindoro; Polillo island; Leyte and Panay islands in the Visayas; and throughout Mindanao. It is widely distributed at elevations up to 300 metres, growing as an epiphyte. In Oriental Mindoro on Mindoro this species grows on mangrove trees at the edges of creeks in bright light, sometimes full sun, with ***Dendrobium crumenatum*** and a ***Thrixspermum*** species for company. It also grows in a swampy region in quite deep shade. Plants from the swamp usually have many small plants along old inflorescences. Plants from both localities flower freely. Blooms appear every couple of weeks and last three to four days.

Notes: Plants from Guam and the Mariana Islands usually have white flowers. See colour photograph on page 99 of *Ferns and Orchids of the Mariana Islands* by Lynn Raulerson and Agnes F. Rinehart.

Dendrobium phillipsii

Ames and Quisumbing

Section: ***Calcarifera*** J.J. Smith

Professor Oakes Ames and Dr. Eduardo Quisumbing named this species in *The Philippine Journal of Science* in 1935. The specific name honours L.H. Phillips of Bukidnon, Mindanao, who collected many orchid species from this region in the 1930s.

Growth habit: semi-pendulous, sympodial. **Pseudobulbs:** terete, up to 60 cm long by 6 mm in diameter. **Leaves:** lanceolate, up to 8 cm long by 2 cm wide. **Inflorescences:** short, bear up to three flowers 2 cm in diameter. **Flower colour:** white; spur and throat of labellum yellow. **Dorsal sepal:** broadly ovate-lanceolate, 1.8 cm long by 1 cm wide. **Petals:** broadly oblong, rounded at tips, 1.7 cm long by 7 mm wide. **Lateral sepals:** triangular-ovate, 4 cm long by 1.3 cm wide. **Labellum:** ovate to oblanceolate, narrowing at base, 3.6 cm long by 1.5 cm wide.

Habitat and Distribution: ***Dendrobium phillipsii*** is endemic to the Philippines and has only been recorded from the province of Bukidnon on Mindanao. It grows as an epiphyte at elevations of around 800 metres.

Photo: Dr. Mark Clements

Dendrobium phillipsii

Dendrobium platycaulon Rolfe

Section: ***Platycaulon*** Schltr.

Robert Rolfe described this plant in 1892 in the *Kew Bulletin.* The specific name refers to the flattened pseudobulbs of this species.

Growth habit: semi-pendulous, sympodial. A most remarkable plant because of the flattened pseudobulbs that are a feature of section ***Platycaulon***. **Pseudobulbs:** up to 20 cm long by 3 cm wide, but only 3 mm thick as if flattened. **Leaves:** deciduous, up to 10 cm long, 2 cm wide. **Inflorescences:** pendulous, up to four from the tip of the pseudobulb; bear up to five blooms about 2.5 cm in diameter. **Flower colour:** pale green to yellow. **Dorsal sepal:** ovate, up to 1.3 cm long, 5 mm wide. **Petals:** ovate, up to 1.3 cm long, 7 mm wide. **Lateral sepals:** triangular, 2.3 cm long, 8 mm wide, joined at base to form a spur about 1.3 cm long. **Labellum:** three-lobed, 2.5 cm long by 1.5 cm wide (when flattened); lateral lobes triangular, enclosing the column; mid-lobe semicircular. There are three ridges running toward the base of the labellum.

Habitat and Distribution: ***Dendrobium platycaulon*** is endemic to the Philippines and has been found in the provinces of Benguet, Laguna and Mountain Province on Luzon; Bohol island in the Visayas; and Surigao del Norte, Surigao del Sur, Zamboanga del Norte and Zamboanga del Sur provinces on Mindanao. It grows as an epiphyte at elevations of about 500 metres. The plants of this species that I have seen growing in the wild have all been in shady locations.

Photo: David Titmuss

Dendrobium platycaulon

Photo: David Banks

Dendrobium polytrichum

Dendrobium polytrichum Ames

Section: ***Crumenata*** Pfitz.

Professor Oakes Ames named this species in the *Philippine Journal of Science* in 1907. The specific name refers to the fimbriated labellum of the flower. Elmer D. Merrill collected the type specimens in May 1904 on Mount Mariveles in Bataan.

Growth habit: upright, sympodial. **Pseudobulbs:** up to 35 cm tall, lowermost nodes thickened, uppermost nodes terete. **Leaves:** terete, 7 cm long by 1.5 mm in diameter. **Inflorescences:** bear solitary flowers up to 2.5 cm wide from near the tips of the stems. **Flower colour:** pure white with a red, orange or yellow blotch on the labellum. **Dorsal sepal:** oblong-lanceolate, up to 1.5 cm long by 4 mm wide. **Petals:** oblong-lanceolate, up to 1.4 cm long by 3.5 mm wide. **Lateral sepals:** lanceolate, up to 1.5 cm long by 5 mm wide, join at the base to form a spur about 8 mm long. **Labellum:** three-lobed, up to 2.2 cm long by 1.2 cm wide (when flattened); lateral lobes orange and rounded; mid-lobe deeply fringed. The coloured blotch on the labellum is hairy.

Habitat and Distribution: *Dendrobium polytrichum* is endemic to the Philippines where it is found in the provinces of Aurora, Bataan, Cagayan, Isabela, Laguna, Quezon and Rizal on Luzon and Oriental Mindoro on Mindoro island. It normally grows as an epiphyte, often in full sun, at elevations to 500 metres.

Dendrobium pristinum Ames

Section: ***Breviflores*** Hooker *f.*

Professor Oakes Ames named this species in 1915 in Fascicle V of his *Studies in the Family Orchidaceae.* The specific name refers to the beauty of the flower.

Photo: David Banks

Dendrobium pristinum

Growth habit: upright, sympodial. **Pseudobulbs:** narrowed at the base, thickening to 7 mm toward the upper portion; overall about 40 cm long. **Leaves:** shiny, leathery, up to 8 cm long by 1 cm wide. **Inflorescences:** semi-pendulous, appear from the leafless pseudobulbs, carry up to five flowers 1.5 cm diameter. **Flower colour:** white to cream with a yellow blotch on the labellum. **Dorsal sepal:** oblong-lanceolate, 7 mm long by 3.5 mm wide. **Petals:** oblanceolate, 7 mm long by 3 mm wide, finely notched. **Lateral sepals:** triangular, 8 mm long by 4 mm wide, form a small spur. **Labellum:** three-lobed, about 1 cm long, finely fringed.

Habitat and Distribution: *Dendrobium pristinum* is endemic to the Philippines where it has only been found in the province of Rizal on Luzon. It grows as an epiphyte at elevations around 1,000 metres. Undoubtedly this species is of wider distribution in the Philippines.

Notes: *Dendrobium pristinum* is closely related to the widely distributed species ***Dendrobium sphegidoglossum***, which was described by Reichenbach *f.* Further investigation may place ***Dendrobium pristinum*** into the synonymy of ***Dendrobium sphegidoglossum***.

Photo: David Titmuss

Dendrobium profusum

Dendrobium profusum

Reichenbach *f.*

Section: ***Calcarifera*** J.J. Smith

Professor Heinrich Reichenbach named this lovely species in 1884 in the *Gardeners' Chronicle*. The specific name refers to its profuse flowering.

Growth habit: pendulous, sympodial. **Pseudobulbs:** up to 50 cm long by 6 mm in diameter. **Leaves:** deciduous, lanceolate, up to 7.5 cm long by 2 cm wide. **Inflorescences:** pendulous, bearing up to 15 flowers 2.5 cm in diameter. **Flower colour:** cream to yellow with pink striping; outside of flower mottled with reddish pink; petals have pinkish red stripe running their length. **Dorsal sepal:** ovate, 1.2 cm long by 6 mm wide. **Petals:** ovate, 1.4 cm long by 7 mm wide. **Lateral sepals:** triangular, 1.4 cm long by 8 mm wide forming a spur about 2.5 cm long. **Labellum:** three-lobed, 2 cm long by 9 mm wide (when flattened), fiddle-shaped; mid-lobe inverted and heart-shaped; side lobes semicircular.

Habitat and Distribution: ***Dendrobium profusum*** is endemic to the Philippines and has been found in Albay, Camarines Norte, Camarines Sur and Quezon provinces on Luzon. It grows as an epiphyte at elevations of 100 to 500 metres in brightly lit habitats.

Photo: Dr. Howard Wood

Dendrobium ramosii

Dendrobium ramosii Ames

Section: ***Pedilonum*** (Blume) Lindley

Professor Oakes Ames described this plant in the *Philippine Journal of Science* in 1912. He named it after a Mr. Ramos, who collected its first specimens.

Growth habit: upright, sympodial. **Pseudobulbs:** very slender, up to 60 cm long, 3 mm wide. **Leaves:** up to 10 cm long, 1 cm wide. **Inflorescences:** short, appear from leafless nodes, carry up to 15 flowers about 1.5 cm in diameter. **Flower colour:** pale purple. **Dorsal sepal:** lanceolate, up to 1 cm long, 4 mm wide. **Petals:** lanceolate, up to 7 mm long, 3 mm wide. **Lateral sepals:** ovate, joined at base to form a spur about 1.3 cm long. **Labellum:** simple, up to 1.2 cm long, 4 mm wide.

Habitat and Distribution: ***Dendrobium ramosii*** is one of the few terrestrial species of the genus. It is endemic to the mountains of Benguet, Mountain Province and Quezon on Luzon where it grows at elevations over 1,200 metres.

Dendrobium sanderae Rolfe

Section: ***Formosae*** (Bentham and Hooker *f.*) Hooker *f.*

Robert Rolfe named this plant in the *Gardeners' Chronicle* in 1909, after Elizabeth Sander, wife of the great orchid patron and nurseryman Frederick Sander.

Growth habit: upright, sympodial. **Pseudobulbs:** can reach over 1 metre in length by 1 cm in diameter. **Leaves:** very stiff and leathery, up to 7 cm long by 2 cm wide. **Inflorescences:** short, appear from the leaf axils, carry up to four flowers 6 cm in diameter. **Flower**

Photo: Lloyd Bradford

Dendrobium sanderae

Photo: David Banks

Dendrobium sanderae var. *major*

Photo: David Banks

Dendrobium sanderae var. *luzonicum*

Photo: David Banks

Dendrobium sanderae var. *parviflorum*

Photo: Jim Cootes

Dendrobium sanderae var. *surigaense*

colour: pure white; lateral lobes of labellum greenish with purple markings. **Dorsal sepal:** lanceolate, up to 4 cm long by 1 cm wide. **Petals:** broadly obovate, up to 3.5 cm long by 2.5 cm wide. **Lateral sepals:** oblong-lanceolate, keeled, up to 3.5 cm long by 1.4 cm wide. **Labellum:** three-lobed; lateral lobes surround the column; mid-lobe flattened and ovate with purplish lines running toward the base of the column.

There are a number of very attractive varieties:

1) ***Dendrobium sanderae*** var. ***luzonicum*** Tiu: up to 50 cm tall; inflorescence bears up to 10 flowers 4.5 cm in diameter; flowers pure white; labellum white with green and purple markings in the base; anther cap tinged with purple.

2) ***Dendrobium sanderae*** var. ***major*** Hort. similar to type species; flowers much larger, up to 8 cm across; petals and labellum much broader.

3) ***Dendrobium sanderae*** var. ***parviflorum*** Anschutz *ex* Quis. grows to a similar size as the previous two varieties; flowers smaller, up to 4 cm in diameter; labellum narrow and straight.

4) ***Dendrobium sanderae*** var. ***surigaense*** Quis. reaches 50 cm tall, with no more than two flowers per inflorescence. **Dorsal sepal:** lanceolate, up to 3 cm long, 1 cm wide, reflexes. **Petals:** elliptic-oblong, up to 3 cm long, 2 cm wide, crinkled edges, reflex. **Lateral sepals:** oblong-lanceolate, tapering to a point, up to 3 cm long, 1.2 cm wide, reflex. **Labellum:** three-lobed; lateral lobes circle column; edge of mid-lobe notched, edges crinkled.

Habitat and Distribution: ***Dendrobium sanderae*** and its varieties are endemic to the Philippines where they have been found in Benguet, Ifugao, Mountain Province, Nueva Vizcaya and Rizal provinces on Luzon and on the southwest island of Palawan.

The plants grow as epiphytes at elevations of over 1,000 metres. The variety ***surigaense*** occurs only in the lowlands of Surigao del Norte and Surigau del Sur provinces on Mindanao. The variety ***luzonicum*** has been recorded from the Calayan Islands and Cagayan province in northern Luzon, where it grows as an epiphyte at altitudes to 300 metres.

Notes: The varieties ***luzonicum*** and ***surigaense*** require a minimum winter temperature of 12°C in more temperate climates.

Photo: David Banks

Dendrobium schuetzei

Dendrobium schuetzei Rolfe

Section: ***Formosae*** (Bentham and Hooker *f.*) Hooker *f.*

Robert Rolfe named this plant in *The Orchid Review* in 1911, possibly after a Mr. Schütze who collected with Wilhelm Micholitz, for Sander, in the Philippines.

Growth habit: upright, sympodial. **Pseudobulbs:** up to 30 cm long, 1.5 cm wide. **Leaves:** very leathery, greyish tinge, up to 6 cm long, 2 cm wide. **Inflorescences:** short, appear from near the top of the pseudobulbs, bear up to four blooms about 6.5 cm in diameter. **Flower colour:** pure white; labellum has green markings at the base. **Dorsal** and **lateral sepals:** lanceolate, up to 3 cm long, 1 cm wide. **Petals:** broadly ovate, up to 3.5 cm long, 3 cm wide. **Labellum:** three-lobed; lateral lobes circle the column; front lobe almost circular, up to 3 cm across. There is a ridge in the centre of the labellum going toward the column.

Habitat and Distribution: ***Dendrobium schuetzei*** is endemic to the provinces of Surigao del Norte and Surigao del Sur on the island of Mindanao where it grows as an epiphyte at elevations to 300 metres.

Dendrobium secundum

(Blume) Lindley

Section: ***Pedilonum*** (Blume) Lindley

Carl Blume originally named this plant ***Pedilonum secundum*** in 1825. This generic name is, at best, considered a section of the vast genus ***Dendrobium***. Lindley gave this species its currently accepted name in the *Botanical Register* in 1829. The specific name refers to the flowers, which all face in one direction.

Growth habit: upright to semi-pendulous, sympodial. **Pseudobulbs:** up to 1 metre long by 1.5 cm wide. **Leaves:** up to 10 cm long by 2.5 cm wide. **Inflorescences:** semi-pendulous, appear from the upper nodes, bear up to 50 flowers about 7 mm across

Photo: David Banks

Dendrobium secundum

the petals. **Flower colour:** variable, ranges from purplish pink to white. Professor Heinrich G. Reichenbach described the white-flowered form in 1882 as ***Dendrobium secundum*** var. ***niveum***. Labellum is orange for the pink species and yellowish for the white forms. **Dorsal sepal:** lanceolate, up to 1 cm long by 3 mm wide. **Petals:** oblong-lanceolate, up to 6 mm long by 2 mm wide. **Lateral sepals:** ovate, join at the base to form a short spur about 1 cm long. **Labellum:** ovate, edges curl up toward the column, up to 1.2 cm long by 5 mm wide.

Habitat and Distribution: ***Dendrobium secundum*** is widely distributed, being found in the Himalayas, Myanmar (Burma), Thailand, Indochina, Peninsular Malaysia, Sumatra, Java and Borneo. In the Philippines it has been found in the provinces of Bataan, Camarines Norte, Camarines Sur and Rizal on Luzon; on the island of Negros (Negros Occidental province) in the Visayas; and the provinces of Agusan del Norte, Agusan del Sur, Davao del Norte, Davao del Sur, Lanao del Norte, Lanao del Sur, Surigao del Norte and Surigao del Sur on Mindanao. It grows as an epiphyte at elevations to 500 metres.

Photo: David Banks

Dendrobium secundum var. *niveum*

Photo: David Banks

Dendrobium serratilabium

Photo: Jim Cootes

Dendrobium serratilabium fma. *album*

Dendrobium serratilabium

L.O. Williams
Section: ***Calcarifera*** J.J. Smith

Louis O. Williams named this plant in 1937 in the *Botanical Museum Leaflets of Harvard University*. The specific name refers to the serrated labellum.

Growth habit: pendulous, sympodial. **Pseudobulbs:** can attain lengths of over 40 cm. New growth often appears midway along the pseudobulb. **Leaves:** up to 8 cm long by 2 cm wide, deciduous. **Inflorescences:** pendulous, appear from the leafless nodes, carry up to three flowers about 3 cm in diameter. **Flower colour:** cream; outside of dorsal and lateral sepals striped with varying degrees of red. **Dorsal sepal:** lanceolate, up to 2 cm long by 5 mm wide. **Petals:** elliptic-lanceolate, up to 1.5 cm long by 4 mm wide. **Lateral sepals:** oblong-lanceolate, join at the base to form a spur about 2.5 cm long. **Labellum:** spathulate, deeply toothed around the edge.

Dendrobium serratilabium fma. *album* Hort.

Section: ***Calcarifera*** J.J. Smith

This variety has not been formally described. Plant as per species, but the flowers are pure yellow or green.

Habitat and Distribution: *Dendrobium serratilabium* and its variety are endemic to the mountains in the provinces of Aurora, Laguna, Quezon and Rizal on Luzon where they grow as epiphytes in fairly heavy shade at elevations between 500 and 1,200 metres.

Dendrobium setigerum

Ames *ex* M.A. Clements
Section: ***Latouria*** (Blume) Miquel

Professor Oakes Ames had ***Dendrobium setigerum*** as a name on a herbarium specimen for this species, but the name was never used. Dr. Mark Clements published the name in the *Australian Orchid Review* Vol. 64 (6) (Dec. 1999/Jan. 2000). The specific name refers to the bristles on the back of the dorsal and lateral sepals.

Growth habit: upright, sympodial. **Pseudobulbs:** club-shaped, up to 25 cm tall, bear up to three leaves. **Leaves:** leathery, up to 10 cm long by 4.5 cm wide. **Inflorescences:** upright, can have up to eight flowers about 3 cm in diameter. **Flower colour:** yellow with purplish spots on the back of the segments; labellum spotted and striped with purple; back of dorsal sepal

Photo: David Titmuss

Dendrobium setigerum

and lateral sepals covered with bristles. **Dorsal sepal:** triangular, up to 1.5 cm long by 1 cm wide. **Petals:** lanceolate, up to 1.5 cm long by 6 mm wide. **Lateral sepals:** triangular, 1.6 cm long by 1 cm wide. **Labellum:** three-lobed, 1.3 cm long by 2.5 cm wide (when flattened), lobes nearly equal; lateral lobes erect, incurved, broadly wedge-shaped, encircling the column; mid-lobe oval, tipped with a small point.

Habitat and Distribution: ***Dendrobium setigerum*** is endemic to the Philippines and has been found in the provinces of Bataan, Laguna, Mountain Province, Pampanga, Quezon and Rizal on Luzon. It grows as an epiphyte, usually at elevations to 400 metres.

Notes: This species is part of the ***Dendrobium macrophyllum*** A. Rich complex, currently under revision.

Dendrobium spurium (Blume) J.J. Smith

Section: ***Fugacia*** J.J. Smith

Carl Blume originally named this plant ***Dendrobium spuria,*** but this is incorrect. Johannes J. Smith rectified the error in *Die Orchideen von Java* in 1905. The specific name means false and refers to the original description as a false ***Dendrocolla***. This name (***Dendrocolla***) is generally accepted as a section of the vandaceous genus ***Thrixspermum***.

Growth habit: pendulous, sympodial. **Pseudobulbs:** can reach 30 cm in length, thin (1.5 mm) at the base and thicken toward the terminal end (1 cm). **Leaves:** up to 14 cm long by 2 cm wide. **Inflorescences:** short, appear from tips of the pseudobulbs; flowers short-lived, appear singly, very showy, about 2.5 cm in diameter. **Flower colour:** cream; pink markings in the labellum. **Dorsal sepal:** oblong to elliptic, up to 1.6 cm long by 7 mm wide. **Petals:** oblong-ovate, up to 1.4 cm long by 3.5 mm wide. **Lateral sepals:** triangular, up to 1.5 cm long by 8.5 mm wide. **Labellum:** up to 1.3 cm

Photo: David Banks

Dendrobium spurium

long by 1.1 cm wide (when flattened); a short oblong base widens abruptly to an almost round, concave blade, which is slightly three-lobed.

Habitat and Distribution: ***Dendrobium spurium*** is a lowland species found in Peninsular Malaysia, Singapore, Sumatra, Java, Borneo and New Guinea. In the Philippines it has been found in Bataan, Cagayan, Ilocos Norte, Laguna and Sorsogon provinces on Luzon and Davao del Sur province on Mindanao. It grows as an epiphyte on tree trunks in brightly lit habitats.

Dendrobium taurinum Lindley

Section: ***Spatulata*** Lindley

Dr. John Lindley named this plant in the *Botanical Register* in 1843. The specific name refers to the fanciful resemblance of the flower to the head of a bull.

Growth habit: upright, sympodial. **Pseudobulbs:** up to 2 metres long by 2 cm wide. **Leaves:** leathery, up to 10 cm long by 4 cm wide. **Inflorescences:** upright, reach 45 cm, carry up to 30 flowers about 4 cm across the petals. **Flower colour:** dorsal and lateral sepals

Photo: Gary Yong Gee

Dendrobium taurinum

Photo: Andres Golamco Jr.

Dendrobium taurinum fma. ***album***

white with pale green striping; petals and labellum reddish purple. **Dorsal** and **lateral sepals:** oblong-lanceolate, up to 2.5 cm long by 7 mm wide, curl and reflex. **Petals:** narrowly tapered and slightly twisted, up to 3 cm long by 6 mm wide. **Labellum:** obovate, three-lobed, three ridges down the centre.

Habitat and Distribution: ***Dendrobium taurinum*** is endemic to the Philippines and has been found in the provinces of Laguna, Quezon, Rizal and Zambales on Luzon; the island of Palawan; Panay island in the Visayas; Dinagat island north of Mindanao; and in the provinces of Bukidnon and Davao del Sur on Mindanao. It grows as an epiphyte at elevations up to 300 metres. The areas in which it grows have high rainfall most of the year.

Notes: ***Dendrobium taurinum*** is notoriously difficult to cultivate if the plant is collected in the wild; it will seldom prosper, even in its native habitat. If it is going to be grown it should be tried from plants grown from seed in flasks.

Dendrobium taurinum fma. ***album*** Valmayor and Tiu

Section: ***Spatulata*** Lindley

Dr. Helen Valmayor and Danny Tiu described this form of the species in 1983 in the *Philippine Orchid Review*. The habit of the plant is the same as the species except that the flowers are white with occasional lavender spots at the tips of the labellum and petals.

Habitat and Distribution: Endemic to the province of Quezon on Luzon, where it is extremely rare.

Dendrobium uniflorum Griffith

Section: ***Distichophyllum*** Hooker *f.*

William Griffith named this plant in 1851 in *Notulae ad Plantas Asiaticas*. The specific name refers to the single-flowered inflorescences of this species.

Growth habit: upright, sympodial. **Pseudobulbs:** reach lengths of up to 40 cm. **Leaves:** leathery, up to 5 cm long by 2 cm wide, very shiny. **Inflorescences:** produced singly, appear opposite the leaves; blooms up to 2 cm in diameter. **Flower colour:** milky white; labellum pale green with brownish lines. **Dorsal sepal:** lanceolate, up to 1.3 cm long by 6.5 mm wide, reflexes. **Petals:** lanceolate, up to 1.1 cm long by 4 mm wide, reflex. **Lateral sepals:** ovate, 2.2 cm long by 8 mm wide, reflex. **Labellum:** large, three-lobed, 2 cm long

Photo: David Banks

Dendrobium uniflorum

Photo: Jim Cootes

Dendrobium ventricosum

by 2.6 cm wide (when flattened); lateral lobes squarish; mid-lobe circular with a notch in the centre.

Habitat and Distribution: ***Dendrobium uniflorum*** is widely spread, being found in Malaysia, Vietnam, Thailand and Borneo. In the Philippines it has been found in Aurora, Bataan, Ilocos Norte, Laguna, Quezon, Rizal, Sorsogon and Zambales provinces on Luzon; Polillo island; Leyte, Negros (Negros Occidental province), Panay, Samar and Sibuyan islands in the Visayas; Basilan island in the Sulu Archipelago; Dinagat island; and Agusan del Norte, Agusan del Sur, Lanao del Norte, Lanao del Sur, Surigao del Norte and Surigao del Sur provinces on Mindanao. It grows as an epiphyte at altitudes from 100 to 700 metres, often in full sun.

Dendrobium ventricosum Kränzlin

Section: ***Crumenata*** Pfitz.

Dr. Fritz Kränzlin named this dainty species in 1910 in A. Engler (ed.) *Das Pflanzenreich. Regni vegetabilis conspectus* IV 50 II B7. The specific name refers to the swelling on the labellum.

Growth habit: upright, sympodial, up to 60 cm tall. **Stems:** flattened. **Leaves:** flattened, up to 5 cm long, pointed. **Inflorescences:** appear singly from near the tips of the pseudobulbs; blooms about 1.5 cm in diameter. **Flower colour:** white; labellum white with purple markings on the lateral lobes, yellow blotch on the centre. **Dorsal sepal:** lanceolate, up to 1 cm long by 4 mm wide. **Petals:** lanceolate, up to 1 cm long by 5 mm wide. **Lateral sepals:** ovate, up to 1 cm long by 5 mm wide. **Labellum:** oblong, up to 1.5 cm long by 6 mm wide, wavy edges.

Habitat and Distribution: ***Dendrobium ventricosum*** is endemic to the Philippines and has been found in the provinces of Sorsogon and Zambales on Luzon and Agusan del Norte, Agusan del Sur, Davao del Norte, Davao del Sur, Surigao del Norte and Surigao del Sur on Mindanao. It grows as an epiphyte at elevations to 300 metres.

Dendrobium victoriae-reginae

Loher

Section: ***Calcarifera*** J.J. Smith

Augustus Loher named this species in the *Gardeners' Chronicle* in 1897, to honour Queen Victoria, who was about to celebrate her 50th anniversary on the British throne.

Growth habit: pendulous, sympodial. **Pseudobulbs:** reach lengths of 60 cm. **Leaves:** up to 7.5 cm long by 2 cm wide. **Inflorescences:** short, appear from opposite the leaves, bear up to three flowers about 4 cm in diameter. **Flower colour:** ranges in colour from blue to violet to white; wide variation in the amount of striping on segments of different clones. There is also

Photo: David Titmuss

Dendrobium victoriae-reginae

Photo: David Banks

Dendrobium wenzelii

an albino form of this delightful species. **Dorsal sepal:** lanceolate, up to 2.6 cm long by 9 mm wide. **Petals:** lanceolate, up to 2.8 cm long by 1.3 cm wide. **Lateral sepals:** oblong-lanceolate, up to 2.9 cm long by 1 cm wide, unite at the base to form a small spur about 6 mm long. **Labellum:** ovate, up to 3 cm long by 1.4 cm wide.

Habitat and Distribution: ***Dendrobium victoriae-reginae*** is endemic to the high mountains (over 1,300 metres) of Ifugao, Mountain Province, Nueva Vizcaya, Pampanga and Quezon provinces on Luzon; Oriental Mindoro province on Mindoro island; throughout Negros island in the Visayas; Camiguin island, north of Mindanao; and Davao del Sur province on Mindanao. It grows as an epiphyte in the mossy cloud forests in deeply shaded localities where there is plenty of air movement. The roots always grow in damp mosses.

Dendrobium wenzelii Ames

Section: ***Aporum*** (Blume) Lindley

Professor Oakes Ames named this miniature growing species in 1915 in his *Studies in the Family Orchidaceae*, Fascicle V after C.A. Wenzel, collector of the type specimens, found in Leyte in March 1913.

Growth habit: upright, sympodial. **Stems:** flattened. **Leaves:** clasp the stem for the entire length—up to 10 cm; terminal portion of the stem bears two orange bracts from which the flowers emerge. **Inflorescences:** short, appear at the tip of stems; blooms about 8 mm in diameter. **Flower colour:** orange; red striping in the labellum. Only one or two flowers are open at a time per stem. **Dorsal sepal:** lanceolate, up to 5 mm long by 3 mm wide. **Petals:** lanceolate, up to 5 mm long by 2 mm wide. **Lateral sepals:** triangular, up to 6 mm long by 5 mm wide. **Labellum:** oblong, up to 8 mm long by 3.5 mm wide, three-lobed, wavy edges.

Habitat and Distribution: ***Dendrobium wenzelii*** is endemic to the Philippines and has been found in the provinces of Aurora and Quezon on Luzon; the province of Oriental Mindoro on the island of Mindoro; Leyte island in the Visayas; and the provinces of Lanao del Norte, Lanao del Sur, Surigao del Norte and Surigao del Sur on Mindanao. It grows as an epiphyte at elevations to 300 metres.

Dendrobium yeageri

Ames and Quisumbing

Section: ***Calcarifera*** J.J. Smith

Professor Oakes Ames and Dr. Eduardo Quisumbing named this species in 1933 in the *Philippine Journal of Science* after Dr. C.H. Yeager, who was with the group that obtained the type specimens in 1932.

Growth habit: pendulous, sympodial. **Pseudobulbs:** up to 50 cm long by 5 mm wide. **Leaves:** lanceolate, up to 10 cm long by 2.5 cm wide. **Inflorescences:** bear up to six flowers 3 cm in diameter. **Flower colour:** pale purple with darker purple stripes; labellum tip white. **Dorsal sepal:** oblong to lanceolate, up to 1.8 cm long by 8 mm wide. **Petals:** narrowly oblong, up to 1.8 cm long by 7 mm wide. **Lateral sepals:** oblong to lanceolate, up to 1.8 cm long by 8 mm wide, join at the base to form a short spur about 1 cm long. **Labellum:** oblanceolate to oblong, narrow at the base, broader at the tip, up to 2.5 cm long by 1 cm wide.

Habitat and Distribution: ***Dendrobium yeageri*** is endemic to the Philippines where it has been found in the provinces of Benguet and Mountain Province on Luzon. It grows as an epiphyte at elevations over 1,000 metres.

Photo: David Banks

Dendrobium yeageri

Dendrochilum Blume

The genus ***Dendrochilum*** has had an interesting nomenclatural history. Carl Blume first described this group of plants in 1825 in *Bijdragen tot de Flora van Nederlandsch Indie.* He then divided the genus into two sections, one characterised by a lateral inflorescence, the other by a terminal inflorescence. In 1881 Bentham proposed that the section ***Platyclinis*** (terminal inflorescence) become a distinct genus, a line of thought followed by some very eminent botanists and taxonomists. In 1904 Rolfe studied a little known taxon, ***Acoridium tenellum***, and found it to be a ***Platyclinis;*** realising that ***Acoridium*** had priority, he then transferred a number of ***Platyclinis*** to ***Acoridium***. Professor Oakes Ames had changing views on ***Acoridium***, several times he accepted it as a section of ***Dendrochilum*** and on other occasions used ***Acoridium*** as a separate genus. The confusion still continues with these generic names.

The type species for this lovely genus is ***Dendrochilum aurantiacum*** Bl.

The centre of distribution for this genus ***Dendrochilum*** is the Philippines, Sumatra and Borneo. Approximately 270 species are recorded. The majority of the species are plants from the mossy cloud forests where they generally grow as epiphytes (on trees) or lithophytes (on rocks); I know of no terrestrial species. The Philippines has the most members for this genus with some 96 species known. Dr. Valmayor states: *"**Dendrochilum** is a genus with a very high percentage of endemism; its species are confined mainly to the high mountains of Luzon. About 85% of the species are located in Luzon, with about 74% found at altitudes of 1,500 metres and above."* (*Orchidiana Philippiniana* H. Valmayor, 1984). Ninety-four percent of the species recorded from the Philippines are endemic. It is quite remarkable that New Guinea, being such a mountainous country, has only two species, Taiwan one member, Thailand three, Malaysia 12, Java 13, Sumatra 61, Borneo 80 and Mount Kinabalu at least 23 species.

Photo: David Titmuss

Dendrochilum arachnites

Photo: David Banks

Dendrochilum auriculare

Dendrochilum arachnites

Reichenbach *f.*

Section: ***Platyclinis*** (Bentham) Pfitzer

Professor Heinrich G. Reichenbach named this untidy-growing species in the *Gardeners' Chronicle* in 1882. The specific name suggests that the flowers have the appearance of a spider.

Growth habit: upright, sympodial, a bit of a rambler having a distance of about 5 cm between pseudobulbs. **Pseudobulbs:** ovoid, 5 cm tall, 1 cm diameter. **Leaves:** lanceolate, up to 16 cm long, 2.5 cm wide, distinctly veined. **Inflorescences:** arch, appear with the new growth; flowers about 2 cm across lateral sepals, their widest point; each can carry up to 30 blooms. **Flower colour:** yellow. **Dorsal** and **lateral sepals:** linear-lanceolate, extend to acute tips, up to 1 cm long, 2 mm wide. **Petals:** lanceolate, up to 7 mm long, 3 mm wide. **Labellum:** oblong, 4 mm long by 1.5 mm wide; three ridges run the length.

Habitat and Distribution: *Dendrochilum arachnites* is endemic to the Philippines and has been found in Benguet, Ifugao, Mountain Province, Nueva Vizcaya, Pampanga, Quezon and Rizal on Luzon; Oriental Mindoro on Mindoro; Leyte island in the Visayas; and Agusan del Norte, Bukidnon, Davao del Norte, Davao Oriental, Davao del Sur, Misamis Occidental, Zamboanga del Norte and Zamboanga del Sur on Mindanao at elevations between 660 and 2,300 metres. It grows as an epiphyte, frequently in bright light, and is one of the more untidy growing members of this genus.

Dendrochilum auriculare Ames

Section: ***Eurybrachium*** Carr *ex* J.J. Wood, H.Æ. Pedersen and J.B. Comber

Professor Oakes Ames named this species in 1909 in *The Philippine Journal of Science.* The specific name refers to the ear-like side lobes of the labellum.

Growth habit: upright, sympodial. **Pseudobulbs:** spindle-shaped to tapered, up to 2.5 cm long, 6 mm wide. **Leaves:** petiolate, leathery, lanceolate, up to 12 cm long, 2.5 cm wide. **Inflorescences:** appear with the new growth, arching, bear up to 30 flowers 8 mm in diameter. **Flower colour:** greenish yellow to white. **Dorsal sepal:** elliptic, up to 8 mm long, 3.5 mm wide. **Petals:** elliptic to broadly ovate, up to 7.5 mm long, 5 mm wide. **Lateral sepals:** elliptic, 7.5 mm long, 3.5 mm wide. **Labellum:** three-lobed; side lobes upright; mid-lobe large, squarish.

Habitat and Distribution: *Dendrochilum auriculare* is endemic to the Philippines and has been recorded from the provinces of Benguet, Ifugao, Nueva Vizcaya, Quezon and Rizal on Luzon. It grows as an epiphyte at elevations between 1,500 and 2,500 metres.

Dendrochilum cobbianum

Reichenbach *f.*

Section: ***Platyclinis*** (Bentham) Pfitzer

Professor Heinrich G. Reichenbach named this species in the *Gardeners' Chronicle* in 1880 after Walter Cobb, the first person to flower this plant in Britain.

Growth habit: upright, sympodial. **Pseudobulbs:** tapered, 4 to 8 cm long, unifoliate. **Leaves:** 12 to 18 cm long, up to 3.5 cm wide, prominent mid-rib. **Inflorescences:** upright; flowering portion pendulous, appears with the new growth; normally longer than the foliage with the 12 mm flowers about 7 mm apart. **Flower colour:** extremely variable; some clones have all orange blooms, others have cream sepals and petals

Dendrochilum cobbianum

Dendrochilum cobbianum 'Old Gold'

with bright orange labellum (as per the type specimen description); yet others have icy green flowers. **Dorsal sepal** and **petals:** lanceolate, up to 5 mm long, 2.5 mm wide. **Lateral sepals:** ovate, up to 6 mm long by 3 mm wide. **Labellum:** triangular, up to 7 mm long by 3 mm wide.

Habitat and Distribution: ***Dendrochilum cobbianum*** is endemic to the Philippines and has been found in Benguet, Mountain Province, Nueva Ecija, Nueva Vizcaya and Rizal on Luzon; Catanduanes island; and Agusan del Norte and Davao del Sur on Mindanao. It grows as an epiphyte on mossy trees trunks at elevations over 1,200 metres. In Mountain Province I have observed this species growing on rocks in full sun.

Dendrochilum convallariaeforme Schauer

Section: ***Eurybrachium*** Carr *ex* J.J. Wood, H.Æ. Pedersen and J.B. Comber

Johann Schauer named this pretty species in *Novorum Actorum Academiae Caesareae Leopoldinae-Carolinae Naturae Curiosorum* in 1843. The specific name refers to the flower's resemblance to lily-of-the-valley (***Convallaria*** spp.). Elmer D. Merrill collected the type specimens of ***Dendrochilum bicallosum*** and ***Dendrochilum bicallosum*** var. ***minor*** on Mount Halcon on Mindoro in November 1906. These two species are now considered synonyms of the widely distributed ***Dendrochilum convallariaeforme***.

Growth habit: upright, sympodial, a small grower. **Pseudobulbs:** about 2.5 cm long by 1.5 cm wide. **Leaves:** up to 10 cm long by 2.5 cm wide. **Inflorescences:** appear with the new growth, arch gracefully, carry up to 50 blooms 8 mm in diameter; flowers spiral around the raceme. **Flower colour:** flesh-pink. **Floral segments:** all of similar shape, lanceolate up to 5 mm long by 2.5 mm wide. **Labellum:** ovate to obovate, 3 mm long by 2 mm wide.

Dendrochilum convallariaeforme

Habitat and Distribution: ***Dendrochilum convallariaeforme*** is endemic to the mountains of Aurora, Camarines Sur, Laguna, Quezon, Rizal and Sorsogon on Luzon; Oriental Mindoro on Mindoro island; Masbate in the Visayas; and Bukidnon and Davao del Sur on Mindanao. It grows as an epiphyte, often in shaded areas, at elevations between 300 and 900 metres.

Photo: David Titmuss

Dendrochilum cootesii

Photo: David Titmuss

Dendrochilum curranii

Dendrochilum cootesii

H.Æ. Pedersen
Section: ***Luzonorchis*** H.Æ. Pedersen

Henrik Æ. Pedersen named this species in *Lindleyana* 12 (4) in 1997. It was named in honour of the author of this book, who first recognized that this species was quite different from other related species.

Growth habit: upright, sympodial. **Pseudobulbs:** dark green, wrinkled, sometimes curved, up to 5 cm long by 1 cm wide topped by a single leaf. **Leaves:** leathery, prominent mid-vein, up to 10 cm long by 3 cm wide. **Inflorescences:** appear with the new growth, slightly longer than the foliage, arch gracefully. In their infancy they resemble the tail of a rattlesnake. The flowers twist around the inflorescence. **Flower colour:** cream most common, but also buff to pinkish clones. The fleshy blooms are unique among the ***Dendrochilum*** species known to me. **Dorsal sepal:** bends back toward the inflorescence, about 1 cm long by 2.5 mm wide. **Petals:** bend back sharply for about the final third of the segment, up to 1 cm long by 2.5 mm wide. **Lateral sepals:** arch gracefully forward, up to 1.2 cm long by 3 mm wide. **Labellum:** small, three-lobed; lateral lobes longer than mid-lobe, enclose the column.

Habitat and Distribution: ***Dendrochilum cootesii*** is a very distinct species endemic to the mountains of Ifugao, Nueva Ecija and Nueva Vizcaya provinces on Luzon. It grows at elevations between 1,200 and 2,000 metres on mossy tree trunks and branches on the fringes of cloud forest.

Dendrochilum curranii Ames

Section: ***Convoluta*** H. Æ. Pedersen

Professor Oakes Ames named this charming, small-growing species in 1908 in his *Studies in the Family Orchidaceae* Fascicle III, after H.M. Curran, collector of the type specimens, which were found on Mount Makiling in the province of Laguna in January 1908.

Growth habit: upright, sympodial. **Pseudobulbs:** tapered, up to 3 cm long by 5 mm in diameter, covered by persistent bracts, unifoliate. **Leaves:** up to 9.5 cm long by 1 cm wide, prominent mid-vein, petiole up to 2 cm long. **Inflorescences:** arch gracefully, appear with the new growth, longer than the leaves; carry about 50 flowers 3.5 mm in diameter. **Flower colour:** pale yellow; labellum reddish brown. **Dorsal sepal:** lanceolate, up to 3 mm long by 1 mm wide. **Petals:** linear to lanceolate, 2.5 mm long by .75 mm wide.

Photo: Dr. Henrik Pedersen

Dendrochilum cymbiforme

Lateral sepals: lanceolate, up to 3 mm long by .75 mm wide. **Labellum:** small, three-lobed; lateral lobes longer than the reflexed mid-lobe.

Habitat and Distribution: ***Dendrochilum curranii*** is endemic to the mountains of Albay, Benguet, Laguna, Pampanga, Quezon and Rizal provinces on Luzon and Oriental Mindoro province on Mindoro island where it grows as an epiphyte at elevations between 600 and 2,000 metres.

Dendrochilum cymbiforme Ames

Section: ***Platyclinis*** (Bentham) Pfitzer

Professor Oakes Ames named this species in the *Philippine Journal of Science* in 1911. The specific name refers to the boat-shaped (cymbiform) labellum.

Growth habit: pendulous, sympodial. **Pseudobulbs:** pear-shaped, 1.3 cm long, 6 mm in diameter, unifoliate. **Leaves:** linear to lanceolate, rigid, up to 28 cm long, 2 cm wide. **Inflorescences:** up to 23 cm long, bear many flowers. **Flower colour:** sepals and petals white; labellum reddish to pale brown. **Dorsal sepal:** oblong, boat-shaped, partly embraces the column, up to 3.5 mm long, 1.6 mm wide. **Petals:** obliquely lanceolate, up to 3 mm long, 1.3 mm wide. **Lateral sepals:** point forward, falcate, up to 4.4 mm long by 1.9 mm wide. **Labellum:** boat-shaped, broadly elliptic-orbicular, up to 4.2 mm long, 3.4 mm wide (flattened).

Habitat and Distribution: ***Dendrochilum cymbiforme*** is endemic to the Philippines and has been recorded from Abra, Apayao, Ifugao and Kalinga on Luzon. It grows as an epiphyte at elevations between 1,665 and 2,010 metres. This species always grows on the undersides of thick branches in mossy forest.

Photo: David Titmuss

Dendrochilum filiforme

Dendrochilum filiforme Lindley

Section: ***Platyclinis*** (Bentham) Pfitzer

Dr. John Lindley named this magnificent ***Dendrochilum*** in the *Botanical Register* in 1840. The specific name refers to the long, slender raceme.

Growth habit: upright, sympodial. **Pseudobulbs:** oval, up to 3 cm tall, diameter 1 cm, unifoliate. **Leaves:** up to 18 cm long, 1.5 cm wide. **Inflorescences:** upright; flowering portion pendulous, wiry, bears a two-ranked raceme that appears with the new growth; blooms about 6 mm in diameter and of good shape, seventy flowers per raceme common. **Flower colour:** bright yellowish green. **Dorsal sepal:** oblong-elliptic, up to 3 mm long, 1 mm wide. **Petals:** obovate, up to 3 mm long, 2 mm wide. **Lateral sepals:** lanceolate, up to 3 mm long, 1.5 mm wide. **Labellum:** obscurely three-lobed; lateral lobes small; mid-lobe ovate.

Habitat and Distribution: ***Dendrochilum filiforme*** is endemic to the Philippines where it has been found in Benguet, Cagayan, Laguna, Mountain Province, Quezon and Rizal on Luzon and Negros Oriental on Negros in the Visayas. This species grows as an epiphyte at elevations between 660 and 2,250 metres.

Photo: David Banks

Dendrochilum glumaceum

Photo: David Banks

Dendrochilum javieri

Dendrochilum glumaceum Lindley

Section: ***Platyclinis*** (Bentham) Pfitzer

Dr. John Lindley named this species in 1841 and the name was published in the *Botanical Register*. The specific name refers to the glume-like (husk-like) appearance of the unopened raceme.

Growth habit: upright, sympodial. **Pseudobulbs:** oval, up to 3 cm high by 1.5 cm in diameter, unifoliate. **Leaves:** up to 30 cm long by 4 cm wide. **Inflorescences:** arch gracefully, appear with the new growth; flowers about 2 cm in diameter. **Flower colour:** white sepals and petals; wide colour range in the labellum includes bright orange, bright yellow and brown. **Dorsal sepal:** linear-oblong tapering to a sharp point, about 8 mm long by 2 mm wide. **Petals:** lanceolate, up to 6 mm long by 2 mm wide. **Lateral sepals:** linear-oblong tapering to a sharp point, up to 8 mm long by 2 mm wide. **Labellum:** three-lobed; lateral lobe small; mid-lobe oblong-ovate.

Habitat and Distribution: ***Dendrochilum glumaceum*** has been found in Borneo. In the Philippines it has been found in Aurora, Bataan, Benguet, Cagayan, Camarines Norte, Camarines Sur, Ifugao, Ilocos Norte, Isabela, Laguna, Mountain Province, Nueva Vizcaya, Pampanga, Quezon and Rizal provinces on Luzon; Catanduanes island; Oriental Mindoro province on Mindoro island; Leyte, Negros (Negros Oriental province) and Panay (Capiz province) islands in the Visayas; and Agusan del Norte, Bukidnon, Davao del Norte, Davao Oriental, Davao del Sur, Lanao del Sur and Misamis Occidental provinces on Mindanao. It grows as an epiphyte at elevations to 2,300 metres. In some places it is extremely common.

Dendrochilum javieri

Magrath, Bulmer and Shafer

Section: ***Acoridium*** (Nees and Meyen) Pfitzer and Kränzlin

This is a recent addition to Philippine orchidology, having only been described in 1989 by Lawrence Magrath, Glenn Bulmer and Ingrid Shafer. The name was published in *Lindleyana*. It is named after Elias Javier, owner of an orchid nursery in the Philippines.

Growth habit: upright, sympodial, up to about 20 cm tall. **Pseudobulbs:** tapered, 2 cm long by 4 mm in diameter. **Leaves:** very leathery, about 18 cm long by 1 cm wide. **Inflorescences:** arching, appear with the new growth; flowers about 12 mm across and a little crowded on the top one-third of the inflorescence. **Flower colour:** bright yellowish green. **Dorsal sepal:** oblong-lanceolate, up to 6.5 mm long by 3 mm wide.

Petals: lanceolate, up to 5 mm long by 2.5 mm wide. **Lateral sepals:** oblong-lanceolate, up to 8 mm long by 3 mm wide. **Labellum:** three-lobed; side lobes squarish; mid-lobe recurved into a pointed "v" shape and inflated below into a saccate, spur-like structure.

Habitat and Distribution: ***Dendrochilum javieri*** has limited distribution, being found as an epiphyte in the mountains of Nueva Ecija and Nueva Vizcaya provinces in central Luzon above 1,200 metres. It usually grows in quite bright light and new growth is bronze-coloured.

Notes: This species was originally named ***Dendrochilum javieriense*** but the specific epithet, which alludes to a locality, is incorrect as the plant was named after Elias Javier. Therefore the correct name for this species is ***Dendrochilum javieri***.

Dendrochilum latifolium var. *macranthum* (Schltr.) H.Æ. Pedersen

Section: ***Platyclinis*** (Bentham) Pfitzer

Dr. Rudolf Schlechter first named this large-growing species ***Dendrochilum grandiflorum*** in 1910. Henrik Æ. Pedersen reduced this plant to a variety of ***Dendrochilum latifolium*** in *Opera Botanica* 130 in 1997. The varietal name refers to the large leaves of this species.

Growth habit: upright, sympodial. **Pseudobulbs:** oval-shaped, up to 5 cm tall by 2.5 cm in diameter, unifoliate. **Leaves:** oblong-lanceolate, up to 40 cm long by 5 cm wide; petiole up to 15 cm in length. **Inflorescences:** upright, flowering portion pendulous, as long as (and in some clones longer) the leaves, appear with new growth; flowers about 1.2 cm in diameter. **Flower colour:** open green, turn orange with age; labellum brownish. **Dorsal** and **lateral sepals:** oblong-lanceolate, up to 9 mm long by 3 mm wide. A large bract at the back of each bloom forces them to face down and is a distinguishing feature of this species. **Petals:** of similar shape to sepals but edges irregularly notched. **Labellum:** three-lobed, 4 mm long by 3 mm wide (when flattened); lateral lobes triangular, roughly edged; mid-lobe circular.

Habitat and Distribution: ***Dendrochilum latifolium*** var. ***macranthum*** is endemic to the Philippines. It has been found in the provinces of Apayao, Aurora, Cagayan, Kalinga, Laguna, Rizal and Quezon on Luzon; the Leyte islands; Negros Occidental and Negros Oriental provinces on the island of Negros; and Panay (Antique province) in the Visayas. It grows as an epiphyte above 1,200 metres.

Photo: David Titmuss

Dendrochilum latifolium var. ***macranthum***

Dendrochilum longibulbum

Ames

Section: ***Convoluta*** H.Æ. Pedersen

Professor Oakes Ames named this species in the *Philippine Journal of Science* in 1912. The specific name refers to the length of the pseudobulbs.

Growth habit: upright, sympodial. **Pseudobulbs:** oval, up to 3 cm long by 5 mm in diameter, unifoliate. **Leaves:** lanceolate, 8 cm long by 1 cm wide. **Inflorescences:** upright, flowering portion pendulous, longer than the leaves, appear with new growth, bear about 20 blooms 6 mm in diameter. **Flower colour:** pale yellow. **Dorsal** and **lateral sepals:** narrowly ovate up to 4 mm long by 2.5 mm wide. **Petals:** rounded, up to 4 mm long. **Labellum:** three-lobed; lateral lobes fractionally longer than the mid-lobe.

Photo: David Banks

Dendrochilum longibulbum

Photo: David Titmuss

Dendrochilum longifolium

Habitat and Distribution: ***Dendrochilum longibulbum*** is endemic to the Philippines where it is found in Albay, Benguet, Laguna, Mountain Province, Quezon, Rizal and Sorsogon on Luzon; Antique on Panay island; and Negros Occidental on Negros island in the Visayas. It grows as an epiphyte, among mosses, on tree trunks in bright light at elevations over 950 metres.

Dendrochilum longifolium

Reichenbach *f.*
Section: ***Platyclinis*** (Bentham) Pfitzer

Professor Heinrich G. Reichenbach named this species in 1856 in *Bonplandia.* The specific name refers to the foliage, which can become quite long when plants grow in heavy shade.

Growth habit: upright, sympodial. **Pseudobulbs:** tapered, up to 6 cm long, 1.5 cm in diameter, unifoliate. **Leaves:** up to 20 cm long, 3.5 cm wide. **Inflorescences:** arch gracefully, longer than leaves, appear with the new growth, bear up to 40 flowers about 1 cm in diameter. **Flower colour:** green; labellum green with two brown ridges. **Floral segments:** oblong-lanceolate, about 6 mm long, 2 mm wide. **Labellum:** three-lobed, up to 7.5 mm long, 1.9 mm wide (flattened).

Habitat and Distribution: ***Dendrochilum longifolium*** is the most widely distributed of all ***Dendrochilum*** species. It is found in Myanmar (Burma), Peninsular Malaysia, Sumatra, Java, Borneo and New Guinea. In the Philippines it has been recorded from Lanao del Sur province on Mindanao and Palawan island. It usually grows as an epiphyte, occasionally as a lithophyte, at elevations from 600 to 2,200 metres.

Dendrochilum microchilum

(Schlechter) Ames
Section: ***Acoridium*** (Nees and Meyen) Pfitzer and Kränzlin

Dr. Rudolf Schlechter first named this charming species ***Platyclinis microchila.*** Professor Oakes Ames transferred the generic name to ***Dendrochilum*** in Fascicle II of his *Studies in the Family Orchidaceae*. The specific name refers to the species' small labellum.

Growth habit: upright, sympodial. **Pseudobulbs:** ovate, 1.5 cm long, 5 mm wide. **Leaves:** ovate, 8.5 cm long, 8 mm wide. **Inflorescences:** arch gracefully, appear with new growth, up to 14.5 cm long, bear up to 40 blooms. **Flower colour:** yellowish. **Dorsal sepal:** linear to lanceolate, up to 3 mm long, 1 mm wide. **Petals:** lanceolate, up to 2 mm long, 1 mm wide. **Lateral sepals:** linear to lanceolate, up to 3 mm long, 1 mm wide. **Labellum:** three-lobed; lateral lobes longer than mid-lobe.

Habitat and Distribution: ***Dendrochilum microchilum*** is endemic to the Philippines and has been found in the provinces of Bataan, Benguet, Camarines Sur, Ifugao, Mountain Province and Rizal on Luzon; Oriental Mindoro on Mindoro island; Antique on Panay island; Leyte, Negros Occidental and Negros Oriental on Negros island; on western Samar island in the Visayas; in the provinces of Agusan del Norte, Bukidnon, Davao del Sur and Misamis Occidental on Mindanao; and Camiguin island, north of Mindanao. This plant grows as an epiphyte between 500 and 2,400 metres.

Photo: David Titmuss

Dendrochilum microchilum

Dendrochilum niveum Ames

Section: ***Platyclinis*** (Bentham) Pfitzer

Professor Oakes Ames named this lovely miniature species in Fascicle VI of *Studies in the Family Orchidaceae* in 1920. The specific name refers to the white blooms. George Boettcher collected the type specimens in Apayao and Kalinga on Luzon in March 1918.

Growth habit: upright, sympodial. **Pseudobulbs:** oval, up to 3 cm tall, 1 cm wide, unifoliate. **Leaves:** lanceolate, up to 17 cm long including petiole (up to 3 cm), 2.5 cm wide; mid-vein very prominent on underside of leaf. **Inflorescences:** arch gracefully, appear with new growth, longer than a new leaf, bear up to 30 flowers about 5 mm in diameter. **Flower colour:** white; labellum pale yellow. **Dorsal** and **lateral sepals:** lanceolate, about 5 mm long, 1.5 mm wide. **Petals:** of similar shape to sepals but a little shorter. **Labellum:** three-lobed; lateral lobes triangular, irregularly notched; mid-lobe circular and irregularly notched.

Photo: David Banks

Dendrochilum niveum

Habitat and Distribution: ***Dendrochilum niveum*** is found in the mountains of Apayao, Kalinga and Rizal provinces on Luzon at about 1,000 metres. It grows as an epiphyte on mossy tree trunks and branches.

Notes: This species is closely related to ***Dendrochilum glumaceum*** Lindley but is smaller in all its floral and vegetative parts. The perfume is also different, being much sweeter than the scent of ***Dendrochilum glumaceum***, which is somewhat spicy.

Dendrochilum pallideflavens Blume

Section: ***Dendrochilum*** Bentham

Carl Blume named this species in 1825 in *Bijdragen tot de Flora van Nederlandsch Indie.* The specific name refers to the pale colour of the flowers.

Growth habit: upright, sympodial. **Pseudobulbs:** fusiform (wider in diameter at the middle than at the ends), spaced at 4 to 6 cm intervals on the rhizome, 4 cm tall, up to 8 mm wide, unifoliate. **Leaves:** lanceolate, up to 12 cm long, 2.5 cm wide. **Inflorescences:** arch gracefully, appear from the new growth, about equal in length to the leaves, bear up to 50 flowers about 6 mm in diameter. **Flower colour:** pale greenish to white. **Dorsal sepal:** linear, shortly pointed tip, 3 mm long, 1.2 mm wide. **Petals:** same as dorsal sepal but shorter, 2 mm long, 0.75 mm wide. **Lateral sepals:** linear to narrowly ovate, 3 mm long, 1.2 mm

Photo: David Banks

Dendrochilum pallideflavens

Photo: David Titmuss

Dendrochilum parvulum (all white)

wide. **Labellum:** tongue-shaped; two short ridges run lengthwise basally.

Habitat and Distribution: ***Dendrochilum pallideflavens*** is one of the most widely distributed ***Dendrochilum*** species known to science. It is found in Myanmar (Burma), Thailand, Malaysia, Sumatra, Java and Borneo. In the Philippines it has been recorded from western Samar island in the Visayas; Agusan del Norte and Zamboanga del Norte provinces on Mindanao; and Basilan, Sulu and Tawitawi islands in the Sulu Archipelago where it grows as an epiphyte at elevations between 80 and 1,400 metres.

Notes: In cultivation this species grows away from its host, producing air roots. These then grow toward the mount and upon reaching it branch freely. ***Dendrochilum pallideflavens*** is quite an untidy plant, unusual for members of this genus.

Dendrochilum parvulum

(Ames) Pfitzer
Section: ***Acoridium*** (Nees and Meyen) Pfitzer and Kränzlin

Professor Oakes Ames originally named this plant ***Acoridium parvulum***. Pfitzer transferred the genus to the more acceptable ***Dendrochilum*** in 1907 in A. Engler (ed.) *Das Pflanzenreich Regni Vegetabilis Conspectus* IV 50 II B7. The specific name refers to the small size of the plant. Elmer D. Merrill collected the type specimens in October 1903 in Bataan, Luzon.

Growth habit: upright, sympodial. **Pseudobulbs:** oval, up to 6 mm long, 2.5 mm in diameter, unifoliate. **Leaves:** up to 3 cm long, 5 mm wide. **Inflorescences:** semi-arching, appear with the new growth, carry up to 20 blooms about 5 mm in diameter. **Flower colour:** white with or without red markings on the labellum. **Dorsal sepal:** lanceolate, up to 3 mm long, 1 mm wide. **Petals:** lanceolate tapering to a sharp point, up to 3 mm long, 1 mm wide. **Lateral sepals:** lanceolate, up to 3 mm long, 1 mm wide. **Labellum:** three-lobed; lateral lobes noticeably longer than mid-lobe.

Habitat and Distribution: ***Dendrochilum parvulum*** is endemic to the Philippines where it has been found in the provinces of Apayao, Bataan, Benguet, Cagayan, Kalinga, Laguna, Mountain Province, Nueva Vizcaya, Quezon and Rizal on Luzon and Lanao del Sur and Misamis Occidental on Mindanao. It grows at elevations over 1,200 metres as an epiphyte.

Dendrochilum pulcherrimum

(Ames) L.O. Williams
Section: ***Heterantha*** H.Æ. Pedersen

Professor Oakes Ames first described this orchid as ***Acoridium pulcherrimum***. L.O. Williams gave it its currently accepted name in the *Philippine Journal of Science*, in 1951. The specific name refers to the beauty of this species when it is in flower. Maximo

Photo: David Titmuss

Dendrochilum parvulum (red lip)

Photo: David Titmuss

Dendrochilum pulcherrimum

Photo: David Banks

Dendrochilum quadrilobum

Ramos and G. Edaño collected the type specimens in Benguet province on Luzon in September 1921.

Growth habit: upright, sympodial. **Pseudobulbs:** oval, up to 3 cm long, 1 cm in diameter. **Leaves:** quite leathery, about 10 cm long, 1 cm wide. **Inflorescences:** heteranthous (terminal internode of shoot from which inflorescence sprouts produces neither pseudobulb nor leaf at any stage), arching, bear up to 50 flowers about 4 mm in diameter. **Flower colour:** bright orange. **Dorsal sepal:** oblong, up to 3 mm long, 1 mm wide. **Petals:** lanceolate, up to 2.5 mm long, 1 mm wide, edges minutely toothed. **Lateral sepals:** oblong-elliptic, up to 3 mm long, 1.5 mm wide. **Labellum:** three-lobed; lateral lobes just a little longer than the mid-lobe.

Habitat and Distribution: ***Dendrochilum pulcherrimum*** is endemic to the mountains of Benguet, Ifugao, Mountain Province, Nueva Ecija and Nueva Vizcaya in central Luzon. It grows at elevations over 1,200 metres as an epiphyte, often in bright light, on the outer branches of trees.

Dendrochilum quadrilobum Ames

Section: ***Acoridium*** (Nees and Meyen) Pfitzer and Kränzlin

Professor Oakes Ames originally named this plant ***Acoridium quadrilobum.*** He changed the name to the more acceptable ***Dendrochilum quadrilobum*** in 1915 in *Studies in the Family Orchidaceae* Fascicle V. The specific name refers to the four-lobed labellum. R.C. McGregor collected the type specimens in the province of Ifugao, Luzon, in February 1913.

Photo: David Banks

Dendrochilum rhombophorum

Photo: David Titmuss

Dendrochilum saccolabium

Growth habit: upright, sympodial. **Pseudobulbs:** oval, up to 2.5 cm long by 5 mm in diameter, covered with sheaths, unifoliate. **Leaves:** linear, up to 6 cm long by 5 mm wide. **Inflorescences:** arching, longer than the leaves, appear with the new growth, carry up to 25 thin, segmented, sweetly perfumed blooms 6 mm in diameter. **Flower colour:** pure white. **Dorsal** and **lateral sepals:** linear-lanceolate, up to 3 mm long by 1.2 mm wide. **Petals:** linear, up to 2 mm long by 0.5 mm wide. **Labellum:** small, vaguely four-lobed; lateral lobes shorter than the mid-lobe.

Habitat and Distribution: ***Dendrochilum quadrilobum*** is endemic to the Philippines. It has been found in Apayao, Ifugao and Kalinga on Luzon where it has limited distribution at elevations over 1,000 metres. It grows as an epiphyte in mossy cloud forests.

Dendrochilum rhombophorum

(Reichenbach *f.*) Ames

Section: ***Acoridium*** (Nees and Meyen) Pfitz. and Krzl.

Professor Heinrich G. Reichenbach first named this plant ***Coelogyne rhombophora***, then changed his mind and transferred the generic name to ***Pholidota rhombophora***. Professor Oakes Ames gave this plant its currently accepted name in 1908 in his *Studies in the Family Orchidaceae*, Fascicle III. The specific name refers to the oval pseudobulbs.

Growth habit: upright, sympodial. **Pseudobulbs:** oval, up to 3 cm tall by 5 mm in diameter, unifoliate. **Leaves:** linear-lanceolate, 10 cm long by 1 cm wide. **Inflorescences:** arching, longer than leaf, appear with the new growth, carry up to 30 flowers 5 mm in diameter. **Flower colour:** creamy yellow to green. **Dorsal sepal:** ovate, up to 2.5 mm long by 1 mm wide. **Petals:** broadly ovate, up to 2.5 mm long by 1.75 mm wide. **Lateral sepals:** ovate, up to 3 mm long by 2 mm wide. **Labellum:** vaguely three-lobed, oblong in shape; mid-lobe triangular, barely noticeable.

Habitat and Distribution: ***Dendrochilum rhombophorum*** is endemic to the Philippines and has been recorded from in Albay, Bataan, Laguna, Quezon and Rizal on Luzon and in Surigao del Norte on Mindanao. It grows as an epiphyte above 1,000 metres.

Dendrochilum saccolabium

Kränzlin

Section: ***Acoridium*** (Nees and Meyen) Pfitzer and Kränzlin

Dr. Fritz Kränzlin named this species in 1916 in *Annalen des K.K. Naturhistorischen Hofmuseums*. The specific name refers to the sac-like labellum.

Growth habit: upright, sympodial. **Pseudobulbs:** tapered, up to 2 cm long by 5 mm in diameter, unifoliate. **Leaves:** very leathery, up to 30 cm long by 7 mm wide. **Inflorescences:** semi-arching, appear with the new growth, carry up to 40 flowers 6 mm in diameter. The spotted bracts behind each flower prevent it from opening fully. **Flower colour:** dull red. **Dorsal** and **lateral sepals:** very rounded—a distinguishing feature of this species, up to 3.5 mm in diameter. **Petals:** small, oblong with rounded tips, up to 2 mm long by 1 mm wide. **Labellum:** three-lobed; mid-lobe is a sac-like structure.

Habitat and Distribution: ***Dendrochilum saccolabium*** is endemic to the Philippines and has been found in Laguna, Quezon and Rizal provinces on Luzon. It grows as an epiphyte at elevations above 1,200 metres.

Dendrochilum smithianum

(Ames) L.O. Williams
Section: ***Heterantha*** H.Æ. Pedersen

Professor Oakes Ames first named this plant ***Acoridium smithianum*** in 1937. Louis O Williams transferred the generic name to ***Dendrochilum*** in the *Philippine Journal of Science* in 1951. The generic name honours a Mr. Smith.

Photo: David Banks

Dendrochilum smithianum

Growth habit: upright, sympodial. **Pseudobulbs:** elliptical, up to 4 cm long by 1.2 cm in diameter. **Leaves:** leathery, linear to lanceolate, up to 12.5 cm long by 2.3 cm wide. **Inflorescences:** arching, bear up to 30 flowers about 4 mm in diameter. Flowers start to open from the proximal end (nearest to the plant) of the inflorescence. **Flower colour:** sepals and petals white to cream; labellum pink. **Dorsal sepal:** linear to lanceolate, up to 4 mm long by almost 1 mm wide. **Petals:** lanceolate, up to 3 mm long by up to 1 mm wide; edges jagged, particularly near the tips. **Lateral sepals:** lanceolate, up to 4 mm long by 1 mm wide. **Labellum:** three-lobed; side lobes pointed, longer than mid-lobe, which is rounded.

Habitat and Distribution: ***Dendrochilum smithianum*** is endemic to the Philippines and has been recorded from the provinces of Aurora, Quezon and Rizal on Luzon. It grows as an epiphyte at elevations about 500 metres.

Notes: ***Dendrochilum smithianum*** is closely related to ***Dendrochilum pulcherrimum*** but both species are readily separable from each other. The flowers of ***Dendrochilum smithianum*** are a paler colour, the inflorescences are much longer and the blooms open much more widely. The tips of the floral segments do not curve inward as much as those of ***Dendrochilum pulcherrimum***. The plants of ***Dendrochilum smithianum*** are also more robust and bigger-growing.

Dendrochilum stenophyllum

L.O. Williams
Section: ***Acoridium*** (Nees and Meyen) Pfitzer and Kränzlin

Louis O. Williams named this orchid in the *Philippine Journal of Science* in 1951. The specific name refers to the narrow leaves.

Growth habit: upright, sympodial. **Pseudobulbs:** tapered, up to 3 cm long by 4 mm in diameter, unifoliate. **Leaves:** up to 23 cm long by 3 mm wide, grooved along their length as if pleated. **Inflorescences:** semi-arching, appear with the new growth, bear up to 40 flowers 4 mm in diameter. **Flower colour:** cream. **Dorsal sepal** and **petals:** lanceolate, up to 2.5 mm long by 1 mm wide. **Lateral sepals:** lanceolate, up to 3 mm long by 1 mm wide. **Labellum:** three-lobed; lateral lobes crescent-shaped.

Photo: David Banks

Dendrochilum stenophyllum

Habitat and Distribution: ***Dendrochilum stenophyllum*** is of limited distribution in the mountains of Aurora, Quezon and Rizal provinces on Luzon. It is epiphytic at elevations about 1,000 metres.

Dendrochilum tenellum

(Nees and Meyen) Ames

Section: ***Acoridium*** (Nees and Meyen) Pfitzer and Kränzlin

Christian Nees and Carl Meyen originally described this species as ***Acoridium tenellum***. The generic name ***Acoridium*** is now considered by most botanists and taxonomists to be a section of ***Dendrochilum***. Professor Oakes Ames transferred the generic name to ***Dendrochilum*** in 1907 in the *Philippine Journal of Science*. The specific name refers to the delicate nature of this species.

Growth habit: upright, sympodial. **Pseudobulbs:** tapered, only 2 cm long by 2 mm in diameter. **Leaves:** grass-like, up to 30 cm long by 1 mm wide. **Inflorescences:** semi-arching, develop about three-quarters of the way along the leaf, appear with the new growth, carry up to 30 flowers about 3 mm in diameter. **Flower colour:** cream; labellum in some clones orange to red. **Dorsal sepal:** oblong, up to 2 mm long by 1 mm wide. **Petals:** elliptic, up to 2 mm long by 1 mm wide. **Lateral sepals:** oblong-lanceolate, up to 2 mm long by 1 mm wide. **Labellum:** three-lobed; lateral lobes a little longer than the mid-lobe.

Habitat and Distribution: ***Dendrochilum tenellum*** is the most widely spread ***Dendrochilum*** species in the Philippines. This species has been found in the provinces of Apayao, Bataan, Benguet, Cagayan, Camarines Sur, Ifugao, Ilocos Norte, Kalinga, Laguna, Mountain Province, Nueva Ecija, Nueva Vizcaya, Pampanga, Pangasinan, Quezon, Rizal, Sorsogon and Zambales on Luzon; the islands of Babuyan and the province of Oriental Mindoro on the island of Mindoro; Leyte island in the Visayas; Camiguin island, north of Mindanao; and the province of Davao del Norte on Mindanao. It is extremely common at elevations between 1,000 and 1,500 metres where it grows as an epiphyte.

Photo: David Titmuss

Dendrochilum tenellum

Dendrochilum tenuibulbum

(Ames) L.O. Williams

Section: ***Convoluta*** H.Æ. Pedersen

Professor Oakes Ames first named this species ***Acoridium tenuibulbum***. Louis O. Williams transferred the generic name to ***Dendrochilum*** in the *Philippine Journal of Science* in 1951. The specific name refers to the slender pseudobulbs of this plant.

Growth habit: upright, sympodial. **Pseudobulbs:** oval, up to 3 cm long by 5 mm in diameter, unifoliate.

Dendrochilum tenuibulbum

Leaves: lanceolate, 8 cm long by 1 cm wide. **Inflorescences:** upright, flowering portion pendulous, longer than the leaves, appear with the new growth, bear about 20 blooms 6 mm in diameter. **Flower colour:** bright yellow. **Dorsal** and **lateral sepals:** narrowly ovate, up to 4 mm long by 2.5 mm wide. **Petals:** rounded, up to 4 mm long. **Labellum:** three-lobed; lateral lobes fractionally longer than the mid-lobe.

Habitat and Distribution: *Dendrochilum tenuibulbum* is endemic to the Philippines where it is found in Albay and Rizal provinces on Luzon and Antique province on Panay island in the Visayas. It grows as an epiphyte, among mosses, on tree trunks in quite bright light at elevations over 950 metres.

Dendrochilum uncatum

Reichenbach *f.*
Section: ***Platyclinis*** (Bentham) Pfitzer

Professor Heinrich G. Reichenbach named this plant in 1855 in *Bonplandia.* The specific name refers to the inwardly bent arms on the column.

Growth habit: upright, sympodial. **Pseudobulbs:** oval, about 2.5 cm high by 1 cm in diameter, unifoliate. **Leaves:** lanceolate, 10 cm long by 1 cm wide. **Inflorescences:** arching, longer than the leaves, appear with the new growth, carry up to 30 flowers about 1 cm in diameter. **Flower colour:** green; labellum has brown markings. **Dorsal sepal:** ovate, up to 5 mm long by 2.5 mm wide. **Petals:** elliptic, up to 5 mm long by 3 mm wide. **Lateral sepals:** oblong-lanceolate, up to 5 mm long by 2.5 mm wide. **Labellum:** three-lobed with two ridges running toward the column.

Habitat and Distribution: *Dendrochilum uncatum* is found in Taiwan. In the Philippines it has been found

Dendrochilum uncatum

in Albay, Aurora, Bataan, Benguet, Cagayan, Ifugao, Laguna, Nueva Vizcaya, Pampanga, Quezon, Rizal, Sorsogon and Zambales provinces on Luzon; Antique and Capiz provinces on Panay island in the Visayas; and Agusan del Norte province on Mindanao. It grows as an epiphyte at elevations over 500 metres.

Notes: This species is closely related to ***Dendrochilum filiforme*** Lindl.

Dendrochilum vanoverberghii

Ames
Section: ***Convoluta*** H.Æ. Pedersen

Professor Oakes Ames named this species in the *Philippine Journal of Science* in 1912, after Father Morice Vanoverbergh, a Belgian priest, who did much work with the native tribes of central Luzon.

Growth habit: upright, sympodial. **Pseudobulbs:** elliptic, up to 2 cm long by 4 mm in diameter. **Leaves:** very leathery, linear, up to 8 cm long by 3 mm wide. **Inflorescences:** arching, bear up to 20 flowers about 3 mm in diameter. **Flower colour:** sepals and petals yellowish to white; labellum red. **Dorsal sepal:** oblong,

Photo: David Titmuss

Dendrochilum vanoverberghii

up to 2.5 mm long by 1 mm wide. **Petals:** oblanceolate, up to 2.3 mm long by .75 mm wide. **Lateral sepals:** ovate, up to 2.4 mm long by 1 mm wide. **Labellum:** three lobed; side lobes shorter than the mid-lobe; mid-lobe oblong, central part a short point.

Habitat and Distribution: *Dendrochilum vanoverberghii* is endemic to the Philippines and has been reported from the mountains of Ifugao, Mountain Province and Nueva Vizcaya provinces in central Luzon. It is an epiphyte at elevations over 2,000 metres, usually on mossy tree trunks or branches.

Dendrochilum wenzelii Ames

Section: ***Acoridium*** (Nees and Meyen) Pfitzer and Kränzlin

Professor Oakes Ames named this species after C.A. Wenzel, an orchid collector, in his *Studies in the Family Orchidaceae*, Fascicle V (1915). Maximo Ramos collected the type specimens on Mount Canayan in Camarines, Luzon, in December 1913.

Growth habit: upright, sympodial. **Pseudobulbs:** tapered, about 2.5 cm long by 5 mm in diameter, unifoliate. **Leaves:** very leathery, up to 30 cm long by 5 mm wide. **Inflorescences:** semi-arching, appear with the new growth; flowers are crowded, with up to 30 about 1 cm across the lateral sepals. **Flower colour:** extremely variable; clones have dull red, brown, orange, or greenish yellow flowers. **Dorsal sepal:** oblong, up to 5 mm long by 1.5 mm wide. **Petals:** oblong-lanceolate, up to 3 mm long by 1 mm wide. **Lateral sepals:** broadly lanceolate, up to 5 mm long by 2 mm wide. **Labellum:** three-lobed; mid-lobe forms a sac-like structure.

Photo: David Banks

Dendrochilum wenzelii (red)

Photo: David Banks

Dendrochilum wenzelii (yellow)

Photo: David Titmuss

Dendrochilum wenzelii (brown)

Photo: David Banks

Dendrochilum williamsii

Habitat and Distribution: ***Dendrochilum wenzelii*** is of limited distribution in the mountains of Albay, Apayao, Camarines Norte, Camarines Sur, Kalinga and Nueva Vizcaya provinces on Luzon; Leyte island in the Visayas; and Bukidnon province on Mindanao. It is epiphytic at elevations above 1,000 metres.

Dendrochilum williamsii (Ames) Pfitz.

Section: ***Acoridium*** (Nees and Meyen) Pfitzer and Kränzlin

Professor Oakes Ames originally named this species ***Acoridium williamsii*** in 1906. Pfitzer transferred the genus in 1907 in A. Engler (ed.) *Das Pflanzenreich. Regni Vegetabilis Conspectus* IV 50 II B7. The plant was named after R.S. Williams, who collected the type specimens in Benguet in September 1904.

Growth habit: upright, sympodial. **Pseudobulbs:** almost nonexistent. **Leaves:** grass-like, about 20 cm long, 2 mm wide, quite thick and flattened. **Inflorescences:** arching, appear about two-thirds of the way along the new leaf, develop with the new growth, about 5 cm long, bear up to 25 blooms about 4 mm in diameter. **Flower colour:** pale brown to pinkish. **Dorsal sepal:** ovate, up to 2.5 mm long, 1.5 mm wide. **Petals:** linear, up to 3 mm long, 1 mm wide. **Lateral sepals:** ovate to lanceolate, up to 2.5 mm long, 1.5 mm wide. **Labellum:** three-lobed; lateral lobes much longer than the mid-lobe.

Habitat and Distribution: ***Dendrochilum williamsii*** is a plant of limited distribution in the mossy cloud forests of the high mountains of Benguet, Ifugao, Nueva Vizcaya and Rizal provinces on Luzon. It grows as an epiphyte on the moss-covered branches of trees at elevations over 1,200 metres.

Dendrochilum yuccaefolium

L.O. Williams

Section: ***Platyclinis*** (Bentham) Pfitzer

Photo: David Titmuss

Dendrochilum yuccaefolium

Louis O. Williams named this plant in 1937 in the *Botanical Museum Leaflets of Harvard University*. The specific name refers to the yucca-like foliage.

Growth habit: upright, sympodial. **Pseudobulbs:** up to 4 cm long, diameter 1.5 cm, unifoliate. **Leaves:** up to 20 cm long, 1.5 cm wide, fairly stiff. **Inflorescences:** semi-arching, short, appear with the new growth, bear up to 30 flowers about 1.5 cm in diameter. **Flower colour:** creamy white. **Floral segments:** linear, up to 1 cm long by 1.5 mm wide. **Labellum:** lanceolate, two small ridges running lengthwise toward the column.

Habitat and Distribution: ***Dendrochilum yuccaefolium*** is of limited distribution, being found in the provinces of Aurora, Rizal and Quezon on Luzon. It grows as an epiphyte at elevations over 500 metres on the trunks and lower branches of trees in areas of good air movement and high constant humidity.

Dipodium R. Brown

Robert Brown established this genus in 1810 in *Prodromus Florae Novae Hollandiae et Insula van Diemen.* The generic name refers to the two stipes of the pollinia.

The type species for this genus is ***Dipodium punctatum*** J.E. Smith.

There are about 20 species in this genus and they are found in China, Laos, Cambodia, Vietnam, Thailand, Peninsular Malaysia, Indonesia, New Guinea, the Philippines, New Caledonia and Australia. The Philippines has only one species of this beautiful genus.

Photo: David Banks

Dipodium paludosum

Dipodium paludosum

(Griffith) Reichenbach *f.*

William Griffith originally named this species ***Grammatophyllum paludosum*** in 1851. Professor Heinrich G. Reichenbach transferred the generic name in *Xenia Orchidacea* in 1862. The specific name means that the plant grows in marshes.

Growth habit: upright, sympodial. **Stem:** up to 1 meter long by 1 cm in diameter. **Leaves:** about 15 cm long by 3 cm wide, clasp the stem. **Roots:** appear from anywhere along the stem. **Inflorescences:** upright, about 20 cm long, carry up to 10 flowers about 5 cm in diameter. **Flower colour:** sepals and petals pale yellow spotted with dull red; labellum white, marked with brown. **Dorsal** and **lateral sepals:** oblong with rounded ends, up to 2.5 cm long, 6 mm wide. **Petals:** oblong-lanceolate, up to 2.5 cm long by 6 mm wide. **Labellum:** narrow, up to 2 cm long by 6 mm wide; hairy ridge runs along the centre.

Habitat and Distribution: *Dipodium paludosum* is widely spread throughout Thailand, Vietnam, Cambodia, Peninsular Malaysia, Sumatra and Borneo. In the Philippines it has been recorded from Albay and Sorsogon provinces on Luzon; Leyte, Panay, Samar and Sibuyan in the Visayas; and Surigao del Sur province on Mindanao. It grows at elevations to 600 metres.

Notes: This species appears to begin life as a terrestrial. As the plant grows longer, it begins to climb the trunk of any nearby tree. The base of the plant will then rot and ***Dipodium paludosum*** lives on as an epiphyte.

Epicrianthes Blume

Carl Blume established this genus in 1825 in *Bijdragen tot de Flora van Nederlandsch Indie.* The generic name refers to the appendages on the petals.

The type species for this genus of most unusual plants is ***Epicrianthes javanica*** Bl.

There are approximately 20 species in this genus, the majority of which are found in Borneo. Other species are found in Laos, Myanmar (Burma), Thailand, Peninsular Malaysia, Indonesia, New Guinea, New Caledonia, the Solomon Islands and the Philippines. At least two species occur in the Philippines, one that is not described.

Epicrianthes flavofimbriata

(J.J. Smith) Garay and Kittredge
Synonym: ***Bulbophyllum flavofimbriatum*** J.J. Smith

Johannes J. Smith first described this beautiful and unusual species as ***Bulbophyllum flavofimbriatum*** in 1931. Doctors Garay and Kittredge transferred the name to ***Epicrianthes*** in 1985 in *Botanical Museum Leaflets of Harvard University.* The specific name refers to the yellow, fringe-like appendages of the petals that are present in the Borneo plant.

Growth habit: pendulous, up to 70 cm long, sympodial. **Rhizome:** about 1 mm in diameter with persistent bracts that also cover the pseudobulbs. **Pseudobulbs:** 1.5 cm long by 5 mm in diameter, deeply furrowed, unifoliate. **Leaves:** leathery, ovate, up to 5.5 cm long by 2 cm wide, edges curl under. **Inflorescences:** short, appear from the base of the pseudobulbs; more than one inflorescence is common; flowers about 1 cm in diameter. **Flower colour:** dorsal and lateral sepals yellowish green on the inside, spotted with brown on the outside; petals cream at the base and the appendages are black; labellum red with greenish stripes. **Dorsal** and **lateral sepals:** ovate, up to 8 mm long by 3.5 mm wide. **Petals:** unique with up to 10 appendages 6 mm long by 1 mm wide that are attached to the base of the petal by small filaments. **Labellum:** very fleshy, about 3 mm long by 1.5 mm wide, highly mobile.

Habitat and Distribution: ***Epicrianthes flavofimbriata*** has been recorded from Borneo (Sabah and Kalimantan). In the Philippines the author has found

Photo: David Titmuss

Epicrianthes flavofimbriata

this species in the province of Aurora on the east coast of central Luzon. It grows as an epiphyte in brightly lit habitats where there is plenty of air movement and high humidity at elevations about 500 metres.

Notes: This genus has not previously been recorded from the Philippines. Alcohol-preserved flowers are held at the Centre for Plant Biodiversity Research, Canberra, Australia, under the number MAC 9195.

There is also another species (not described) from the Philippines that was collected by G.E. Edaño in 1948 (No. PNH 5404). This plant was collected in the region of Sibulan, Negros Oriental province, Negros island, in the Visayas, at an elevation of about 800 metres. My thanks to Doctor Leslie A. Garay, who brought this Visayan species to my attention.

Epigeneium Gagnepain

This genus has been known by a variety of names over the years, firstly ***Sarcopodium***, then ***Dendrobium*** or ***Bulbophyllum***, later ***Katherinea*** and now ***Epigeneium***. Francois Gagnepain proposed the current name in 1932 in *Bulletin du Museum d'Histoire Naturelle, Paris.* The generic name refers to the position of the segments in relation to the column.

The type species for this magnificent genus is ***Epigeneium fargesii*** (Finet) Gagn.

There are approximately 35 species that are found in India, China, Taiwan, Vietnam, Thailand, Indonesia, Borneo, New Guinea and the Philippines. There are only three species in the Philippines.

Epigeneium acuminatum

(Rolfe) Summerhayes

Professor Oakes Ames first named this plant ***Dendrobium acuminatum*** in 1908; Rolfe then changed the genus to ***Sarcopodium***. Summerhayes gave this plant its currently accepted name in the *Kew Bulletin* in 1957. The specific name refers to the sharply pointed segments of the flower. T.E. Borden collected the type specimens in Bataan, Luzon, in May 1904.

Photo: Dr. Howard Wood

Epigeneium acuminatum

Growth habit: upright, sympodial. **Pseudobulbs:** distinctly four-angled, about 4 cm long, 2.5 cm high, bifoliate. **Leaves:** very leathery, up to 12 cm long, 3 cm wide. **Inflorescences:** upright, carry up to 10 blooms 5 cm in diameter. **Flower colour:** sepals and petals creamy white; labellum marked with brown at the base. In February 1996 I saw clones with a beautiful pink blush at the back of the dorsal and lateral sepals. **Dorsal** and **lateral sepals:** linear, tapering to a sharp point, about 2.5 cm long, 6 mm wide. **Petals:** linear, tapering to a point, up to 2.5 cm long, 7 mm wide. **Labellum:** three-lobed, up to 2 cm long, 1.6 cm wide (flattened); lateral lobes triangular-ovate; mid-lobe ovate.

Habitat and Distribution: ***Epigeneium acuminatum*** is endemic to the island of Luzon and is found in Apayao, Aurora, Bataan, Benguet, Kalinga, Quezon and Rizal. It grows as an epiphyte at elevations over 500 metres, often in full sun.

Epigeneium treacherianum

(Reichenbach *f. ex* Hooker *f.*) Summerhayes
Synonyms: ***Dendrobium lyonii*** Ames, ***Epigeneium lyonii*** (Ames) Summerh.

Photo: David Titmuss

Epigeneium treacherianum

Professor Reichenbach named this glorious plant ***Dendrobium treacherianum*** in 1881, after W.H. Treacher, a colonial secretary in Labuan island, near Sabah. Summerhayes transferred this species to its currently accepted genus in the *Kew Bulletin* in 1957.

Growth habit: upright, sympodial. **Pseudobulbs:** four-angled, about 4 cm square by 3 cm high, flattened over the stout rhizome. **Leaves:** very leathery, up to 15 cm long by 3 cm wide; pseudobulbs have two sharp thorn-like projections when the leaves have fallen. **Inflorescences:** arch gracefully, carry up to 20 flowers 7.5 cm in diameter. **Flower colour:** white at the tips darkening to purplish red in the centre; labellum the same colour as the other segments. There is also an albino form. **Dorsal** and **lateral sepals:** linear, tapering to a point, 5 cm long by 1 cm wide. **Petals:** linear-lanceolate, pointed, 4.5 cm long by 1 cm wide. **Labellum:** three-lobed, 3 cm long by 2.5 cm wide (when flattened); lateral lobes obtuse; mid-lobe ovate, drawn out to a point.

Habitat and Distribution: ***Epigeneium treacherianum*** is found in Borneo. In the Philippines it has been found in Apayao, Bataan, Kalinga, Laguna, Nueva Ecija, Nueva Vizcaya and Rizal provinces on Luzon; Oriental Mindoro province on Mindoro island; and Samar island in the Visayas. It inhabits elevations from 600 to 1,200 metres, growing as an epiphyte and occasionally as a lithophyte, often in bright-lit situations.

Eria Lindley

Dr. John Lindley described this genus in 1825 in the *Botanical Register*. Erias are closely related to the genus ***Dendrobium***; the major difference is that ***Eria*** has eight pollinia and ***Dendrobium*** only four. The generic name comes from the Greek word *erion* meaning woolly, in reference to the outer surface of the sepals in some species in this remarkable genus.

The type species for this genus is ***Eria javanica*** (Sw.) Bl.

Members for this genus are found throughout India, China, tropical Asia, New Guinea, Australia and the islands of the Pacific Ocean. Current estimates have about 500 members in ***Eria*** of which approximately 69 species occur in the Philippines. I believe many more await description.

Eria aporoides Lindley

Section: ***Aporodes*** Lindley
Synonym: ***Dendrobium aporoides*** (Lindl.) Merr.

Dr. John Lindley described this plant in 1859 and the name was published in the *Journal of the Linnean Society*. Its specific name refers to its resemblance to members of the ***Aporum*** section of ***Dendrobium***.

Growth habit: upright, sympodial, up to 20 cm tall. **Leaves:** two-ranked, clasp the stem alternately, very leathery, resemble teeth of a saw. **Inflorescences:** short, numerous bright orange bracts; blooms 6 mm in diameter, fragrant. **Flower colour:** white sepals and petals; labellum pale yellow. **Dorsal sepal** and **petals:** ovate, up to 4 mm long, 2.5 mm wide. **Lateral sepals:** triangular, 6 mm long, 3 mm wide, join at base to form small spur. **Labellum:** three-lobed, up to 5 mm long, 2 mm wide.

Habitat and Distribution: ***Eria aporoides*** is endemic to the Philippines and has been recorded from Albay, Aurora, Cagayan, Camarines Norte, Camarines Sur, Ilocos Norte, Laguna, Nueva Vizcaya, Quezon, Rizal and Zambales on Luzon; Polillo island; Leyte island in the Visayas; Camiguin island, north of Mindanao; Agusan del Norte, Agusan del Sur, Davao del Norte, Davao Oriental, Davao del Sur, Surigao del Norte, Surigao del Sur, Zamboanga del Norte and Zamboanga del Sur on Mindanao; and Basilan island. It grows as a lithophyte and an epiphyte at elevations to 1,000 metres.

Photo: Jim Cootes

Eria aporoides

Photo: David Banks

Eria brachystachya

Eria brachystachya Reichenbach *f.*

Section: ***Cylindrolobus*** Blume

Professor Heinrich G. Reichenbach described this plant in 1855 in *Bonplandia*. The specific name refers to the short inflorescences.

Growth habit: upright, sympodial. **Pseudobulbs:** at least 70 cm long by 5 mm in diameter. **Leaves:** up to

Photo: Andres Golamco Jr.

Eria bractescens

Photo: Jim Cootes

Eria floribunda

12.5 cm long, 8 mm wide, light green. **Inflorescences:** short, appear from opposite leaves, up to 5 cm long, carry lovely yellow bracts; blooms sweetly fragrant, about 2.5 cm in diameter. **Flower colour:** pure white; brown markings on labellum. **Dorsal sepal:** ovate, rounded tip, up to 1 cm long, 6 mm wide. **Petals:** oblong to ovate, similar dimensions to other segments. **Lateral sepals:** oblong, rounded tips, up to 1 cm long, 5 mm wide. **Labellum:** three-lobed, oblong.

Habitat and Distribution: ***Eria brachystachya*** has been found in Apayao, Camarines Norte, Camarines Sur, Kalinga, Quezon, Rizal, Sorsogon and Tarlac on Luzon; the Babuyan islands; Oriental Mindoro on Mindoro; Leyte island in the Visayas; Camiguin island, north of Mindanao; and Agusan del Norte, Agusan del Sur and Bukidnon on Mindanao. It grows as an epiphyte at elevations to 1,000 metres.

Eria bractescens Lindley

Section: ***Hymenaria*** Lindley

Dr. John Lindley named this plant in the *Botanical Register* in 1841. The specific name refers to the prominent floral bracts.

Growth habit: upright, sympodial. **Pseudobulbs:** ovate, up to 7.5 cm long, 3 cm in diameter. **Leaves:** up to four, oblong-lanceolate, up to 20 cm long, 3 cm wide. **Inflorescences:** upright, up to 15 cm long, several per pseudobulb; bottom portion has numerous cream bracts about 2 cm long, 5 mm wide; flowers up to 2 cm in diameter. **Flower colour:** cream. **Dorsal sepal:** elliptic, up to 1 cm long, 4 mm wide. **Petals:** similar to dorsal sepal. **Lateral sepals:** sickle-shaped, up to 1 cm long, 4 mm wide. **Labellum:** three-lobed, wedge-shaped, about 1 cm long, 6 mm wide.

Habitat and Distribution: ***Eria bractescens*** is a widely distributed species found in Nepal, Sikkim, India, Myanmar (Burma), Thailand, Laos, Cambodia, Vietnam, Malaysia, Indonesia and New Guinea. In the Philippines it has been found in Bataan and Rizal on Luzon; on Samar island in the Visayas; in Zamboanga del Norte and Zamboanga del Sur on Mindanao; and on Basilan island. It grows as an epiphyte at elevations up to 500 metres.

Eria floribunda Lindley

Section: ***Urostachya*** Lindley

Dr. John Lindley described this species in 1843 in the *Botanical Register*. The specific name refers to the plant's floriferous nature.

Growth habit: upright, sympodial. **Pseudobulbs:** up to 30 cm tall, 1 cm in diameter. **Leaves:** up to 15 cm long, 2.5 cm wide. **Inflorescences:** arching, 10 cm long, up to five from each new pseudobulb, carry up to 100 blooms about 5 mm in diameter. **Flower colour:** white; dark red anther cap. **Dorsal sepal:** ovate, rounded tip, up to 4 mm long by 3 mm wide. **Petals:** oblong, up to 3 mm long by 1.5 mm wide. **Lateral sepals:** triangular, up to 4 mm long by 4 mm wide. **Labellum:** about 4 mm long; base forms a scoop with two forward pointing lobes; mid-lobe fan-shaped.

Habitat and Distribution: ***Eria floribunda*** is found throughout Myanmar (Burma), Thailand, Vietnam,

Indonesia and Borneo. In the Philippines it has been found in Apayao, Kalinga, Nueva Ecija and Rizal provinces on Luzon; Palawan island; and Agusan del Norte, Agusan del Sur, Bukidnon and North and South Cotabato provinces on Mindanao. It grows as an epiphyte at elevations to 1,300 metres.

Eria gigantea Ames

Section: ***Mycaranthes*** Lindley

Professor Oakes Ames named this plant in 1908 in *Studies in the Family Orchidaceae*, Fascicle II. The specific name refers to the species' large growth habit. D. LeRoy Topping collected the type specimens in Bataan, Luzon, in 1904.

Growth habit: upright, sympodial. **Pseudobulbs:** up to 1 metre long, 2 cm in diameter. **Leaves:** 30 cm long, 3 cm wide, enclose the entire pseudobulb. **Inflorescences:** arching, up to four at each flowering, covered with small white hairs, appear from top of pseudobulbs, up to 20 cm long; blooms up to 7 mm in diameter. **Flower colour:** pale yellow with purple spots. **Dorsal sepal:** oblong-ovate, up to 4 mm long, 3 mm wide. **Petals:** oblong, 3 mm long, 1.5 mm wide. **Lateral sepals:** triangular-ovate, up to 4 mm long, 2.5 mm wide. **Labellum:** three-lobed, fairly large callus in the centre.

Habitat and Distribution: ***Eria gigantea*** is endemic to the Philippines and has been found in Bataan, Isabela, Laguna, Nueva Ecija and Rizal on Luzon and Agusan del Norte, Agusan del Sur, Lanao del Norte, Lanao del Sur and Misamis Occidental on Mindanao. It usually grows as a lithophyte and occasionally as a terrestrial at elevations between 600 and 1,300 metres.

Photo: David Titmuss

Eria gigantea

Photo: David Banks

Eria javanica

Eria javanica (Swartz) Blume

Section: ***Eria*** Lindley

Eria javanica was first named ***Dendrobium javanicum*** in 1800 by the Swedish botanist Olof Swartz. In 1836 Carl Blume gave this plant its currently accepted name in *Rumphia*. The specific name refers to Java, one of the islands where this plant is found.

Growth habit: upright, sympodial; plant variable in size. **Pseudobulbs:** 3 to 12 cm high, 2 to 4 cm in diameter, occasionally four-angled. **Leaves:** shiny, leathery, 15 to 60 cm long, 2.5 to 6 cm wide. **Inflorescences:** upright, up to 60 cm long; several can appear from the pseudobulb, up to 50 star-shaped flowers per inflorescence. **Flower colour:** cream, sometimes with red striping along segments and lateral lobes of labellum; blossoms up to 5 cm in diameter but more usually about 2.5 cm across the petals. **All segments:** oblong-lanceolate, about 1.5 cm long, 4 mm wide. **Labellum:** three-lobed, about 1 cm long, 4 mm wide.

Photo: David Banks

Eria longissima

Photo: Jim Cootes

Eria ornata

Habitat and Distribution: ***Eria javanica*** is widely spread throughout Asia, being found in India, Laos, Myanmar (Burma), Thailand, Peninsular Malaysia, Indonesia and New Guinea. In the Philippines it has been found in Albay, Benguet, Camarines Sur, Laguna, Mountain Province and Rizal on Luzon and Lanao del Norte, Lanao del Sur, Zamboanga del Norte and Zamboanga del Sur on Mindanao. It is found at elevations up to 1,200 metres and normally grows as an epiphyte.

Eria longissima Ames and Quisumbing

Section: ***Cylindrolobus*** Blume

Professor Oakes Ames and Dr. Eduardo Quisumbing named this plant in 1931 in the *Philippine Journal of Science*. The specific name alludes to its height.

Growth habit: upright, sympodial, a large grower. **Pseudobulbs:** readily attain heights over 1 metre. **Leaves:** up to 20 cm long, 3 cm wide. **Inflorescences:** short, appear opposite leaves, carry up to four flowers 2.5 cm in diameter. **Flower colour:** sepals and petals white, orange spotting on outside of small spur; labellum has cream centre, white margin, brownish markings on mid and lateral lobes. Each inflorescence has numerous yellow-green bracts. **Dorsal sepal:** oblong, up to 1.5 cm long, 6 mm wide. **Petals:** elliptic-oblong, about 1.2 cm long, 5 mm wide. **Lateral sepals:** triangular-lanceolate, up to 1.5 cm long, 6 mm wide. **Labellum:** three-lobed, 1 cm long, 8 mm wide (flattened); three ridges run most of the length.

Habitat and Distribution: ***Eria longissima*** is endemic to the Philippines and is found in the provinces of Laguna and Rizal on Luzon. It grows as an epiphyte at elevations about 1,000 metres.

Eria ornata (Blume) Lindley

Section: ***Dendrolirium*** Lindley

Carl Blume first named this species ***Dendrolirium ornatum*** in 1825. ***Dendrolirium*** is now considered to be a section of the genus ***Eria***. Lindley gave this plant its currently accepted name in 1830 in *Genera and Species of Orchidaceous Plants*. The specific name refers to the showiness of this orchid.

Growth habit: upright, sympodial. **Pseudobulbs:** flattened, up to 8 cm high, 3 cm wide, spaced at about 7 cm intervals along a rhizome 6 mm in diameter. **Leaves:** up to three, leathery, about 15 cm long, 3 cm wide. **Inflorescences:** upright, up to 30 cm long, on the upper half. Each inflorescence can carry up to 50 flowers about 2.5 cm in diameter. **Flower colour:** from salmon pink to dull brown; bright orange bract of variable size behind each flower; outside of lateral and dorsal sepals covered with little brown hairs. **Dorsal sepal:** lanceolate, up to 1.5 cm long, 4.5 mm wide. **Petals:** lanceolate, up to 1.3 cm long, 3 mm wide, bend forward to cover the column. **Lateral sepals:** triangular, up to 1.5 cm long, 8 mm wide. **Labellum:** three-lobed, 1.4 cm long, 6 mm wide; several ridges are present in the basal half.

Habitat and Distribution: ***Eria ornata*** is a widely spread species, being found in Thailand, Peninsular Malaysia, Sumatra, Java, Borneo and Sulawesi. In the Philippines it is found in Benguet, Ifugao, Mountain Province, Nueva Vizcaya, Rizal and Zambales provinces on Luzon and Agusan del Norte, Agusan del Sur, Bukidnon, Davao del Sur, Zamboanga del Norte and Zamboanga del Sur provinces on Mindanao. It grows as an epiphyte at elevations to 1,450 metres.

Eria ovata Lindley

Section: ***Urostachya*** Lindley

Dr. John Lindley named this species in the *Botanical Register* in 1844. The specific name possibly refers to the shape of the labellum.

Growth habit: upright, sympodial. **Pseudobulbs:** up to 20 cm long, 1.2 cm in diameter, bear up to four leaves. **Leaves:** linear-oblong to oblong-lanceolate, up to 15 cm long, 2.1 cm wide, unequally bilobed at the tip. **Inflorescences:** upright, up to 23 cm long, appear from top of pseudobulbs, bear up to 45 flowers about 1.3 cm in diameter; flowers have a most unpleasant perfume. **Flower colour:** pale yellow; labellum pale yellow at front, very dark red at the base; inflorescence green and shiny; each flower has a green bract 1 cm long, 3 mm wide that bends backward. **Dorsal sepal:** ovate, 1 cm long, 4 mm wide. **Petals:** lanceolate, pointed, 8 mm long, 3 mm wide. **Lateral sepals:** triangular, pointed, 1 cm long, 4 mm wide. **Labellum:** small, ovate, pointed, 5 mm long, 4 mm wide.

Habitat and Distribution: ***Eria ovata*** is found in Taiwan and the Ryukyu Islands. In the Philippines it has been found in Albay, Bataan, Benguet, Cagayan, Ifugao and Laguna provinces on Luzon; the Leyte islands and Negros Oriental province on Negros island in the Visayas; and Agusan del Norte and Bukidnon provinces on Mindanao. It grows as an epiphyte at elevations between 500 and 2,550 metres.

Photo: David Titmuss

Eria ovata

Eria ovata var. *retroflexa*

(Lindley) Garay and Sweet

Section: ***Urostachya*** Lindley

Dr. John Lindley first named this species ***Eria retroflexa***. Dr. Leslie Garay and Herman Sweet gave the plant varietal status in 1974 in *Orchids of the Southern Ryukyu Islands.* The varietal name possibly refers to the bent back bracts on the inflorescence.

Growth habit: upright, sympodial. **Pseudobulbs:** up to 20 cm long by 1.2 cm in diameter, bear up to four leaves. **Leaves:** linear-oblong to oblong-lanceolate, up to 17 cm long by 2.5 cm wide. **Inflorescences:** upright, up to four, appear from along the pseudobulbs, bear up to 25 flowers about 8 mm in diameter; flowers do not open widely. **Flower colour:** pale yellow and pink; inflorescence pink and shiny. Each flower has a pink and green bract, 6 mm long by 2 mm wide, which bends backward. **Dorsal** and **lateral sepals:** triangular, pointed, 8 mm long by 2 mm wide. **Petals:** lanceolate, pointed, 6 mm long by 2 mm wide. **Labellum:** small, ovate, pointed, bright yellow with purple at the base.

Photo: David Titmuss

Eria ovata var. ***retroflexa***

Habitat and Distribution: ***Eria ovata*** var. ***retroflexa*** is a widely distributed species found in Taiwan, the Ryukyu Islands, Borneo, Sulawesi (Celebes) and New Guinea. In the Philippines it has been found in Rizal and Sorsogon provinces on Luzon; Oriental Mindoro province on Mindoro island ; Bohol and Leyte islands in the Visayas; and Surigao del Norte province on Mindanao. It grows as an epiphyte at elevations to 1,200 metres.

Photo: David Banks

Eria polyura

Eria polyura Lindley

Section: ***Polyura*** Schlechter

Dr. John Lindley named this species in 1841 in the *Botanical Register.* The specific name is from the Greek *poly* (many) *oura* (tail), referring to the many tail-like inflorescences that appear simultaneously.

Growth habit: semi-pendulous, sympodial. **Pseudobulbs:** about 20 cm long by 1 cm in diameter, somewhat club-shaped. This orchid has an interesting growth habit. Each new growth begins about halfway along the previous year's pseudobulb, making the plant longer each year. I have seen this species growing in its native habitat, where it starts its life a metre or so up a tree trunk. As the plant ages it grows happily as a semi-terrestrial along the top of leaves on the forest floor. Quite frequently plants in the wild attain a length of several metres. **Leaves:** up to five, lanceolate, shiny, up to 15 cm long by 2 cm wide. **Inflorescences:** pendulous, about 10 cm long, appear from the upper portion of the plant, bear up to 40 flowers 1.5 cm in diameter. **Flower colour:** creamy white; labellum reddish. **Dorsal sepal** and **petals:** oblong, up to 7.5 mm long by 4 mm wide. **Lateral sepals:** obovate, with rounded tips about 7.5 mm long by 4 mm wide. **Labellum:** heart-shaped to ovate, about 1 cm long.

Habitat and Distribution: ***Eria polyura*** is endemic to the Philippines. It has been found in Albay, Aurora, Bataan, Benguet, Cagayan, Ilocos Norte, Laguna, Mountain Province, Quezon and Rizal provinces on Luzon; Palawan island; Leyte, Panay and Sibuyan islands in the Visayas; and Lanao del Norte, Lanao del Sur, Surigao del Norte and Surigao del Sur provinces on Mindanao. It grows as an epiphyte and sometimes as a semi-terrestrial at elevations to 2,400 metres.

Photo: David Banks

Eria profusa

Eria profusa Lindley

Section: ***Hymenaria*** Lindley

Dr. John Lindley named this plant in 1842 in the *Botanical Register.* The specific name refers to the profusion of blooms produced by this species.

Growth habit: upright, sympodial. **Pseudobulbs:** oval-shaped, up to 7.5 cm tall, 4.5 cm in diameter. **Leaves:** two to four, lanceolate, up to 27 cm long, 4.5 cm wide. **Inflorescences:** arching, up to 37 cm long, bear up to 70 blooms covered with short brown hairs; flowers 1 cm in diameter, outside of dorsal and lateral sepals have small brownish hairs, anther cap dark red. **Flower colour:** pale yellow; labellum bright yellow. **Dorsal sepal:** elliptic-oblong, up to 5 mm long, 3 mm wide. **Petals:** oblong, up to 5 mm long, 2.5 mm wide. **Lateral sepals:** triangular, up to 5 mm long, 5 mm wide. **Labellum:** three-lobed; mid-lobe almost round.

Habitat and Distribution: ***Eria profusa*** is endemic to the Philippines where it has been found in Apayao, Isabela, Kalinga and Rizal provinces on Luzon; Negros Oriental province on Negros island in the Visayas; Agusan del Norte and Surigao del Sur provinces on Mindanao; and islands of the Sulu Archipelago. It grows as an epiphyte at elevations to 1,000 metres.

Eria rhodoptera Reichenbach *f.*

Section: ***Hymenaria*** Lindley

Professor Heinrich G. Reichenbach named this plant in 1882 in the *Gardeners' Chronicle*. The specific name refers to the rosy-coloured petals of the flower.

Growth habit: upright, sympodial. **Pseudobulbs:** up to 10 cm tall, diameter 3 cm. **Leaves:** leathery, up to five, about 18 cm long by 3 cm wide at the top of the pseudobulb. **Inflorescences:** upright, up to 20 cm long, bear approximately 30 blooms 1.5 cm in diameter. **Flower colour:** sepals white to cream; petals crimson; labellum white; lateral lobes crimson. Each bloom has a cream coloured bract about 2 cm long, 3 mm wide. **Dorsal sepal:** oblong, up to 1.5 cm long, 4 mm wide. **Petals:** oblong, point forward over the column. **Lateral sepals:** oblong-lanceolate, up to 1.5 cm long, 5 mm wide. **Labellum:** three-lobed, 1 cm long, 4 mm wide.

Habitat and Distribution: ***Eria rhodoptera*** is endemic to the Philippines and has been found in Rizal province on Luzon and on Busuanga and Palawan islands. It grows as an epiphyte at elevations around 600 metres.

Photo: David Titmuss

Eria rhodoptera

Eria tridens Ames

Section: ***Urostachya*** Lindley

Professor Oakes Ames named this beautiful species in *Elmer's Leaflets of Philippine Botany* in 1912. The specific name refers to the shape of the labellum, which resembles a trident.

Growth habit: upright, sympodial. **Pseudobulbs:** carry two leaves, up to 5 cm tall by 1.8 cm in diameter. **Leaves:** 18 cm long by 2.5 cm wide, unequally lobed at the tips. **Inflorescences:** covered in short white hairs, arching, up to three appear from each pseudobulb with about 35 flowers 8 mm in diameter. **Flower colour:** pinkish white; back of each flower has a greenish yellow bract 7 mm long by 2.5 mm wide; backs of dorsal and lateral sepals have minute white hairs. **Dorsal sepal:** oblong to elliptic, up to 6 mm long by 2 mm wide. **Petals:** ovate, up to 5 mm long by 2 mm wide. All these segments have pointed tips. **Lateral sepals:** broadly triangular, up to 8 mm long by 6 mm wide. **Labellum:** distinctly tapered, 5 mm long by 3.5 mm wide; apex has three teeth.

Habitat and Distribution: ***Eria tridens*** is endemic to Mount Apo in Davao del Sur province on Mindanao. It grows as an epiphyte at elevations around 1,000 metres.

Photo: David Banks

Eria tridens

Photo: David Banks

Eria vanoverberghii

Eria vanoverberghii Ames

Section: ***Mycaranthes*** Lindley

Professor Oakes Ames named this interesting species in 1912 in the *Philippine Journal of Science.* The specific name honours Father Vanoverbergh, who collected the type specimens in the Mountain Province in central Luzon.

Growth habit: upright to semi-pendulous, sympodial. **Pseudobulbs:** up to 1 metre long by 1 cm in diameter. **Leaves:** strap-shaped, pointed at the tip, leathery, up to 25 cm long by 3 cm wide. **Inflorescences:** up to 25 cm long; up to six appear from the tip of the pseudobulb, bear many small flowers about 1 cm in diameter. **Flower colour:** white; labellum white with very dark red markings; back of flower has a large and conspicuous white bract. **Dorsal sepal:** oblong to lanceolate, hairy on the back, up to 7 mm long by 3 mm wide. **Petals:** oblong to almost elliptic, up to 5 mm long by 2 mm wide. **Lateral sepals:** oblong, acute, hairy at the back, up to 7 mm long by 3 mm wide. **Labellum:** three-lobed; side lobes triangular; mid-lobe nearly square with a pointed tip 3 mm long by 4.5 mm wide.

Habitat and Distribution: ***Eria vanoverberghii*** is endemic to the Philippines and has been recorded from the provinces of Apayao, Kalinga, Mountain Province, Nueva Vizcaya, Quezon and Rizal on Luzon and has been collected in the province of Oriental Mindoro on the island of Mindoro. It grows as an epiphyte at elevations between 300 and 1,550 metres.

Photo: Jim Cootes

Eria xanthocheila

Eria xanthocheila Ridley

Section: ***Hymenaria*** Lindley

Henry Ridley named this species in 1907 in *Material for a Flora of the Malay Archipelago.* The specific name refers to the bright yellow labellum.

Growth habit: upright, sympodial; a creeping rhizome as thick as the pseudobulbs, about 1 cm. **Pseudobulbs:** up to 30 cm tall, carry up to four leaves. **Leaves:** 12 cm long, 2.5 cm wide. **Inflorescences:** arching, up to three appear from each pseudobulb with about 40 flowers 1.5 cm in diameter. **Flower colour:** yellowish sepals and petals; labellum bright yellow; lateral lobes of labellum light purple, in some clones yellow; behind each flower there is a creamy yellow bract 1.2 cm long, 6 mm wide, which bends backward. **Dorsal sepal:** ovate, up to 1 cm long, 5 mm wide; tip of segment rounded. **Petals:** ovate, up to 8 mm long, 4 mm wide; tip of segment rounded. **Lateral sepals:** broadly ovate, up to 1 cm long by 8 mm wide; tip of segment rounded. **Labellum:** distinctly three-lobed; lateral lobes upright; mid-lobe almost circular.

Habitat and Distribution: ***Eria xanthocheila*** is found in the lowlands of Thailand, Malaysia, Sumatra and Java where it grows as an epiphyte at elevations to 1,200 metres. The plants from the Philippines are from an unknown locality. It is possible that this species is not described.

Euanthe Schlechter

Dr. Rudolf Schlechter named this genus, of only one species, in 1914 in *Die Orchideen.* It is closely related to ***Vanda,*** its original name. Differences in the labellum separate the two genera. The generic name refers to the showy inflorescences.

This magnificent genus is endemic to the island of Mindanao.

Euanthe sanderiana

(Reichenbach *f.*) Schlechter
Synonym: ***Vanda sanderiana*** Rchb. *f.*

Professor Heinrich G. Reichenbach originally named this species ***Vanda sanderiana*** in 1882 in the *Gardeners' Chronicle*, in honour of Henry F. Sander, famous nurseryman and patron of orchids, of St. Albans, England. Dr. Rudolf Schlechter created the generic name, *Euanthe* in *Die Orchideen* in 1914 and transferred ***Vanda sanderiana*** to his new genus.

Growth habit: upright, monopodial, up to 1 metre tall. **Leaves:** arch gracefully, up to 40 cm long by 3 cm wide, tips unequally notched. **Inflorescences:** upright, carry up to 10 flowers 10 cm in diameter. **Flower colour:** dorsal sepal and petals pale pink with some dark spotting toward the centre; lateral sepals greenish brown with darker brown tessellations; labellum purplish brown. **Dorsal and lateral sepals:** broadly elliptic, up to 4 cm long by 2.5 cm wide. **Petals:** broadly elliptic, up to 3.5 cm long by 2 cm wide. **Labellum:** three-lobed, about 2.5 cm long by 1.75 cm wide; side lobes rounded, upright; mid-lobe oblong to circular; three ridges. Has no spur, which is the main feature separating ***Euanthe*** from ***Vanda***.

Habitat and Distribution: ***Euanthe sanderiana*** is endemic to Mindanao island where it is found in North and South Cotabato, Davao del Norte, Davao del Sur, Zamboanga del Norte and Zamboanga del Sur. It grows as an epiphyte at elevations to 500 metres.

Notes: In my opinion, undoubtedly the finest and most important orchid species from the Philippines.

Photo: Jim Cootes

Euanthe sanderiana

Photo: Jim Cootes

Euanthe sanderiana var. ***albata***

Euanthe sanderiana var. *albata*

Reichenbach *f.*

Professor Heinrich G. Reichenbach described this in the *Gardeners' Chronicle* in 1887. The habit of the plant and the flower size are just a little smaller than those of the species. The flowers are green and white.

Habitat and Distribution: Reported from Davao del Sur and South Cotabato on Mindanao provinces where it grows as an epiphyte at elevations to 500 metres. This variety is extremely rare in its natural habitat.

Notes: The orchid world is fortunate for the dedication of growers in Hawaii and Thailand who have done wonderful work in propagating both the albino and normal-coloured forms of this glorious species. Vastly superior forms are now readily available to any orchid grower who can provide the winter minimum temperature of 15 degrees Celsius that this species requires.

Flickingeria Hawkes

Alex Hawkes established this genus of plants, with short-lived flowers, in 1965 in *The Orchid Weekly.* The generic name honours Edward Flickinger, an orchid journal editor.

The type species for this genus of peculiar plants is ***Flickingeria angulata*** (Bl.) Hawkes.

There are approximately 50 species found throughout mainland Asia, Indonesia, Borneo, New Guinea, the Philippines, the Pacific Islands and the north of Australia. There are at least nine species described from the Philippines.

Flickingeria fimbriata (Blume) Hawkes

Section: ***Plicatiles*** Seidenf.

This widespread species was first named in 1825 by Carl Blume, but the generic name ***Desmotrichum*** was already used for a genus of algae. Dr. John Lindley then transferred the generic name to ***Dendrobium*** in 1830. The new generic name, ***Flickingeria,*** was applied to this group of species in 1961 by Alex Hawkes in *The Orchid Weekly.* The specific name is from the Latin *fimbria,* meaning thread or shred, and refers to the decomposed remnants of the sheaths of the branching stems.

Photo: Hans Fessel

Flickingeria fimbriata

Growth habit: upright, sympodial. **Rhizome:** up to 6 mm in diameter. **Stems:** appear about every sixth node and carry a pseudobulb. **Pseudobulb:** up to 6 cm long by 2 cm in diameter, unifoliate. **Leaves:** leathery, terminal on the pseudobulb, elliptic to obtuse in shape, up to 7 cm long by 3 cm wide. **Inflorescences:** short, single flowered, perfumed; blooms, about 3 cm in diameter, in the axil of the leaf. **Flower colour:** cream with brownish red spotting to pinkish with darker spots; labellum pale yellow with brownish red streaks to creamy white. **Dorsal sepal:** lanceolate, up to 2 cm long by 5 mm wide. **Petals:** similar in size and shape to the other segments. **Lateral sepals:** linear to lanceolate, 2 cm long by 5 mm wide. **Labellum:** three-lobed, up to 2 cm long by 1 cm wide at the tip; mid-lobe ruffled; side lobes triangular. Three ridges run the length of the labellum and once past the side lobes they become wavy.

Habitat and Distribution: ***Flickingeria fimbriata*** is a widespread species found in China, Vietnam, Thailand, the Nicobar Islands, Malaysia, Sumatra, Java, Sulawesi (Celebes) and Borneo. In the Philippines it has been found in Aurora, Bataan, Cagayan, Isabela, Laguna, Pampanga, Quezon, Rizal and Tarlac provinces on Luzon; Dinagat island; Catanduanes and Palawan islands; Leyte, Panay and Samar islands in the Visayas; and Agusan del Norte, Lanao del Sur and Surigao del Sur provinces on Mindanao. It grows as an epiphyte at elevations to 800 metres.

Flickingeria interjecta (Ames) Hawkes

Section: ***Bilobulatae*** Seidenf.

Professor Oakes Ames first named this plant ***Dendrobium interjectum***. The new generic name was applied to this group of species in 1961 by Alex Hawkes in *The Orchid Weekly.* The specific name refers to the space between the pseudobulbs.

Growth habit: upright, sympodial. **Rhizome:** up to 3 mm in diameter. **Stems:** appear about every sixth node, carry a pseudobulb. **Pseudobulb:** up to 2.5 cm long by 7 mm in diameter, unifoliate. **Leaves:** leathery, terminal on the pseudobulb, elliptic to obtuse in shape, up to 4 cm long by 9 mm wide. **Inflorescences:** short, single-flowered; blooms about 8 mm in diameter, behind the leaf. **Flower colour:** pale yellow; labellum yellow for the apical third, purplish the basal two-thirds. **Dorsal sepal:** broadly ovate, up to 4.5 mm long by 3.5 mm wide. **Petals:** linear, 4.5 mm long by 2 mm wide. **Lateral sepals:** triangular, 9 mm long by 5 mm wide. **Labellum:** three-lobed, 1.25 cm long by 6 mm wide (when flattened); mid-lobe divided.

Habitat and Distribution: ***Flickingeria interjecta*** is endemic to the Philippines and has been found in Aurora and Quezon provinces on Luzon and Leyte and Panay (Capiz province) islands in the Visayas. It grows as an epiphyte, usually in bright light at elevations between 100 and 700 metres.

Flickingeria interjecta

Flickingeria junctiloba

Flickingeria junctiloba

Fessel and Lückel
Section: ***Bilobulatae*** Seidenf.

Hans Fessel and Emil Lückel named this species in the September-October 1998 issue of *Die Orchidee.* The specific name refers to the conspicuous connection of the side lobes and the callus on the labellum.

Growth habit: upright, sympodial. **Internodes:** 7 mm to 1.3 cm long, 1 mm in diameter; each internode has a number of hair-like bracts. **Pseudobulbs:** consist of two to three internodes with upper internode about 3.5 mm in diameter, bear a single leaf; new growth emerges from previous year's pseudobulb. **Leaves:** leathery, lanceolate, 4 cm long by 1.5 cm wide. **Inflorescences:** single-flowered, appear from a bract beneath the leaf. **Flower colour:** sepals and petals greenish yellow; labellum white; anther cap yellow; column has purple spotting at the base. **Dorsal sepal:** broadly lanceolate, 5 mm long, 2 mm wide. **Petals:** linear, 6 mm long, 1.5 mm wide. **Lateral sepals:** ovate, 6 mm long, 4 mm wide. **Labellum:** three-lobed; mid-lobe parallel ending in bi-lobed tip 2 mm long; overall length of labellum about 8 mm, 2 mm wide (at mid-lobe). Conspicuous, conical callus on labellum near column.

Habitat and Distribution: ***Flickingeria junctiloba*** is endemic to Leyte island in the Visayas. It grows as an epiphyte in the mountains on the west of the island.

Flickingeria luxurians

Flickingeria luxurians

(J.J. Smith) Hawkes
Section: ***Plicatiles*** Seidenf.

Johannes J. Smith named this pretty species as a ***Dendrobium*** in 1921. The generic name was transferred by Alex Hawkes in *The Orchid Weekly* in 1961. The specific name may refer to the luxuriant growth of some clones of this species.

Growth habit: upright, sympodial. **Rhizome:** up to 7 mm in diameter, creeps along the host plant. **Stems:**

Photo: David Banks

Flickingeria scopa

Photo: Hans Fessel

Flickingeria xantholeuca

appear at intervals of about every 6 to 7 cm, up to 7 cm high, bear a pseudobulb. **Pseudobulbs:** about 4 cm high by 2 cm in diameter, unifoliate. **Leaves:** lanceolate, leathery, up to 10 cm long by 3.5 cm wide. **Inflorescences:** short, single-flowered, perfumed; blooms about 3 cm in diameter, appear in the leaf axil. **Flower colour:** white with reddish spots and blotches toward the centre. **Floral segments:** oblong to lanceolate, up to 2 cm long by 6 mm wide. **Labellum:** three-lobed, about 2 cm long by 1 cm wide at the tip.

Habitat and Distribution: ***Flickingeria luxurians*** is found in Thailand, Malaysia, Java and Borneo. In the Philippines it has been recorded from Surigao del Norte and Surigao del Sur provinces on Mindanao and the nearby island of Dinagat. It grows as an epiphyte at elevations up to 300 metres.

Flickingeria scopa (Lindley) Brieger

Section: ***Flickingeria*** Seidenf.
Synonym: ***Dendrobium scopa*** Lindl.

Dr. John Lindley first named this species ***Dendrobium scopa*** in 1842 in the *Botanical Register.* Alex Hawkes transferred this plant to the genus ***Flickingeria*** in 1961 in *The Orchid Weekly.* The specific name refers to the bristly hairs at the end of the labellum.

Growth habit: upright, becomes semi-pendulous as the plant becomes older and longer, sympodial. **Pseudobulbs:** consist of three to four nodes of varying lengths with the uppermost one being the longest at up to 6 cm long by 1.8 cm in diameter, unifoliate; new growth arises from the base of the uppermost node and gets progressively smaller. The plant can attain a length of over 1 metre. **Leaves:** ovate, leathery, up to 15 cm long by 6 cm wide. **Inflorescences:** short, single-flowered, appear from bracts at the junction of the leaf and the top of the pseudobulb; blooms about 2 cm in diameter and short-lived, barely lasting a day. **Flower colour:** sepals and petals cream to milky white; labellum yellow and sometimes spotted with purple in the centre. **Dorsal** and **lateral sepals:** lanceolate, up to 1 cm long by 3 mm wide. **Petals:** narrower, up to 1 cm long by 2 mm wide. **Labellum:** three-lobed; lateral lobes small; mid-lobe divided into about 50 wavy, hair-like segments.

Habitat and Distribution: ***Flickingeria scopa*** is found in New Guinea, Sulawesi, Taiwan and Guam. In the Philippines it has been found in the provinces of Aurora, Laguna, Quezon, Rizal and Tarlac on Luzon; Oriental Mindoro on Mindoro island; and North and South Cotabato on Mindanao. It grows as an epiphyte, occasionally as a lithophyte, at elevations to 500 metres.

Notes: The larger growing species of ***Flickingeria*** are commonly sold at roadside nurseries in the Philippines as Cattleya, which they sometimes resemble when the upper growth is removed.

Flickingeria xantholeuca

(Rchb. *f.*) Hawkes
Section: ***Bilobulatae*** Seidenf.

Professor Heinrich G. Reichenbach first named this species ***Dendrobium xantholeucum*** in *Xenia Orchidacea* in 1862. The new generic name was applied to this group of species in 1961 by Alex Hawkes in *The Orchid Weekly.* The specific name refers to the yellowish white flowers.

Growth habit: upright, sympodial. **Pseudobulbs:** slightly flattened, up to 3 cm long by 1 cm in diameter, unifoliate. **Leaves:** ovate-acute, up to 7.5 cm long by 2.5 cm wide. **Inflorescences:** single-flowered, appear from bracts at front of leaf. **Flower colour:** sepals and petals pale greenish yellow; labellum yellow; anther cap bright green. **Dorsal sepal:** broadly ovate, up to 5 mm long by 4 mm wide. **Petals:** triangular, up to 4 mm long by 2.5 mm wide. **Lateral sepals:** triangular, up to 6 mm long by 5 mm wide. **Labellum:** three-lobed; mid-lobe deeply cleft; side lobes short, triangular; two ridges run the length of the labellum, front half of ridges wavy; a third ridge runs half the length of the labellum from the front.

Habitat and Distribution: ***Flickingeria xantholeuca*** is a widely distributed species and has been recorded from Thailand, Malaysia, Sumatra, Java and Borneo. In the Philippines it is only known from Leyte in the Visayas and the province of Davao del Sur in Mindanao. It grows as an epiphyte at elevations up to 500 metres.

Goodyera R. Brown

Robert Brown established this genus of terrestrial orchids in 1813 in *W. Aiton et W.T. Aiton, Hortus Kewensis.* The generic name honours John Goodyer, an English botanist of the 1600s.

The type species for this genus is ***Goodyera repens*** (L.) R. Br.

The genus consists of about 40 species that are found throughout the warmer parts of Europe, the Middle East, Madagascar, North America, Mexico, the islands of the Pacific, northern Australia and throughout Asia. The Philippines has nine species.

Goodyera clausa

(A.A. Eaton *ex* Ames) Schlechter

A.A. Eaton first named this species ***Epipactis clausa*** in *Studies in the Family Orchidaceae*, published by Professor Oakes Ames 1908. Rudolf Schlechter changed the generic name in *Engler's Botanische Jahrbücher* in 1911. The specific name refers to the flowers, which were not fully open when the original description was made.

Growth habit: terrestrial, sympodial. **Stems:** upright, up to 20 cm tall. **Leaves:** three, 2 to 3 cm apart, ovate-lanceolate, up to 3.5 cm long by 1.3 cm wide, dark green with white or yellow veins. **Inflorescence:** about 4 cm long, hairy, up to 20 blooms. **Flower colour:** white. **Dorsal sepal:** ovate to ovate-lanceolate. **Petals:** not symmetrical, obtuse, widen into a round or axe-shaped plate above claw. **Lateral sepals:** ovate to ovate-lanceolate. **Labellum:** saccate at base, very broadly ovate.

Habitat and Distribution: ***Goodyera clausa*** is endemic to the Philippines and has been recorded from Benguet, Ifugao and Quezon provinces on Luzon and Negros Oriental province on Negros island in the Visayas. It grows as a terrestrial in damp, shaded localities at elevations about 2,200 metres.

Photo: Dr. Henrik Pedersen

Goodyera clausa

Grammatophyllum Blume

Carl Blume established this genus in 1825 in *Bijdragen tot de Flora van Nederlandsch Indie.* The generic name refers to the spotted floral segments, which in some species are very pronounced.

The type species for this genus is the magnificent ***Grammatophyllum speciosum*** Bl.

There are about 12 members in this genus and they are found in Myanmar (Burma), Thailand, Laos, Vietnam, Peninsular Malaysia, Indonesia, Borneo, New Guinea, the Philippines and the Solomon Islands. Eight species of this extremely attractive genus are recorded from the Philippines.

Photo: Jim Cootes

Grammatophyllum elegans

Grammatophyllum elegans

Reichenbach *f.*

Professor Heinrich G. Reichenbach named this species in 1882 in the *Gardeners' Chronicle*. The specific name refers to the elegant nature of the flowers.

Growth habit: upright, sympodial. **Pseudobulbs:** large, up to 20 cm tall, diameter 5 cm. **Leaves:** up to four, up to 30 cm tall, 6 cm wide. **Inflorescences:** arch, up to a metre long, carry up to 50 flowers 7 cm in diameter. **Flower colour:** brown. **Dorsal sepal:** oblanceolate, up to 3.5 cm long, 1.5 cm wide. **Petals:** similar shape to dorsal sepal but marginally smaller. **Lateral sepals:** oblong, up to 3.5 cm long, 1.5 cm wide. **Labellum:** small, three-lobed, up to 2 cm long, 1 cm wide.

Habitat and Distribution: ***Grammatophyllum elegans*** is endemic to Albay on Luzon. It is also reported from Fiji, but is most likely to be a colour variation of the widespread ***Grammatophyllum scriptum*** (L.) Bl. It grows as an epiphyte at elevations to 300 metres.

Grammatophyllum martae

Quisumbing *ex* Valmayor and Tiu

Dr. Eduardo Quisumbing named this species after Marta Rivilla, an active member of the Philippine Orchid Society in the 1970s, but did not publish the name. Dr. Helen Valmayor and Danny Tiu published it in the *Philippine Orchid Review* in 1983.

Growth habit: upright, sympodial. **Pseudobulbs:** large, up to 22 cm tall, 10 cm in diameter. **Leaves:** up to four, about 60 cm long, 10 cm wide. **Inflorescences:** arching, reach 1.5 metres in length, bear up to 80 blooms 7.5 cm in diameter. **Flower colour:** blotched with dark brown, green edging around sepals and petals, of extremely good shape. **Dorsal sepal:** oblanceolate, up to 4 cm long, 2.5 cm wide. **Petals:** spathulate, up to 4 cm long, 2 cm wide. **Lateral sepals:** obovate, up to 4 cm long, 2 cm wide. **Labellum:** three-lobed, small.

Habitat and Distribution: ***Grammatophyllum martae*** is endemic to Negros Occidental on Negros island in Visayas where it grows as an epiphyte at elevations up to 300 metres.

Notes: ***Grammatophyllum martae*** could be a tetraploid form of ***Grammatophyllum elegans***.

Grammatophyllum measuresianum Weathers

John Weathers described this species in 1889 in the *Garden and Forest*. The specific name honours a Mr. Measures, who was first to flower this species.

Growth habit: upright, sympodial. **Pseudobulbs:** ovoid, to 25 cm tall, 6 cm in diameter. **Leaves:** four to six, at top of pseudobulbs, broadly lanceolate, up to 60 cm long, 10 cm wide. **Inflorescences:** upright to semi-arching, up to 1.5 metres long, bear up to 80 flowers about 5 cm in diameter. **Flower colour:** base colour whitish to pale yellow, spotted in varying degrees of dark chocolate brown. **Dorsal sepal:** broadly lanceolate, up to 3.8 cm long, 1.8 cm wide. **Petals:** oblanceolate, up to 3.5 cm long, 1.4 cm wide. **Lateral sepals:** broadly lanceolate, up to 3.8 cm long, 1.8 cm wide. **Labellum:** three-lobed, 2 cm long, 2 cm wide (flattened) side lobes encircle column, square at front; mid-lobe rectangular with rounded tip 6 mm long,

Photo: Jim Cootes

Grammatophyllum martae

Photo: David Titmuss

Grammatophyllum measuresianum

5 mm wide; rear of mid-lobe hairy. All floral segments reflex for the apical third.

Habitat and Distribution: ***Grammatophyllum measuresianum*** is endemic to the Philippines and has been recorded from Batangas on Luzon; Oriental Mindoro on Mindoro; and Palawan island. It grows as an epiphyte at elevations up to 500 metres.

Photo: Jim Cootes

Grammatophyllum multiflorum var. ***tigrinum***

Grammatophyllum multiflorum var. *tigrinum* Lindley

Dr. John Lindley first named this variety in 1842 in the *Botanical Register*. The specific name refers to the multi-flowered inflorescences, and the varietal name alludes to the markings on the floral segments.

Growth habit: upright, sympodial. **Pseudobulbs:** up to 15 cm long, 4 cm in diameter. **Leaves:** oblanceolate, up to 4 per pseudobulb, up to 30 cm long, 6 cm wide. **Inflorescences:** arching, up to 1.5 metres long, carry up to 100 flowers about 5 cm in diameter. **Flower colour:** variable, usually olive green with chocolate brown blotches. **Dorsal sepal:** lanceolate, up to 2.5 cm long, 1 cm wide. **Petals:** oblanceolate, up to 2.5 cm long, 7 mm wide. **Lateral sepals:** oblanceolate, up to 2.5 cm long, 1 cm wide. **Labellum:** three-lobed; side lobes upright, rounded tips; mid-lobe small.

Habitat and Distribution: ***Grammatophyllum multiflorum*** var. ***tigrinum*** is endemic to the Philippines and has been found in Albay, Camarines Norte, Camarines Sur, Cavite, Laguna, Quezon and Sorsogon on Luzon; Catanduanes and Palawan islands; and Leyte and Samar islands in the Visayas. Interestingly there are no records from Mindanao of this species.

Notes: ***Grammatophyllum multiflorum*** var. ***tigrinum*** is confused with ***Grammatophyllum scriptum*** (Linn.) Bl., a species that does not occur in the Philippines. The shape of the side lobes differs as does the apex of the mid-lobe of the labellum.

Photo: David Banks

Grammatophyllum multiflorum fma. ***citrinum***

Grammatophyllum multiflorum fma. *citrinum* Hort.

The albinistic form of the species has not been formally described under this name. It was, however, described in 1983 by Professor Valmayor and Danny Tiu as ***Grammatophyllum scriptum*** fma. ***citrinum***, a species that does not occur in the Philippines.

This is the albino variety of the species in which all the segments are yellowish green with no coloured blotching. The labellum is green and white.

Habitat and Distribution: The albino form is only found in the southern province of Albay on Luzon where it is rare. Both species grow as epiphytes at elevations up to 300 metres.

Photo: Jack Jannese

Grammatophyllum speciosum

Photo: David Banks

Grammatophyllum stapeliiflorum

Grammatophyllum speciosum

Blume

Carl Blume named this fabulous species in 1825 in *Bijdragen tot de Flora van Nederlandsch Indie.* The specific name refers to the very showy flowers.

Growth habit: upright, sympodial, one of the largest-growing orchid species known. **Pseudobulbs:** up to 3 m long, 5 cm wide. **Leaves:** up to 40 cm long, 2 cm wide. **Inflorescences:** upright, attain a length of 2 metres (several inflorescences can be produced during the flowering season), bear about 100 flowers up to 7 cm in diameter. Usually the first three or four blooms have no column or labellum and only four floral segments. **Flower colour:** yellow or cream with maroon spotting. **Dorsal sepal:** lanceolate, up to 3.5 cm long by 1.5 cm wide. **Petals:** elliptic, up to 3.5 cm long by 2.2 cm wide. **Lateral sepals:** lanceolate, up to 3.5 cm long by 1.8 cm wide. **Labellum:** three-lobed, small.

Habitat and Distribution: ***Grammatophyllum speciosum*** is widespread, being found in Vietnam, Myanmar, Thailand, Peninsular Malaysia, Indonesia, Borneo and Papua New Guinea. In the Philippines it is found in Camarines Norte, Camarines Sur, Nueva Ecija, Quezon and Sorsogon on Luzon; Oriental Mindoro on Mindoro island; and Davao del Sur and Surigao del Sur on Mindanao. It grows as an epiphyte, high in old trees, in almost full sun at elevations to 300 metres.

Notes: Because of the size of this species it is difficult to maintain in a small glass house and is really only a plant for tropical areas where it can be grown outdoors in a garden or in the low fork of a tree. According to the literature available to me this species can be reluctant to flower away from its native habitat.

Grammatophyllum stapeliiflorum

(Teijsmann and Binnendijk) J.J. Smith
Synonym: ***Grammangis stapeliiflora*** Schltr.

Johannes Teijsmann and Simon Binnendijk originally named this plant ***Cymbidium stapeliiflorum*** in 1862. Johannes J. Smith gave this species its currently accepted name in 1905 in *Die Orchideen von Java in Flora von Buitenzorg.* The specific name refers to the resemblance of the blooms to some species of cactus (***Stapelia*** spp.).

Growth habit: upright, sympodial. **Pseudobulbs:** up to 12 cm long by 5 cm in diameter. **Leaves:** two to three, about 30 cm long by 5 cm wide. **Inflorescences:** pendulous, bearing about 10 flowers 6 cm in diameter.

Flower colour: dorsal and lateral segments' base colour pale brown heavily spotted with dark purple; petals almost solid, dark purple; labellum similarly marked to the dorsal and lateral sepals. **Dorsal sepal:** ovate, up to 3 cm long by 2 cm wide. **Petals:** lanceolate, up to 3 cm long by 1.2 cm wide. **Lateral sepals:** oblanceolate to obovate, up to 3 cm long by 1.5 cm wide. **Labellum:** three-lobed; lateral lobes upright; mid-lobe small, almost round.

Habitat and Distribution: ***Grammatophyllum stapeliiflorum*** is fairly widely distributed, being found in Malaysia, Papua New Guinea, Sumatra, Java and Sulawesi (Celebes). In the Philippines it has been found in Surigao del Norte and Surigao del Sur provinces on Mindanao, but is undoubtedly more widely distributed. It grows as an epiphyte at low elevations up to 500 metres.

Grastidium Blume

Carl Blume created this genus in 1825 in *Bijdragen tot de Flora van Nederlandsch Indie.* The generic concept did not gain acceptance by the botanists of those times; it was mainly accepted as a section of the very large genus ***Dendrobium***. In 1981 F.G. Brieger reinstated ***Grastidium,*** but his system of classification was not supported by modern botanists. Recently, Dr. Mark Clements and David Jones (Centre for Plant Biodiversity Research, Canberra, Australia) did some preliminary phylogenetic analysis of the group and studies into the morphology, reproductive biology and general biology of these plants. These studies have provided clear support for part of what Brieger has proposed. The generic name refers to the grass-like growth habit, or grass-like leaves, of many of the species.

Type species for this genus is ***Grastidium salaccense*** Bl.

There are about 200 species in the genus ***Grastidium*** found throughout Asia, Sri Lanka, Taiwan, Australia, New Caledonia and as far east as Tahiti. In the Philippines there are at least seven species in this genus. I feel certain many more await description.

Grastidium luzonense

(Lindley) M.A. Clements and D.L. Jones

Dr. John Lindley named this plant ***Dendrobium luzonense*** in 1844 in the *Botanical Register.* Dr. Mark Clements and David Jones transferred the generic name to ***Grastidium*** in *Lasianthera* Vol. 1, issue 2. The specific name refers to Luzon island, where the original plants of the species were collected.

Growth habit: upright, sympodial. **Pseudobulbs:** up to 60 cm tall, 3 to 4 mm in diameter. **Leaves:** shiny green, of a thin texture, about 10 cm long by 1 cm wide. **Inflorescences:** short; flowers are borne in pairs from the opposite side of the leaves; spurs of blooms face each other (one of the distinguishing features of this genus); blooms about 2 cm in diameter. **Flower colour:** sepals and petals olive-green; labellum reddish brown with a single cream-coloured ridge running toward base of the column. **Dorsal sepal:** lanceolate, up to 1.4 cm long by 5 mm wide. **Petals:** lanceolate, 1.5 cm long by 6 mm wide. **Lateral sepals:** oblong-lanceolate, up to 1.5 cm long by 8 mm wide. **Labellum:** three-lobed, 9 mm long by 4 mm wide; a single ridge runs the length of the labellum.

Photo: David Titmuss

Grastidium luzonense

Habitat and Distribution: ***Grastidium luzonense*** is endemic to the Philippines and has been found in Apayao, Benguet, Kalinga, Pangasinan and Sorsogon provinces on Luzon; the islands of Catanduanes; Oriental Mindoro province on Mindoro island; Leyte island in the Visayas; and Agusan del Norte, Davao del Sur and Lanao del Sur provinces on Mindanao. It grows as an epiphyte at elevations up to about 1,300 metres.

Photo: David Titmuss

Grastidium ornithoflorum

Grastidium ornithoflorum

(Ames) M.A. Clements and D.L. Jones

Professor Oakes Ames named this species ***Dendrobium ornithoflorum*** in 1908 in Fascicle III of his *Studies in the Family Orchidaceae.* Dr. Mark Clements and David Jones transferred the generic name to ***Grastidium*** in *Lasianthera* Vol. 1, issue 2. The specific name refers to the buds of the blooms, which have a bird-like appearance.

Growth habit: upright, becomes semi-pendulous on older, mature plants, sympodial. **Pseudobulbs:** up to 70 cm long, 3 mm in diameter, branching. **Leaves:** thin-textured, linear, up to 10 cm long, 5 mm wide, tip unevenly bilobed. **Inflorescences:** appear from small scales opposite the leaves, two-flowered; mentums face each other. **Flower colour:** pale yellow; labellum pale yellow with green ridges. **Dorsal sepal:** triangular, up to 1.4 cm long, 4 mm wide. **Petals:** linear to acute, up to 1.3 cm long by 1.5 mm wide. **Lateral sepals:** triangular, up to 1.6 cm long by 6 mm wide. **Labellum:** three-lobed, 1.3 cm long by 5 mm wide (when flattened); lateral lobes very pointed, much shorter than mid-lobe; mid-lobe pointed, unevenly edged; three ridges run lengthwise toward the short column.

Habitat and Distribution: ***Grastidium ornithoflorum*** is endemic to the Philippines and has been found in Quezon on Luzon; Negros Oriental on Negros island in the Visayas; and Misamis Occidental on Mindanao. It grows as an epiphyte on deciduous trees at elevations up to 2,000 metres.

Photo: Dr. Mark Clements

Grastidium verruculosum

Grastidium verruculosum

(Ames) M.A. Clements and D.L. Jones

Professor Oakes Ames first named this plant ***Dendrobium verruculosum*** in 1914. Dr. Mark Clements and David Jones transferred the generic name to ***Grastidium*** in *Lasianthera* Vol. 1, issue 2. The specific refers to the small wart-like growths at the base of the leaves.

Growth habit: upright, sympodial. **Pseudobulbs:** terete, up to 60 cm long. **Leaves:** grass-like, linear to lanceolate, up to 12 cm long by 8 mm wide. **Inflorescences:** two-flowered, appear from dried bracts opposite the leaves. **Flower colour:** greenish yellow. **Dorsal sepal:** lanceolate, 1 cm long by 3 mm wide. **Petals:** acutely triangular, 1 cm long by 1.5 mm wide. **Lateral sepals:** lanceolate 1 cm long by 4 mm wide. **Labellum:** three-lobed; 8 mm long by 3 mm wide; side lobes obtuse; mid-lobe pointed with three ridges running the length of the labellum.

Habitat and Distribution: ***Grastidium verruculosum*** is endemic to the Philippines and has been recorded from the provinces of Ilocos Norte, Quezon, Rizal and Sorsogon on Luzon and the island of Leyte in the Visayas. It grows as an epiphyte, sometimes on mangroves, at elevations up to 300 metres.

Grosourdya Reichenbach *f.*

Professor Heinrich G. Reichenbach established this genus in 1864 in *Botanische Zeitung* and named it after Dr. von Grosourdy, a specialist in medicinal plants.

The type species for this small genus is ***Grosourdya elegans*** Rchb. *f.*

This is a genus of approximately 10 species that are found throughout Asia. The Philippines has two representatives.

Grosourdya appendiculata

(Blume) Reichenbach *f.*

This miniature-growing species has had a host of names. Carl Blume first named this species ***Dendrocolla appendiculata*** in 1825. In 1867 Reichenbach gave the plant its currently accepted name in *Xenia Orchidacea.* The specific name refers to appendages, possibly hairs or crests.

Growth habit: semi-pendulous, monopodial, similar to Australia's ***Sarcochilus falcatus*** but the flowers are very different. **Flower colour:** pale yellow heavily spotted with brownish red; labellum white. **Inflorescences:** semi-pendulous, usually several produced with the flowers opening one or two at a time; blooms about 1.5 cm in diameter, last one day, are most attractive. **Dorsal sepal:** oblong, up to 7.5 mm long by 2.5 mm wide. **Petals:** oblong, of similar dimensions to the dorsal sepal. **Lateral sepals:** ovate, up to 7.5 mm long by 3.5 mm wide. **Labellum:** three-lobed; lateral lobes upright; mid-lobe triangular.

Photo: David Titmuss

Grosourdya appendiculata

Habitat and Distribution: ***Grosourdya appendiculata*** is widely distributed, being found in the Andaman Islands, Myanmar, Thailand, Peninsular Malaysia, Java and Borneo. In the Philippines it has been found in the province of Nueva Ecija on Luzon and the southwest island of Palawan. Normally it grows as an epiphyte on the tips of branches in quite shady positions at elevations to 1,200 metres.

Hippeophyllum Schlechter

Dr. Rudolf Schlechter proposed this genus in *Nachtrage zur Flora der Deutschen Schutzgebiete in der Südsee* in 1905. The generic name refers to the equitant habit of the leaves of the species.

The type species for this genus is ***Hippeophyllum micranthum*** Schltr.

There are four members of this genus, which are found in Peninsular Malaysia, Sumatra and New Guinea. The Philippines has one endemic species.

Hippeophyllum wenzelii Ames

Professor Oakes Ames named this species in 1913 in *The Philippine Journal of Science.* The specific name honours C.A. Wenzel, who collected the type specimens.

Growth habit: upright, sympodial. **Rhizome:** diameter 5 mm; internodes about 3 mm long, produce many roots; plants at 4 cm intervals. **Pseudobulbs:** nonexistent. **Leaves:** leathery, tapered, up to 12 cm long, 6 mm wide, upright, leaf bases overlap. **Inflorescences:** upright, much longer than leaves, covered in short

Hippeophyllum wenzelii

bracts that rest against stem. **Flower colour:** dull green. **Dorsal sepal:** oblong to lanceolate, up to 3 mm long, 1 mm wide. **Petals:** narrowly spathulate, up to 2.5 mm long, 0.75 mm wide. **Lateral sepals:** similar to dorsal sepal. **Labellum:** three-lobed; lateral lobes upright, pointed; mid-lobe circular, up to 2.5 mm long, 1.2 mm wide.

Habitat and Distribution: ***Hippeophyllum wenzelii*** is endemic to the Philippines and is only known from Leyte island in the Visayas. It grows as an epiphyte at elevations about 60 metres.

Notes: This genus appears to be related to ***Oberonia*** but the extended rhizome between the plants is not found in that genus.

Kingidium P.F. Hunt

P.F. Hunt established this genus in 1970 in the *Kew Bulletin.* It was named after Sir George King, an author in the 1800s.

The type species for this genus is ***Kingidium deliciosum*** (Rchb. *f.*) Sweet.

There are approximately five species, which are found throughout Sri Lanka, northern India, China, Myanmar (Burma), Thailand, Malaysia, Laos, Cambodia, Vietnam, Indonesia, the Philippines, Sulawesi (Celebes) and Borneo. There is only one species found in the Philippines.

Kingidium philippinense

(Ames) Gruß and Röllke

Professor Oakes Ames first named this species ***Doritis philippinense*** in 1908. In 1995 Olaf Gruß and Lutz Röllke made the new combination of ***Kingidium philippinense*** in *Die Orchidee.* The specific name refers to the Philippines.

Growth habit: semi-pendulous, monopodial, resembles a small ***Phalaenopsis***. **Leaves:** two to three, thinly textured, up to 12 cm long, 5 cm wide. **Inflorescences:** upright, branch on adult plants; flowers up to 2 cm in diameter. **Flower colour:** pale pink sepals and petals; labellum dark pink. **Dorsal** and **lateral sepals:** oblong, up to 1 cm long by 4 mm wide. **Petals:** oblong, up to 9 mm long by 3 mm wide. **Labellum:** 1 cm long by 6 mm wide, three-lobed; mid-lobe triangular.

Kingidium philippinense

Habitat and Distribution: ***Kingidium philippinense*** is endemic to the Philippines where it is found in the provinces of Bataan, Bulacan, Cagayan and Rizal on Luzon; the islands of Bancalan and Palawan; Leyte island in the Visayas; and the provinces of Agusan del Norte, Agusan del Sur, Bukidnon, North and South Cotabato, Davao del Norte and Davao del Sur on Mindanao. It usually grows in quite heavy shade as an epiphyte at elevations to 800 metres.

Liparis L.C. Richard

Louis C. Richard established this genus in 1818 in *Memoires du Museum d'Histoire Naturelle, Paris.* The name refers to the greasy or shiny appearance of the leaves of some members for the genus.

The type species for this large genus is ***Liparis loeselii*** (L.) L.C. Rich.

There are approximately 250 species in this genus, which are found throughout the tropical and temperate areas of the world. The Philippines has about 36 species.

Liparis caespitosa (Thouars) Lindley

Thouars first named this miniature-growing species ***Malaxis caespitosa***. Dr. John Lindley transferred the generic name to ***Liparis*** in the *Botanical Register* in 1825. The specific name refers to the clumping habit of this pretty little species.

Growth habit: upright, sympodial. **Pseudobulbs:** ovate, up to 5 mm tall by 3 mm in diameter. **Leaves:** lanceolate, up to 3.5 cm long by 6.5 mm wide. **Inflorescences:** upright, appear from top of pseudobulb, bear up to 15 small blooms. **Flower colour:** green. **Dorsal sepal:** lanceolate, up to 3 mm long by 1 mm wide, reflexes. **Petals:** linear, 2.5 mm long by 0.5 mm wide, reflex. **Lateral sepals:** lanceolate, up to 3 mm long by 1 mm wide, reflex. **Labellum:** triangular, 4 mm long by 3 mm wide.

Habitat and Distribution: ***Liparis caespitosa*** is a widely distributed species that has been recorded from Africa, Madagascar, Mauritius, the Mascarene Islands, India, Sri Lanka, Myanmar (Burma), Thailand, Indochina, Indonesia, New Guinea and Taiwan. In the Philippines it has been recorded from Bataan, Benguet, Mountain Province and Rizal on Luzon; Leyte island in the Visayas; and North and South Cotabato, Davao del Sur and Lanao del Sur on Mindanao. It grows as an epiphyte and lithophyte at elevations between 800 and 1,600 metres.

Photo: David Titmuss

Liparis caespitosa

Photo: Jim Cootes

Liparis condylobulbon

Liparis condylobulbon
Reichenbach *f.*

Professor Heinrich G. Reichenbach named this small-flowered species in 1862 in *Hamburger Garten-Blumenzeitung.* The specific name refers to the shape of the pseudobulbs.

Growth habit: upright, sympodial. **Rhizome:** stout, about 4 mm diameter, carry the club-shaped pseudobulbs at intervals of up to 7 cm. **Pseudobulbs:** up to 14 cm high by 2 cm in diameter at their widest point, bifoliate. **Leaves:** about 15 cm long by 2.5 cm wide. **Inflorescences:** upright, about 20 cm long, bear many flowers 3 mm in diameter. **Flower colour:** olive-green. **Dorsal sepal:** lanceolate, up to 3 mm long by 1 mm wide. **Petals:** oblong, up to 3 mm long by 0.75 mm wide. All these segments reflex toward the pedicel. **Lateral sepals:** lanceolate, up to 3 mm long by 1 mm wide. **Labellum:** obovate, up to 4 mm long by 2.5 mm wide, reflexes for its top third.

Habitat and Distribution: ***Liparis condylobulbon*** is widely spread throughout Thailand, Indonesia, Sulawesi (Celebes), Borneo, Taiwan, Australia (limited distribution), the Solomon Islands, Samoa and Fiji. In the Philippines it has been found in Albay, Apayao, Benguet, Kalinga, Laguna, Mountain Province, Nueva Vizcaya, Quezon, Rizal and Sorsogon provinces on

Photo: David Banks

Liparis elegans

Photo: David Titmuss

Liparis latifolia

Luzon; the Catanduanes islands; Mindoro (Oriental Mindoro province); Palawan; Leyte and Samar islands in the Visayas; Agusan del Norte, Agusan del Sur, Davao del Norte, Davao Oriental, Davao del Sur, Lanao del Norte, Lanao del Sur, Zamboanga del Norte and Zamboanga del Sur provinces on Mindanao; and on Basilan island. It grows as a lithophyte or epiphyte at elevations up to 1,500 metres.

Liparis elegans Lindley

Dr. John Lindley named this species in 1830 in *Genera and Species of Orchidaceous Plants.* The specific name refers to the elegant nature of this plant.

Growth habit: upright, sympodial. **Pseudobulbs:** up to 3.5 cm tall, 1.5 cm in diameter, unifoliate. **Leaves:** 10 cm long, 2 cm wide. **Inflorescences:** upright, about 10 cm long, up to 100 flowers 6 mm in diameter. **Flower colour:** dull green; labellum pale brown with tiny white hairs along the edges. **Dorsal** and **lateral sepals:** oblong, up to 4 mm long, 2 mm wide. **Petals:** linear, up to 4 mm long, 1 mm wide. **Labellum:** tapered, 6 mm long, 3 mm wide (when flattened).

Habitat and Distribution: ***Liparis elegans*** is widely distributed, being found in Thailand, Malaysia, Indonesia and Borneo. In the Philippines it has been found in Agusan del Sur on Mindanao and on Basilan island. It grows as an epiphyte at elevations to 400 metres.

Liparis latifolia (Blume) Lindley

Carl Blume first named this species ***Malaxis latifolia*** in 1825. Dr. John Lindley transferred the generic name to ***Liparis*** in his *Genera and Species of Orchidaceous Plants* published in 1830. The specific name refers to the wide leaves of this plant.

Growth habit: upright, sympodial. **Pseudobulbs:** up to 7 cm tall, 3 cm wide, appear flattened, usually unifoliate but occasionally bifoliate. **Leaves:** up to 20 cm long, 7 cm wide. **Inflorescences:** upright; up to 80 blooms 1 cm in diameter. **Flower colour:** reddish brown. Plants of this species that have been growing in bright light have a pinkish tinge to pseudobulbs and foliage. **Dorsal** and **lateral sepals:** oblong, up to 6 mm long by 2 mm wide. **Petals:** linear, up to 6 mm long by 1 mm wide. **Labellum:** tapered, front edged toothed.

Habitat and Distribution: ***Liparis latifolia*** has been recorded from China, Malaysia, Thailand, Indonesia, Borneo, Timor and New Guinea. It appears to be endemic to Mindanao island (no provinces recorded), growing as an epiphyte at elevations to 1,100 metres.

Notes: ***Liparis latifolia*** was recorded by Professor Oakes Ames in Fascicle V of his *Studies in the Family Orchidaceae* (page 80) and is mentioned in later fascicles as occurring in the Philippines. Professor Valmayor, in *Orchidiana Philippiniana*, does not record this species. Study of the drawings of this species in Vol. 1 of *Orchids of Borneo* (page 174) convinces me that this plant occurs in the Philippines. ***Liparis latifolia*** is sold, and frequently cultivated, as ***Liparis nutans***, a totally different orchid from the section ***Distichon.***

Liparis merrillii (Ames) Schlechter

Professor Oakes Ames first named this species as ***Cestichus merrillii*** in 1905. Dr Rudolf Schlechter transferred the generic name to ***Liparis*** in 1911 in *Fedde, Repertorium Specierum Novarum, Beihefte.* The specific name honours Elmer D. Merrill, collector of the type specimens.

Growth habit: upright, sympodial. **Pseudobulbs:** conical, small, up to 1 cm high, 8 mm in diameter, unifoliate, spaced at intervals of about 5 cm on a rhizome 2 mm in diameter. **Leaves:** linear, up to 20 cm long, 1 cm wide. **Inflorescences:** upright, flattened, appear from the top of pseudobulbs. **Flower colour:** yellowish orange to red. **Dorsal sepal:** ovate to lanceolate, reflexes, up to 7 mm long, 3 mm wide. **Petals:** linear, up to 6 mm long, 1 mm wide. **Lateral sepals:** ovate to lanceolate, not reflexing, up to 7 mm long, 3 mm wide. **Labellum:** wedge-shaped, bi-lobed at the apex, up to 8 mm long, 6 mm wide; the base has two rounded calli (small lumps).

Habitat and Distribution: ***Liparis merrillii*** is endemic to the Philippines and has been found in Bataan, Ifugao, Mountain Province, Nueva Vizcaya, Nueva Ecija, Quezon and Rizal on Luzon; Oriental Mindoro on Mindoro island; Leyte island in the Visayas; and Misamis Occidental and Zamboanga del Norte on Mindanao. It grows as an epiphyte, often in deep mosses, at elevations between 300 and 2,900 metres.

Photo:: Jim Cootes

Liparis merrillii

Photo: Jim Cootes

Liparis parviflora

Liparis parviflora (Blume) Lindley

Carl Blume first named this plant ***Malaxis parviflora*** in 1825. Dr. John Lindley transferred the generic name in his *Genera and Species of Orchidaceous Plants* in 1830. The specific name refers to the small flowers.

Growth habit: upright, sympodial. **Pseudobulbs:** ovate, up to 6 cm long, 2.8 cm wide, somewhat flattened, bear one or two leaves. **Leaves:** elliptic, up to 24 cm long, 6 cm wide, folded at base. **Inflorescences:** semi-pendulous, up to 50 cm long, bear over 130 flowers; numerous bracts at the base. **Flower colour:** sepals and petals cream to pale green; labellum dull red. **Dorsal sepal:** linear, 5 mm long by 1.5 mm wide. **Petals:** linear, 5.5 mm long by 1 mm wide. **Lateral sepals:** linear, 5 mm long by 2 mm wide. **Labellum:** oblong, 6 mm long by 3 mm wide (when flattened), distinctly bent in middle, cleft at apex, minute white hairs around the edge.

Habitat and Distribution: ***Liparis parviflora*** has been found in Malaysia, Thailand, Indonesia, Sulawesi (Celebes) and Borneo. In the Philippines it has been recorded from Apayao, Benguet, Kalinga, Laguna, Mountain Province, Nueva Ecija, Pangasinan, Quezon and Rizal provinces on Luzon; Oriental Mindoro province on Mindoro island; Busuanga island; Leyte island in the Visayas; Agusan del Norte, Agusan del Sur, North and South Cotabato and Davao del Norte provinces on Mindanao; and Basilan island, south of Zamboanga province. It grows as an epiphyte at elevations to 2,000 metres. In Pangasinan province I have observed this plant growing in full sun.

Notes: The colour of this species is extremely variable, some clones are all cream to green, others have pink tips to the segments and others are totally pink. A very attractive orchid in any of its colour variations.

Liparis philippinensis

Liparis viridiflora

Liparis wenzelii

Liparis philippinensis

(Ames) Schlechter

Professor Oakes Ames first named this species ***Cestichus philippinensis*** in 1905. Dr. Rudolf Schlechter transferred the generic name to ***Liparis*** in 1911 in *Fedde, Repertorium Specierum Novarum, Beihefte.* The specific name refers to the Philippines.

Growth habit: upright, sympodial. **Pseudobulbs:** short, conical, unifoliate, 2 cm tall by 2 cm in diameter. **Leaves:** linear, up to 20 cm long by 2 cm wide. **Inflorescences:** upright, flattened, appear from the top of the pseudobulbs. **Flower colour:** bright green; column darker green. **Dorsal sepal:** linear, reflexes back, 1.3 cm long by 1 mm wide. **Petals:** linear, reflex back; 1 cm long by 0.5 mm wide. **Lateral sepals:** similar to dorsal sepal. **Labellum:** broadly lanceolate, apex pointed, bent down; 1 cm long by 4 mm wide; two rounded protuberances at the base.

Habitat and Distribution: *Liparis philippinensis* is endemic to the Philippines and has been recorded from Abra, Bataan, Benguet, Cagayan, Laguna, Mountain Province, Nueva Vizcaya, Pampanga and Rizal provinces on Luzon; Oriental Mindoro province on Mindoro island; Panay (Antique province), Leyte and Negros (Negros Oriental province) islands in the Visayas; and Agusan del Norte, Agusan del Sur, Bukidnon and North and South Cotabato provinces on Mindanao. It grows as an epiphyte at elevations between 300 and 2,900 metres. I have observed this species growing in central Luzon and it is always living with deep mosses around the roots.

Liparis viridiflora (Blume) Lindley

Carl Blume first named this species ***Malaxis viridiflora*** in 1825. Dr. John Lindley gave the plant its currently accepted name in 1830 in *Genera and Species of Orchidaceous Plants.* The specific name alludes to the green flowers.

Growth habit: upright, sympodial. **Pseudobulbs:** tapered, up to 9 cm long by 2 cm wide, bifoliate. **Leaves:** shiny, up to 23 cm long by 3 cm wide. **Inflorescences:** upright, up to 12 cm long, bear many flowers 3 mm diameter. **Flower colour:** pale green. **Dorsal** and **lateral sepals:** linear, up to 3 mm long by 1 mm wide. **Petals:** linear, up to 3 mm long by .5 mm wide; all segments reflex back toward the pedicel. **Labellum:** tapered and bent for half its length (2 mm).

Habitat and Distribution: *Liparis viridiflora* is widespread, being found throughout India, Sri Lanka, China, Taiwan, Myanmar (Burma), Thailand, Indochina, Malaysia, Indonesia, Samoa and Fiji. In the Philippines it has been found in Benguet, Ilocos Norte, Laguna, Mountain Province, Quezon and Rizal provinces on Luzon; Oriental Mindoro province on Mindoro island; and Zamboanga del Sur province on Mindanao. It grows as an epiphyte at elevations to 1,650 metres.

Liparis wenzelii Ames

Professor Oakes Ames named this species in 1915 in *Studies in the Family Orchidaceae*, Fascicle V. The specific name honours C.A. Wenzel, collector of the type specimens.

Growth habit: upright, sympodial. **Pseudobulbs:** short, conical, unifoliate, 2 cm tall, 2 cm in diameter. **Leaves:** oblong-lanceolate, up to 20 cm long, 2 cm wide. **Inflorescences:** upright, appear from the top of pseudobulbs. **Flower colour:** dull green; labellum pale brown. **Dorsal sepal:** oblong, reflexes back, 6 mm long, 2 mm wide. **Petals:** linear, reflex back; 5 mm long, 1 mm wide. **Lateral sepals:** similar to dorsal sepal. **Labellum:** triangular; apex bi-lobed, bent down; 8 mm long, 4 mm wide.

Habitat and Distribution: ***Liparis wenzelii*** is endemic to the Philippines and has only been recorded from the island Leyte in the Visayas. It grows as an epiphyte at elevations around 500 metres.

Luisia Gaudichaud

Charles Gaudichaud named this genus in 1826 in *Voyage sur l'Uranie et la Physicienne.* It was named after Don Luis de Torres, a botanist of those times.

The type species for this genus is ***Luisia teretifolia*** Gaud.

There are about 36 species in the genus and they are found in India, Sri Lanka, Myanmar (Burma), Thailand, Peninsular Malaysia, Indonesia, Borneo, New Guinea, Australia and Polynesia. Four species are represented among the islands of the Philippines.

Luisia cordatilabia

Ames and Quisumbing

Professor Oakes Ames and Dr. Eduardo Quisumbing named this species in 1933 in the *Philippine Journal of Science*. The specific name refers to the heart-shaped labellum of the flower.

Growth habit: upright, monopodial, up to 1 metre long, stems occasionally branch. **Leaves:** terete, up to 12 cm long by about 3 mm wide. **Inflorescences:** short, appear from the upper leaf nodes, carry up to 10 flowers about 2 cm across the petals. **Flower colour:** sepals and petals yellowish green; labellum maroon, almost black. **Dorsal sepal:** lanceolate-oblong, up to 1 cm long by 5 mm wide. **Petals:** oblanceolate, up to 1 cm long by 5 mm wide. **Lateral sepals:** boat-shaped, up to 1 cm long by 4 mm wide. **Labellum:** three-lobed; lateral lobes upright; mid-lobe heart-shaped, up to 1 cm across.

Photo: David Titmuss

Luisia cordatilabia

Habitat and Distribution: ***Luisia cordatilabia*** is endemic to the Philippines, in the mountains of Benguet, Mountain Province, Nueva Ecija and Nueva Vizcaya on Luzon. It grows as an epiphyte, usually in bright light at elevations above 1,200 metres.

Macropodanthus L.O. Williams

Louis O. Williams established this small genus of monopodial orchids in 1938 in the *Botanical Museum Leaflets of Harvard University*. The generic name refers to the long column foot.

The type species for this genus is ***Macropodanthus philippinensis*** L.O. Wms.

There are approximately six species in the genus, and they are found throughout Southeast Asia. There is only one species found in the Philippines.

Macropodanthus philippinensis L.O. Williams

The specific name refers to the Philippines where this unusual species is found.

Growth habit: semi-pendulous, monopodial; a small-growing vandaceous species with stems about 3 mm in diameter. **Leaves:** leathery, about 15 cm long by 3 cm wide. **Inflorescences:** pendulous, carry up to 15 blooms about 1.5 cm in diameter; segments reflex after blooms are open for a day or two. **Flower colour:** pale olive green; labellum white. **Dorsal sepal:** elliptic and concave, up to 1.2 cm long by 6 mm wide. **Petals:** narrowly obovate, up to 9 mm long by 4 mm wide. **Lateral sepals:** broadly ovate, up to 1.3 cm long by 6 mm wide, reflex after the flowers have been open for a couple of days. **Labellum:** vaguely five-lobed; lateral lobes small; mid-lobe rounded.

Photo: David Banks

Macropodanthus philippinensis

Habitat and Distribution: ***Macropodanthus philippinensis*** is endemic to the Philippines where it has been found in Aklan province on Panay island in the Visayas. It grows as an epiphyte at elevations up to 500 metres.

Malleola J.J. Smith and Schlechter

Johannes J. Smith and Dr. Rudolf Schlechter established this genus in 1913 in *Fedde, Repertorium Specierum Novarum, Beihefte*. The generic name refers to the hammer-like structure of the column.

The type species for this genus is ***Malleola sphingoides*** J.J. Sm.

There are about 30 species in the genus and they are found in Thailand, Vietnam, Peninsular Malaysia, Indonesia, New Guinea, the Solomon Islands and the Philippines. The two species found in the Philippines are endemic.

Malleola constricta Ames

Professor Oakes Ames named this plant in 1915 in Fascicle V of his *Studies in the Family Orchidaceae*. The specific name refers to the constricted nectary of the flower. C.A. Wenzel collected the type specimens in December 1912 on the island of Leyte.

Growth habit: semi-pendulous, monopodial; a miniature vandaceous species with stems about 3 mm in diameter. **Leaves:** very leathery, have the appearance of velvet, about 8 cm long by 1.5 cm wide. **Inflorescences:** pendulous, carry up to 50 blooms about 5 mm in diameter. **Flower colour:** brown with yellow edging; labellum white with purple spotting. **Dorsal sepal:** elliptic and concave, up to 3 mm long by 2 mm wide. **Petals:** oblong, up to 3 mm long by 1.5 mm wide. **Lateral sepals:** broadly ovate, up to 3 mm long by 3 mm wide. **Labellum:** three-lobed; lateral lobes small; mid-lobe pointed.

Photo: David Banks

Malleola constricta

Habitat and Distribution: ***Malleola constricta*** is endemic to the Philippines where it is found in Laguna and Sorsogon provinces on Luzon; the nearby island of Polillo; Bohol and Leyte islands in the Visayas; Agusan del Sur and North and South Cotabato provinces on Mindanao; and in the Sulu Archipelago. It grows as an epiphyte at elevations to 700 metres.

Megalotus Garay

Dr. Leslie Garay established this genus in 1972 in *Botanical Museum Leaflets of Harvard University*. The generic name refers to the ear-like appendages on the column.

The type species for this small genus is ***Megalotus bifidus*** (Lindl.) Garay.

This genus appears to be endemic to the Philippines and has only one species.

Megalotus bifidus (Lindley) Garay

Dr. John Lindley first named this species ***Saccolabium bifidum***. Dr. Leslie Garay gave the plant its currently accepted name in 1972 in the *Botanical Museum Leaflets of Harvard University*. The specific name refers to the bifid (split) labellum.

Growth habit: semi-pendulous, monopodial. **Leaves:** oblong to elliptic, up to 7 cm long, 2 cm wide; leaf tips unequally bilobed. **Inflorescences:** pendulous, bear up to 50 small blooms about 4 mm in diameter. **Flower colour:** pale green sepals and petals; labellum white with a purple spot. The labellum is split in half for about one third of its length (hence the specific name). **Dorsal sepal:** elliptic-oblong, up to 2 mm long, 1 mm wide. **Petals:** lanceolate, up to 2 mm long, 1 mm wide. **Lateral sepals:** oblong with pointed tips, up to 2 mm long, 1 mm wide. **Labellum:** three-lobed; lateral lobes very small; mid-lobe extended up to 4 mm long with its centre split for about 1.5 mm.

Photo: David Titmuss

Megalotus bifidus

Habitat and Distribution: *Megalotus bifidus* is endemic to the Philippines, where it is found in Apayao, Kalinga, Laguna and Nueva Ecija provinces on Luzon; Oriental Mindoro province on Mindoro island; Leyte island in the Visayas; and Agusan del Sur and North and South Cotabato provinces on Mindanao. It grows as an epiphyte at elevations to 1,200 metres.

Micropera Lindley

Dr. John Lindley named this genus in the *Botanical Register* in 1832. The generic name refers to the small, saccate labellum.

The type species for this genus is ***Micropera pallida*** (Roxb.) Lindl.

There are about 14 species; these are found in India, Thailand, Indochina (Vietnam, Laos, Cambodia), Peninsular Malaysia, Indonesia, New Guinea, Australia, the Solomon Islands and the Philippines. There are three species found in the Philippine archipelago; all are endemic.

Micropera philippinensis (Lindley) Garay

Dr. John Lindley originally named this plant ***Camarotis philippinensis***. Dr. Leslie Garay gave it its currently accepted name in 1972 in the *Botanical Museum Leaflets of Harvard University*. The specific name refers to the Philippines, where this orchid is found.

Growth habit: upright, monopodial. **Stems:** reach 40 cm in height. **Leaves:** very leathery, up to 15 cm long by 1.5 cm wide. **Inflorescences:** upright, carry up to 20 flowers 1.5 cm in diameter. **Flower colour:** white to cream. **Dorsal sepal:** oblong, up to 1 cm long by 3 mm wide. **Petals:** oblong, up to 1 cm long by 3 mm wide. **Lateral sepals:** oblong, up to 1 cm long by 3 mm wide. **Labellum:** three-lobed; S-shaped appendage on the column.

Habitat and Distribution: *Micropera philippinensis* is endemic to the Philippines where it is found in Laguna, Rizal and Sorsogon provinces on Luzon and Bohol and Leyte islands in the Visayas. It occurs at elevations to 850 metres and grows as an epiphyte.

Photo: David Banks

Micropera philippinensis

Photo: David Banks

Micropera utriculosa

Micropera utriculosa

(Ames) Garay

Professor Oakes Ames first named this plant ***Camarotis utriculosa*** in 1915. Dr. Leslie Garay gave this species its currently accepted name in the *Botanical Museum Leaflets of Harvard University* in 1972. The specific name refers to the inflated spur.

Growth habit: upright, monopodial; not as large as ***Micropera philippinensis*** at 30 cm in height. **Leaves:** leathery, up to 15 cm long, 1 cm wide, arch gracefully. **Inflorescences:** upright, carry up to 12 blooms approximately 1 cm in diameter. **Flower colour:** purple. **Dorsal sepal:** oblanceolate, up to 5 mm long by 3 mm wide. **Petals:** oblong-elliptic, up to 5 mm long by 2.5 mm wide. **Lateral sepals:** obovate, up to 5 mm long by 3 mm wide. **Labellum:** three-lobed, fleshy.

Habitat and Distribution: ***Micropera utriculosa*** is endemic to the Philippines where it is found in Albay, Cavite and Quezon on Luzon; Bohol island in the Visayas; and Agusan del Sur and North and South Cotabato on Mindanao. It grows as an epiphyte at elevations to 1,000 metres.

Nephelaphyllum Blume

Carl Blume established this genus of terrestrial orchids in 1825 in *Bijdragen tot de Flora van Nederlandsch Indie.* The generic name is in reference to the haziness of the upper surface of the leaf.

The type species for this genus is ***Nephelaphyllum pulchrum*** Bl.

There are about eight species in the genus and they are found in India, Thailand, Indochina (Laos, Cambodia, Vietnam), Hong Kong and the Philippines. Two species are found in the Philippines.

Nephelaphyllum pulchrum Blume

Carl Blume named this plant in 1825 in *Bijdragen tot de Flora van Nederlandsch Indie.* The specific name means beautiful or lovely.

Growth habit: upright, sympodial, bit of a climber, rhizome about 1 cm in diameter. **Leaves:** heart-shaped, 7.5 cm long by 6 cm wide, silvery green with darker veining; underside purplish. **Inflorescences:** upright, carry about five flowers in which the sepals and petals all reflex badly; blooms about 2 cm across. **Flower colour:** sepals and petals brownish; labellum uppermost, white with yellow markings in the centre. **Floral segments:** linear, taper to a point for their final quarter, up to 1 cm long by 2.5 mm wide. **Labellum:** uppermost, ovate, up to 1 cm long by 6 mm wide.

Habitat and Distribution: ***Nephelaphyllum pulchrum*** is widespread, being found in India, Myanmar (Burma), Thailand, Vietnam, Cambodia, Peninsular Malaysia, Sumatra, Java and Borneo. In the Philippines it has been recorded from the Palawan island. It grows as a terrestrial between 300 and 1,000 metres, usually in heavily shaded localities and often close to creeks.

Nephelaphyllum pulchrum

Oberonia Lindley

Dr. John Lindley established this genus of miniature-flowered orchids in 1830 in his *Genera and Species of Orchidaceous Plants.* The generic name honours the mythical character Oberon, who lived among the branches of trees, alluding to the epiphytic nature of these delightful little plants.

The type species for this genus of miniature-flowered orchids is ***Oberonia iridifolia*** Lindley.

Oberonia, a widely distributed genus with approximately 300 members, has been recorded from East Africa to India, Sri Lanka, throughout Southeast Asia, New Guinea, Australia and the islands of the Pacific Ocean. At least 24 species have been described from the Philippines.

Oberonia setigera Ames

Professor Oakes Ames named this plant in 1912 in the *Philippine Journal of Science.* The specific name refers to the bristle-like bracts behind the flowers.

Growth habit: semi-pendulous, sympodial. **Leaves:** sword-like, reddish, up to 5 cm long by 6 mm wide. **Inflorescences:** pendulous, bear many small flowers about 2 mm in diameter. **Flower colour:** reddish brown. Each bloom has a very prominent bract about 1 cm long. **Dorsal sepal:** lanceolate, 1.2 mm long by 0.5 mm wide. **Petals:** fringed, oblanceolate, 1.2 mm long by 0.3 mm wide. **Lateral sepals:** lanceolate, 1.3 mm long by 0.5 mm wide. **Labellum:** three-lobed; mid-lobe square; side lobes have seven tooth-like projections.

Oberonia setigera

Habitat and Distribution: ***Oberonia setigera*** is endemic to the Philippines and has been recorded from Bataan and Nueva Vizcaya provinces on Luzon. It grows as an epiphyte on tree trunks and branches in shaded locations at elevations around 1,200 metres.

Photo: David Titmuss

Oberonia toppingii

Oberonia toppingii Ames

Professor Oakes Ames named this plant in 1913 in the *Philippine Journal of Science,* honouring D. LeRoy Topping, who collected the type specimen in 1908.

Growth habit: semi-pendulous, sympodial. **Leaves:** sword-like, up to 5 cm long by 9 mm wide. **Inflorescences:** pendulous, up to 8 cm long, bear many small flowers about 1.5 mm in diameter. **Flower colour:** brownish orange; anther cap yellow. Each bloom is backed by a bract about 4 mm long that dries up from the apex. **Dorsal sepal:** lanceolate, 1 mm long by 0.5 mm wide. **Petals:** linear, 1 mm long by 0.3 mm wide. **Lateral sepals:** lanceolate, 1 mm long by 0.5 mm wide. **Labellum:** 2 mm long by 1 mm wide, three-lobed; mid-lobe split; side lobes small, like two arms.

Habitat and Distribution: ***Oberonia toppingii*** is endemic to the Philippines and has only been recorded from Rizal province on Luzon. It grows as an epiphyte in shaded locations at elevations to 300 metres.

Omoea Blume

Carl Blume established this genus in 1825 in his *Bijdragen tot de Flora van Nederlandsch Indie.* The generic name means similar and probably refers to the similarity between ***Omoea*** and ***Ceratochilus***.

Type species for the genus is ***Omoea micrantha*** (Bl.) J.J. Sm.

This genus of monopodial orchids is only known from Java and the Philippines. There are two species in ***Omoea***; one is endemic to the Philippines.

Omoea philippinensis Ames

Professor Oakes Ames named this species in 1915 in his *Studies in the Family Orchidaceae* Fascicle V. The specific name refers to the Philippines.

Growth habit: upright, monopodial. **Stems:** up to 10 cm long, 3 mm in diameter. **Leaves:** very leathery, thick, flattened, grooved top and bottom, up to 3 cm long, 3 mm wide. **Inflorescences:** short, bear up to six blooms. **Flower colour:** petals and sepals green with brown tips; labellum white, purple stripe on side lobes. **Dorsal sepal:** lanceolate, concave, pointed at tip, up to 4 mm long, 1.5 mm wide. **Petals:** linear, point forward, up to 3 mm long, 1 mm wide. **Lateral sepals:** lanceolate, concave, pointed at tip, up to 4 mm long, 1.5 mm wide. **Labellum:** three-lobed with spur; side lobes triangular; mid-lobe heart-shaped.

Photo: David Banks

Omoea philippinensis

Habitat and Distribution: ***Omoea philippinensis*** is endemic to the Philippines. It has been recorded from Bataan, Pampanga, Nueva Ecija and Nueva Vizcaya on Luzon. It grows as an epiphyte at elevations between 800 and 2,000 metres.

Paphiopedilum Pfitzer

Ernst Pfitzer established this genus in 1886. The name was published in *Morphologische Studien ueber die Orchideenbluthe.* Before 1886 ***Paphiopedilum*** species were known as members of the genus ***Cypripedium.*** The generic name refers to the resemblance of the labellum to a slipper.

The type species for this widely grown genus is ***Paphiopedilum insigne*** (Lindl.) Pfitz.

Paphiopedilum are an Asian genus of orchids and are found in Southern China, India, Myanmar (Burma), Thailand, Vietnam, Indonesia, New Guinea, the Solomon Islands and the Philippines. The majority of the species are semi-terrestrial, a number are lithophytes and some are epiphytes. They occur at elevations to about 2,500 metres. There are between 60 and 70 species, depending on whose taxonomy you wish to accept. The Philippines has about 10 species.

Paphiopedilum acmodontum

Schoser *ex* M. Wood
Section: ***Chloroneura*** Hallier

This beautiful species was not described until Mark Wood named it in 1976 in *The Orchid Review.* Dr. Schoser first brought ***Paphiopedilum acmodontum*** to the attention of orchid growers in 1969 at the World Orchid Conference in Sydney, Australia. The specific name refers to teeth-like projections on the labellum.

Paphiopedilum acmodontum

Growth habit: upright, sympodial. **Leaves:** mottled, up to 20 cm long, 3 cm wide. **Inflorescences:** upright, carry a single bloom up to 7 cm across petals. **Flower colour:** dorsal sepal white, striped vertically with purple lines of uneven length; petals pink for half their length, upper half bright green, lower half white blushed with pink, random spotting on the inner two-thirds of the segment; labellum olive green, shiny; synsepalum white, purple and green stripes. **Dorsal sepal:** ovate, pointed at the apex, 3.5 cm long, 3 cm wide. **Petals:** oblong, stand out horizontally, up to 4 cm long, 1.3 cm wide. **Labellum:** 4 cm long, 2.3 cm wide, shiny, three tooth-like projections on upper edge. **Synsepalum:** up to 2.5 cm long, 1.7 cm wide.

Habitat and Distribution: ***Paphiopedilum acmodontum*** is endemic to Negros island in the Visayas. It grows as a terrestrial at elevations to 1,000 metres.

Paphiopedilum adductum Asher

Section: ***Mastigopetalum*** Hallier

This multiflowered species was not described until 1983. How it ever escaped detection until then is quite remarkable. Dr. James Asher named it in *The Orchid Digest.* The specific name refers to the strongly adducted staminode.

Paphiopedilum adductum

Growth habit: upright, sympodial. **Leaves:** plain green, up to 25 cm long by 4 cm wide. **Inflorescences:**

upright, bear up to three blooms. **Flower colour:** dorsal sepal creamy white with dark red vertical striping; petals yellowish with dark red stripes; labellum reddish brown with darker striping; synsepalum, creamy white with dark red vertical striping. **Dorsal sepal:** ovate, pointed tip, up to 5.5 cm long, 3 cm wide. **Petals:** narrowly linear, gradually tapering, drooping, about 10 cm long, up to 8 mm wide. **Labellum:** up to 4.5 cm long, 2 cm wide. **Synsepalum:** oblanceolate, up to 6 cm long, 3 cm wide.

Habitat and Distribution: ***Paphiopedilum adductum*** is endemic to Mindanao island where it grows as an epiphyte at elevations from 1,250 to 1,350 metres.

Paphiopedilum argus

(Reichenbach *f.*) Stein
Section: ***Barbata*** Kränzlin

Professor Heinrich G. Reichenbach described this plant in 1873 as ***Cypripedium argus***. George Stein gave this orchid its currently accepted name in 1892 in *Stein's Orchideenbuch*. The specific name refers to Argus, the hundred-eyed monster of Greek mythology, in reference to the many spotted petals.

Growth habit: upright, sympodial. **Leaves:** prettily mottled, up to 30 cm long, 4 cm wide. **Inflorescences:** upright, carry a single bloom up to 7 cm across petals. **Flower colour:** dorsal sepal white, barred vertically with green and purple stripes; petals white, striped green lengthwise, tipped with purple and covered with dark spots that frequently fuse together; labellum dark brown on top fading to paler brown underneath, veined; synsepalum greenish, striped with darker green. **Dorsal sepal:** ovate, tip pointed, up to 4 cm long by 2.8 cm wide. **Petals:** oblong, up to 5 cm long by 1.3 cm wide, about 20 to 30 degrees below the horizontal; edged with warts. **Labellum:** up to 4 cm long by 2.4 cm wide. **Synsepalum:** up to 4 cm long by 2 cm wide.

Habitat and Distribution: ***Paphiopedilum argus*** is endemic to the mountains of Benguet, Ifugao and Quezon provinces on Luzon; and Bohol island and Negros Oriental province on Negros island in the Visayas where it grows on the forest floor at elevations above 600 metres.

Paphiopedilum argus

Paphiopedilum ciliolare

Paphiopedilum ciliolare

(Reichenbach *f.*) Stein
Section: ***Loripetalum*** Karasawa and Saito

Professor Heinrich G. Reichenbach originally named this plant ***Cypripedium ciliolare*** in 1882. George Stein gave this plant its currently accepted name in 1892 in *Stein's Orchideenbuch*. ***Paphiopedilum ciliolare*** is closely related to ***Paphiopedilum superbiens*** and ***Paphiopedilum curtisii,*** and some taxonomists consider them to be subspecies. The specific name is from the Latin *ciliaris,* meaning fringed with hairs, in reference to the densely hairy margins of the petals.

Growth habit: upright, sympodial. **Leaves:** mottled, up to 15 cm long by 4 cm wide. **Inflorescences:** upright, carry a solitary bloom up to 8 cm across the petals. **Flower colour:** dorsal sepal purplish at base with vertical striping of purple and dark green, outer edge white; petals purple, heavily spotted with black on the inner two-thirds; labellum dull purple; synsepalum white with green vertical veins. **Dorsal sepal:** broadly ovate, pointed at the tip, up to 5 cm long by 3.5 cm wide. **Petals:** oblong, up to 5 cm long by 1.4 cm wide, about 45 degrees below the horizontal. **Labellum:** up to 5 cm long by 3 cm wide. **Synsepalum:** up to 4 cm long by 2 cm wide.

Habitat and Distribution: ***Paphiopedilum ciliolare*** is endemic to the Mountain Province on Luzon; the islands Camiguin and Dinagat north of Mindanao; and the provinces of Agusan del Norte, Agusan del Sur, Surigao del Norte and Surigao del Sur on Mindanao. It grows as a terrestrial, usually at the bases of shrubs, at elevations between 300 and 1,800 metres.

Paphiopedilum fowliei Birk

Section: ***Barbata*** Kränzlin

Lance Birk named this in 1981 in *The Orchid Digest*, in honour of Dr. Jack Fowlie, its long-time editor.

Growth habit: upright, sympodial, a small grower. **Leaves:** mottled, with a distinctly bluish green shade, span seldom over 17.5 cm. **Inflorescences:** upright, a single bloom about 4 cm across the petals. **Flower colour:** dorsal sepal white, vertically striped with green, pink blush at base; petals white, striped lengthwise with green, tipped with pink, black warts on upper and lower edges; labellum pale brown; synsepalum greenish, striped with darker green. **Dorsal sepal:** ovate, pointed at the apex, up to 3 cm long by 2.4 cm wide. **Petals:** linear, up to 3.5 cm long by 6 mm wide, curve at about 45 degrees below the horizontal, reflexing. **Labellum:** up to 4.5 cm long by 2.8 cm wide. **Synsepalum:** up to 2 cm long by 1.5 cm wide.

Habitat and Distribution: ***Paphiopedilum fowliei*** is endemic to Palawan island. It grows as a terrestrial at medium elevations of about 700 metres.

Paphiopedilum fowliei fma. *christianae* Braem

Section: ***Barbata*** (Kränzlin) V.A. Albert and B. Pettersen

Dr. Guido Braem named this form in *Orchidées. Culture and Protection* in 1998 in honour of Christiana Richards of the state of Florida, USA. This is the form of the species that lacks all anthocyanins in the flower, therefore the flowers are green and white only—having no other colour.

Photo: David Banks

Paphiopedilum fowliei

Photo: Dr. Guido Braem

Paphiopedilum fowliei fma. ***christianae***

Habitat and Distribution: ***Paphiopedilum fowliei*** fma. ***christianae*** is found in the same locality as the normally coloured form of the species.

Paphiopedilum haynaldianum

(Reichenbach *f.*) Stein

Section: ***Polyantha*** Pfitzer

Professor Heinrich G. Reichenbach first described ***Paphiopedilum haynaldianum*** as ***Cypripedium haynaldianum*** in 1874. George Stein gave this species its currently accepted name in 1892, in *Stein's Orchideenbuch*. The specific name refers to Cardinal Haynald, the Archbishop of Kaloesa in Hungary.

Growth habit: upright, sympodial. **Leaves:** up to 35 cm long by 4 cm wide, bright green. **Inflorescences:** arched, carry up to five blooms about 12 cm across the petals. **Flower colour:** dorsal sepal white for top third, greenish below, covered with pink, striped and spotted

Photo: David Banks

Paphiopedilum haynaldianum

Photo: Gary Yong Gee

Paphiopedilum haynaldianum fma. ***album***

with brown; petals tipped with pinkish purple, inner halves olive green mottled with brown; labellum light brown and shiny as if varnished; synsepalum yellowish green spotted with purple. **Dorsal sepal:** ovate with bottom half reflexing somewhat, up to 6 cm long by 3 cm wide. **Petals:** narrowly spathulate, up to 7 cm long by 2 cm wide. **Labellum:** up to 6 cm long by 2.5 cm wide. **Synsepalum:** up to 4.5 cm long by 2.5 cm wide.

Habitat and Distribution: ***Paphiopedilum haynaldianum*** is endemic to the Philippines and is found in the provinces of Apayao, Batangas, Benguet, Kalinga, Mountain Province, Nueva Ecija and Rizal on Luzon; Negros Occidental on Negros island in the Visayas; and Surigao del Norte and Surigao del Sur on Mindanao. It usually grows as a lithophyte and frequently as an epiphyte at elevations up to 1,500 metres.

Paphiopedilum haynaldianum

fma. ***album*** (Hort. *ex* Asher) Braem

Section: ***Polyantha*** Pfitzer

The zoologist Asher first described this species as ***Paphiopedilum haynaldianum*** var. ***album***. Dr. Guido Braem transferred it from varietal to forma status in *The Genus Paphiopedilum—Natural History and Cultivation, Vol. 1* in 1998. The forma name refers to the flower's lack of anthocyanins. This rare albino species form has yellowish green and white flowers.

Photo: David Banks

Paphiopedilum hennisianum

Habitat and Distribution: The locality of the albino form of this wonderful species is unknown.

Paphiopedilum hennisianum

(M. Wood) Fowlie

Section: ***Barbata*** Kränzlin

This species has been in cultivation for over 20 years and was first bought to the notice of orchid growers at The World Orchid Conference in Sydney, Australia, in 1969 but was not validly described until 1977. Mark Wood subsequently described the plant and named it as a variety of ***Paphiopedilum barbatum***. Dr. Jack Fowlie raised it to specific rank in 1977 in *The Orchid Digest*. ***Paphiopedilum hennisianum*** was named in honour of Kurt Hennis of Hildesheim, Germany.

Growth habit: upright, sympodial. **Leaves:** up to 18 cm long by 4 cm wide, mottled. **Inflorescences:** upright, carry one flower up to 6 cm across the petals. **Flower colour:** dorsal sepal white, striped vertically with green; petals white with black warts along the edges, striped horizontally with green; labellum pale brown; synsepalum white with green stripes. **Dorsal sepal:** ovate, pointed at the tip, up to 4 cm long by 3 cm wide. **Petals:** linear, curve backward toward the tips, up to 4 cm long by 1 cm wide. **Labellum:** up to 4 cm long by 2 cm wide. **Synsepalum:** up to 3.2 cm long by 2 cm wide.

Habitat and Distribution: ***Paphiopedilum hennisianum*** is endemic to the Philippines and has been found in the province of Bukidnon on Mindanao. It grows terrestrially at medium elevations to 500 metres.

Photo: Olaf Gruss

Paphiopedilum hennisianum fma. *christiansenii*

Paphiopedilum hennisianum fma. *christiansenii* (Gruß and Röth) Gruß and Röth

Section: ***Barbata*** (Kränzlin) V.A. Albert and B. Pettersen

Olaf Gruß and Jürgen Röth first named this orchid as a variety in 1996 in *Die Orchidee*. They then transferred the status from a variety to a form in *Caesiana* no. 12 in 1999. The forma name honours Hans Christiansen, the first person to flower this albinistic form of the species in Denmark. The dimensions of the plant and the size of the flowers are the same as the species. **Flower colour:** bright green and white.

Habitat and Distribution: *Paphiopedilum hennisianum* fma. ***christiansenii*** is found in the same locality as the normally coloured form of the species.

Paphiopedilum parnatanum

Cavestro

Subsection: ***Barbata*** (Kränzlin) V.A. Albert and B. Pettersen

W. Cavestro named this species in 1999 in the French magazine *Orchidees*, in honour of A.S. Parnata, the first person to flower the type specimens.

Photo: Olaf Gruss

Paphiopedilum parnatanum

Growth habit: upright, sympodial. **Leaves:** prettily mottled, up to 22 cm long by 3.5 cm wide. **Inflorescences:** upright, carry a single bloom up to 7 cm across the petals. **Flower colour:** dorsal sepal white, barred vertically with green stripes, numerous black spots at the base; petals green, striped lengthwise with darker green, tipped with white and covered with many dark spots that occasionally fuse together; labellum dark brown on top fading to paler brown underneath, veined with darker brown; synsepalum greenish, striped with darker green, white at the tips. **Dorsal sepal:** ovate, tip pointed, up to 4.5 cm long by 3 cm wide. **Petals:** oblong, up to 5 cm long by 1.5 cm wide, between 30 to 45 degrees below the horizontal; edged with warts. **Labellum:** up to 4 cm long by 2.4 cm wide. **Synsepalum:** up to 3.5 cm long by 1.5 cm wide.

Habitat and Distribution: *Paphiopedilum parnatanum* is endemic to Tarlac province on Luzon where it grows at altitudes of about 700 metres on the forest floor among mosses and grasses. The locality of this species usually has plenty of air movement.

Paphiopedilum philippinense

(Reichenbach *f.*) Stein

Section: ***Mastigopetalum*** Hallier

Professor Heinrich G. Reichenbach named this species in 1862 in *Bonplandia.* George Stein gave this plant its currently accepted name in 1892 in *Stein's Orchideenbuch*. The specific name refers to the country of origin.

Growth habit: upright, sympodial. **Leaves:** up to 30 cm long by 4 cm wide, bright green, shiny. **Inflorescences:** arch, bear up to five flowers 20 cm across the petals. **Flower colour:** dorsal sepal white striped vertically with dark purple; petals yellow edged with purple; labellum yellowish green; synsepalum white

Photo: David Banks

Paphiopedilum philippinense

Photo: Jim Cootes

Paphiopedilum philippinense fma. *alboflavum*

Photo: Ray Hill

Paphiopedilum randsii

striped with dark purple. **Dorsal sepal:** ovate, pointed at the tip, up to 5 cm long by 2.5 cm wide. **Petals:** narrowly linear, up to 12 cm long by about 6 mm wide, taper, almost horizontal to the dorsal sepal. There is great variation in the amount of twisting of the petals. **Labellum:** up to 4 cm long, 2.5 cm wide. **Synsepalum:** up to 5 cm long, 2.5 cm wide.

Habitat and Distribution: ***Paphiopedilum philippinense*** is the most widespread ***Paphiopedilum*** in the Philippines, being found in the southern provinces of Luzon; Bohol and Cebu islands and Negros Occidental on Negros island in the Visayas; Agusan del Norte and Surigao del Norte provinces on Mindanao; and on Palawan. It was recently recorded from some islands north of Borneo and Sabah, eastern Malaysia. It grows at elevations up to 500 metres in fairly bright light.

Paphiopedilum philippinense fma. *alboflavum* Gruß

Section: ***Mastigopetalum*** Hallier

Olaf Gruß established this form of ***Paphiopedilum philippinense*** in the *Die Orchidee* 51(3): 354-355 in the year 2000. The specific name refers to flowers, which are lacking anthocyanin pigments.

Growth habit: upright, sympodial. **Leaves:** up to 30 cm long, 4 cm wide, bright green, shiny. **Inflorescences:** arch, bear up to five flowers 20 cm across petals. **Flower colour:** yellowish green and white (i.e. lacking any colour pigments). **Dorsal sepal:** ovate, pointed at the tip, up to 5 cm long, 2.5 cm wide. **Petals:** narrowly linear, up to 12 cm long, about 6 mm wide, taper, almost horizontal to dorsal sepal. Great variation in the amount of twisting of petals. **Labellum:** up to 4 cm long, 2.5 cm wide. **Synsepalum:** up to 5 cm long, 2.5 cm wide.

Habitat and Distribution: ***Paphiopedilum philippinense*** fma. ***alboflavum*** has only been found on a few occasions. Its locality is a closely guarded secret.

Paphiopedilum randsii Fowlie

Section: ***Mastigopetalum*** Hallier

Dr. Jack Fowlie named this species in 1969 in *The Orchid Digest*. He named it after Ray Rands, the owner of an orchid nursery in California, USA.

Growth habit: upright, sympodial. **Leaves:** up to 30 cm long, 5 cm wide, plain green. **Inflorescences:** upright, carry up to five blooms about 7 cm across the petals. **Flower colour:** dorsal sepal cream-white with vertical maroon stripes; petals creamy white with horizontal maroon stripes; labellum bright yellowish green with a brownish overlay; synsepalum creamy white with vertical maroon stripes. **Dorsal sepal:** ovate, up to 4 cm long, 2.5 cm wide, basal third reflexes. **Petals:** linear, up to 6 cm long, 1.2 cm wide, curve gracefully. **Labellum:** up to 5 cm long, 2.3 cm wide. **Synsepalum:** up to 3 cm long, 2 cm wide.

Habitat and Distribution: ***Paphiopedilum randsii*** is endemic to the provinces of Agusan del Sur, Surigao del Norte and Surigao del Sur on Mindanao. It grows as an epiphyte at low elevations of about 500 metres.

Notes: The habitat of this plant is under threat from the activities of illegal loggers and charcoal producers.

Paphiopedilum roebbelenii

(Reichenbach *f.*) Pfitzer
Section: ***Mastigopetalum*** Hallier

Professor Heinrich G. Reichenbach first named this species ***Cypripedium roebbelenii*** in 1883. Pfitzer transferred the generic name to ***Paphiopedilum*** in *Engler, Botanische Jahbücher* in 1894. The specific name honours Carl Röbbelen, the plant's original collector. ***Paphiopedilum roebbelenii*** is very closely related to ***Paphiopedilum philippinense*** but the two species are readily separable because of the much narrower leaves, the much hairier peduncle and the lack of green markings on the staminode.

Habitat and Distribution: ***Paphiopedilum roebbelenii*** is only found in the province of Rizal on Luzon.

Paphiopedilum roebbelenii

Paphiopedilum urbanianum

Fowlie
Section: ***Chloroneura*** Hallier

Dr. Jack Fowlie named this species in 1981 in *The Orchid Digest* after Jacinta Urban, the owner of an orchid nursery in the Philippines.

Growth habit: upright, sympodial. **Leaves:** up to 20 cm long, 4 cm wide; mottled, very showy. **Inflorescences:** upright, carry one flower about 7 cm across the petals. **Flower colour:** dorsal sepal white with green lines running vertically; petals white and green for the basal third, dark pink for the remainder, central third spotted with black; labellum brown and shiny; synsepalum green with darker markings. **Dorsal sepal:** ovate, pointed at the apex, up to 3.5 cm long, 3.2 cm wide. **Petals:** linear, up to 4.5 cm long, 1.2 cm wide, about 40 to 45 degrees below the horizontal. **Labellum:** up to 4.5 cm long by 2.5 cm wide. **Synsepalum:** up to 3 cm long by 2 cm wide.

Paphiopedilum urbanianum

Habitat and Distribution: ***Paphiopedilum urbanianum*** is endemic to Oriental Mindoro on Mindoro where it grows at elevations between 500 and 800 metres. It lives as a terrestrial, usually in fairly shaded locations.

Phaius Loureiro

Juan Loureiro named this genus in 1790 in his *Flora Cochinchinensis.* The generic name refers to the dark colours of some of the flowers.

The type species for this attractive genus is ***Phaius tancarvilleae*** (Banks *ex* l'Heritier) Bl.

There are about 50 species found throughout tropical Africa, all of Asia and Australia. The plants are mainly terrestrial in habit, and there are nine species to be found in the Philippines.

Phaius flavus (Bl.) Lindley

Carl Blume first named this species ***Limodorum flavum***. Dr. John Lindley transferred the genus to ***Phaius*** in 1830 in his *Genera and Species of Orchidaceous Plants.* The specific name refers to the yellow flowers.

Photo: Lloyd Bradford

Phaius flavus

Growth habit: upright, sympodial. **Pseudobulbs:** conical, up to 10 cm high by 5 cm wide. **Leaves:** lanceolate, up to 50 cm long by 12 cm wide; have yellow spots and a thin texture. **Inflorescences:** upright, up to 70 cm tall, appear from the base of the pseudobulbs, bear up to 20 blooms about 6 cm in diameter. **Flower colour:** sepals and petals yellow; labellum yellow, brown markings. **Dorsal** and **lateral sepals:** elliptic, up to 3 cm long by 1 cm wide. **Petals:** of similar shape but narrower. **Labellum:** encloses the column, edges wavy.

Habitat and Distribution: ***Phaius flavus*** is a widely spread species, being found in Japan, China, India, Sri Lanka, Malaysia, Thailand, Laos, Vietnam, Indonesia, Sulawesi (Celebes), Ceram, New Guinea and Taiwan. In the Philippines it has been found in Benguet, Cagayan, Mountain Province, Rizal and Sorsogon provinces on Luzon; Oriental Mindoro province on Mindoro island; and Panay island in the Visayas. It grows as a terrestrial at elevations to 1,500 metres.

Photo: Jim Cootes

Phaius tancarvilleae

Phaius tancarvilleae

(Banks *ex* l'Heritier) Blume

Sir Joseph Banks first described this species in 1788, in *l'Heritier, Sertum Anglicum* as ***Limodorum tancarvilleae***. In 1856 Carl Blume transferred the genus to ***Phaius*** in *Museum Botanicum Lugdunum Batavium.* The specific name honours the Countess of Tancarville.

Growth habit: upright, sympodial. **Pseudobulbs:** up to 15 cm tall by 6 cm in diameter. **Leaves:** up to 1 metre long by 20 cm wide. **Inflorescences:** upright, up to 2 metres tall, carry up to 15 blooms about 10 cm in diameter. **Flower colour:** whitish on the back, reddish brown on the front; labellum maroon on the inside, whitish on the outside. **Sepals** and **petals:** narrowly elliptic, up to 5 cm long by 1.5 cm wide. **Labellum:** trumpet-shaped, edges wavy.

Habitat and Distribution: ***Phaius tancarvilleae*** is extremely widespread, being found in the northwest Himalayas, Sri Lanka, Thailand, Myanmar (Burma), Laos, Vietnam, China, Japan, Taiwan, Malaysia, Indonesia, New Guinea, Australia, New Caledonia, the Cook Islands, Tonga, Samoa and Fiji. In the Philippines it has been found in the Mountain Province and Quezon province of Luzon. It grows as a terrestrial at elevations up to 1,500 metres.

Phalaenopsis Blume

Carl Blume established this genus in 1825 in *Bijdragen tot de Flora van Nederlandsch Indie.* The name comes from two Greek words that suggest the flowers have a resemblance to moths or butterflies in flight.

The type species for this magnificent genus is ***Phalaenopsis amabilis*** (Linn.) Blume.

Phalaenopsis is an Asian genus found in India, throughout Southeast Asia and the north of Australia. The centre of distribution is in Indonesia and the Philippines. They are mainly plants of low elevations (up to 500 metres), but some species occur at higher elevations (up to 1,500 metres). Normally they grow as epiphytes in close proximity to running water. There are approximately 21 species in the Philippines.

Phalaenopsis amabilis

(Linnaeus) Blume
Section: ***Phalaenopsis*** Bentham

Rumphius, who named the plant ***Angraecum album majus***, published the first description and illustration of this species in 1750 in *Herbarium Amboinense*. The original orchid was found on Ambon island. Two years later Osbech found the same species in Java and sent the plants to Carl Linnaeus who named it ***Epidendrum amabile*** in *Species Plantarum.* Roxburgh named it ***Cymbidium amabile*** in 1814. Carl Blume gave this plant its current name in 1825 in *Bijdragen tot de Flora van Nederlandsch Indie.* This species is the type species for this genus ***Phalaenopsis***. The specific name refers to the beauty of the flowers.

Growth habit: upright, monopodial. **Leaves:** leathery, up to 35 cm long by 10 cm wide, up to three pairs of leaves on old, well-grown plants. **Inflorescence:** arching, reach lengths of over 60 cm, occasionally branch; flowers up to 7.5 cm in diameter (superior or line-bred clones can be bigger). **Flower colour:** milky white; labellum white with yellow and brown markings. **Dorsal sepal:** elliptic, up to 3 cm long by 1.2 cm wide. **Petals:** broadly spathulate, up to 3.5 cm high by 3.5 cm wide. **Lateral sepals:** ovate, up to 3 cm long by 1.5 cm wide. **Labellum:** three-lobed; lateral lobes upright; mid-lobe cross-shaped with two hair like appendages at the tip. There is a large callus in the centre of the labellum, which is fleshier than the other segments.

Habitat and Distribution: ***Phalaenopsis amabilis*** is a widely spread species, being found throughout Indonesia, Borneo, Sulawesi (Celebes), Papua New Guinea and northern Australia. In the Philippines it has been found in the southern islands of Balabac, Bancalan and Lambugan near Palawan; Palawan island; and Sulu and Tawitawi in the Sulu Archipelago. It grows as an epiphyte at elevations to 500 metres.

Photo: Jim Cootes

Phalaenopsis amabilis

Photo: Jim Cootes

Phalaenopsis aphrodite

Phalaenopsis aphrodite

Reichenbach *f.*
Section: ***Phalaenopsis*** Bentham

This species is closely related to ***Phalaenopsis amabilis*** and at different times has been considered a variety of that species. Kamel first described this plant in 1704 as ***Visco-Aloes Luzonis decima quarta***. Due to lack of money this name was never published, but in 1709 an illustration of this orchid was included in *Gazophyllaci Naturae et Artis* by Petiver. Professor

Heinrich G. Reichenbach gave this plant its accepted name in 1862 in *Hamburger Garten-Blumenzeitung*. The specific name refers to Aphrodite, the Greek goddess of love.

Growth habit: upright, monopodial. **Leaves:** leathery, up to 30 cm long by 7 cm wide, purplish tinge under the leaves. **Inflorescences:** upright, branching in older plants; flowers up to 6 cm in diameter. **Flower colour:** milky-white; labellum white with yellow and red markings. **Dorsal sepal:** elliptic, up to 3 cm long by 1.2 cm wide. **Petals:** broadly spathulate, up to 3 cm long by 3 cm wide. **Lateral sepals:** ovate, up to 3 cm long by 1.5 cm wide. **Labellum:** three-lobed; callus, located at the junction of the middle and lateral lobes, is a different shape than that of ***Phalaenopsis amabilis***.

Habitat and Distribution: ***Phalaenopsis aphrodite*** is the most widely spread of all the large-flowered *Phalaenopsis* species. It is found throughout the islands of the Philippines north of Mindanao where it grows at elevations of up to 500 metres. The form of this species found in Taiwan grows as an epiphyte at elevations to 1,500 metres.

Notes: This is the most commonly cultivated ***Phalaenopsis*** species in the Philippines and is frequently seen in gardens throughout the country.

Phalaenopsis cornucervi

(Breda) Blume and Reichenbach *f.*
Section: ***Polychilos*** (Breda) Reichenbach *f.*

Jacob van Breda first described this species as ***Polychilos cornucervi*** in 1827. Carl Blume and Heinrich Reichenbach realised a mistake had been made as far as the genus of this species was concerned and they made the transfer in 1860 in *Hamburger Garten-Blumenzeitung*. The specific name refers to the inflorescence, which resembles the antlers of a deer.

Growth habit: upright, monopodial. **Leaves:** very bright green, shiny, up to 20 cm long by 4 cm wide. **Inflorescences:** upright, remarkable as they are round at the base and flatten laterally where the flowers appear and will occasionally branch; flowers star-shaped, about 2.5 cm in diameter, fleshy with a waxy appearance. This plant can be continually in flower. **Flower colour:** pale yellow blotched and spotted with reddish brown. **Dorsal sepal:** lanceolate, up to 1.5 cm long by 4 mm wide. **Petals:** lanceolate, marginally smaller than the other segments. **Lateral sepals:** elliptic-lanceolate, up to 1.5 cm long by 4 mm wide. **Labellum:** three-lobed; lateral lobes upright; mid-lobe triangular.

Phalaenopsis cornucervi

Habitat and Distribution: ***Phalaenopsis cornucervi*** is quite widespread, being found in the Nicobar Islands, Myanmar (Burma), Thailand, Malaysia, Sumatra, Java and Borneo. In the Philippines it has only been found on Palawan island. It grows as an epiphyte at elevations to 750 metres.

Phalaenopsis equestris

(Schauer) Reichenbach *f.*
Section: ***Stauroglottis*** (Schauer) Bentham

Johann Schauer first described this species as ***Stauroglottis equestris*** in 1843. Professor Heinrich G. Reichenbach gave the plant its current name in 1850 in *Linnaea* 22. The Latin name *equestris* pertains to horses and a proud and handsome rider.

Growth habit: upright, monopodial. **Leaves:** very leathery, up to 20 cm long, 4 cm wide. **Inflorescences:** upright, branch on older plants; flowers star-shaped, up to 4 cm in diameter; dorsal and lateral sepals reflex. **Flower colour:** segments usually pink; trowel-shaped labellum very variable—I have seen clones with labellums in pale pink, rose-purple, brown and orange. **Dorsal sepal:** oblong-elliptic, up to 1.5 cm long by 4 mm wide. **Petals:** elliptic, up to 1.5 cm long by 5 mm wide. **Lateral sepals:** oblong-elliptic, 1.5 cm long by 5 mm wide. **Labellum:** three-lobed; lateral lobes are upright; mid-lobe obovate; callus at the junction of the mid-lobe and lateral lobe.

Photo: David Banks

Phalaenopsis equestris

Habitat and Distribution: ***Phalaenopsis equestris*** is found throughout the Philippine archipelago and the islands of southern Taiwan. It grows as an epiphyte at elevations to 300 metres.

Phalaenopsis equestris var. *alba* (Hort.) Sweet

Section: ***Stauroglottis*** (Schauer) Bentham

Dr. Herman Sweet formally named this variety in the *American Orchid Society Bulletin* in 1967 (vol. 36, page 143). Plant habit as per species, but the flowers are pure white with yellow or purplish markings in the labellum.

Habitat and Distribution: ***Phalaenopsis equestris*** var. ***alba*** has been found in the provinces of Laguna, Quezon and Rizal on Luzon. It grows as an epiphyte at elevations to 300 metres.

Phalaenopsis fasciata Reichenbach *f.*

Section: ***Zebrinae*** Pfitzer

Professor Heinrich G. Reichenbach named this species in the *Gardeners' Chronicle* in 1882. The specific name refers to the broad, parallel stripes of colour on the floral segments.

Growth habit: upright, monopodial. **Leaves:** bright green, up to 20 cm long by 6 cm wide. **Inflorescences:** arching, longer than the leaves, carry a single star-shaped flower up to 4 cm in diameter. A large, healthy plant can have four inflorescences at any one time and remain in bloom for many months. **Flower colour:** creamy yellow background, cinnamon-brown barring; labellum orange at basal end, pink at distal end. **Dorsal sepal:** elliptic, up to 2 cm long by 6 mm wide. **Petals:** elliptic, similar in size to the other segments. **Lateral sepals:** elliptic-ovate, up to 2 cm long by 6 mm wide. **Labellum:** three-lobed; lateral lobes upright; mid-lobe oblong-ovate.

Photo: David Banks

Phalaenopsis equestris var. *alba*

Photo: David Banks

Phalaenopsis fasciata

Habitat and Distribution: ***Phalaenopsis fasciata*** is endemic to the Philippines and is found in Bataan, Albay and Sorsogon provinces on Luzon; Bohol island in the Visayas; and Davao del Sur province on Mindanao. It is epiphytic at elevations to 300 metres.

Phalaenopsis heiroglyphica

(Reichenbach *f.*) Sweet

Section: ***Zebrinae*** Pfitzer

Professor Heinrich G. Reichenbach first named this pretty species ***Phalaenopsis lueddemanniana*** var. ***heiroglyphica*** in 1887 because of the plant's superficial resemblance to ***Phalaenopsis lueddemanniana***. The late Dr. Herman Sweet raised this plant to specific status in the *American Orchid Society Bulletin* in 1969. The specific name refers to the hieroglyphic-like markings on the sepals and petals of the flowers.

Photo: David Titmuss

Phalaenopsis heiroglyphica

Photo: Jim Cootes

Phalaenopsis lindenii 'Peter'

Photo: Elias Javier

Phalaenopsis lindenii var. *alba*

Growth habit: upright, monopodial. **Leaves:** plain green and shiny, up to 25 cm long by 5 cm wide. **Inflorescences:** upright, will occasionally branch; blooms star-shaped, up to 4 cm in diameter. **Flower colour:** sepals and petals cream or white covered with brown or maroon hieroglyphic-like markings; labellum narrow, white or pinkish. **Dorsal** and **lateral sepals:** elliptic-ovate, up to 2 cm long by 5 mm wide. **Petals:** of similar size and shape as the other segments. **Labellum:** three-lobed; lateral lobes upright; mid-lobe oblong, obscurely toothed toward the tip.

Habitat and Distribution: ***Phalaenopsis heiroglyphica*** is endemic to the Philippines and has been found on Palawan and Polillo islands and in Surigao del Norte and Surigao del Sur provinces on Mindanao. It grows at elevations to 500 metres as an epiphyte in very shaded locations.

Phalaenopsis lindenii Loher

Section: ***Stauroglottis*** (Schauer) Bentham

Augustus Loher described this plant in 1895 in the *Journal des Orchidees* and named it after Jean Linden, a French horticulturist of that time.

Growth habit: upright, monopodial. **Leaves:** up to 20 cm long by 4 cm wide, dark green with silvery white mottling. The underside of leaves vary from distinctly purple to bright green. **Inflorescences:** upright, slender, occasionally branching. **Flower colour:** white with pale pink suffusion. In some clones the dorsal and lateral sepals reflex. Labellum darker pink at the apex with radiating darker, rose-coloured lines shading into brown at the base; flowers up to 3 cm in diameter. **Dorsal sepal:** oblong-elliptic, up to 1.5 cm long by 5 mm wide. **Petals:** spathulate, up to 1.5 cm long by 5 mm wide. **Lateral sepals:** oblong-ovate, up to 1.5 cm long by 5 mm wide. **Labellum:** three-lobed; lateral lobes upright; mid-lobe circular.

Phalaenopsis lindenii var. *alba* Hort.

Section: ***Stauroglottis*** (Schauer) Bentham

The plant is similar to the species but the undersides of the leaves are green. All the floral segments are white; the markings on the labellum green.

Habitat and Distribution: ***Phalaenopsis lindenii*** and the ***alba*** variety are high altitude plants from the mountains of Benguet, Isabela, Mountain Province, Nueva Ecija and Nueva Vizcaya provinces in central Luzon. They grow as epiphytes at elevations between 1,000 and 1,500 metres. Their habitat receives over 4 metres of rain per year, the driest time being December and January, the coolest months in the Philippines.

Phalaenopsis lueddemanniana Reichenbach *f.*

Section: ***Zebrinae*** Pfitzer

Professor Heinrich G. Reichenbach named this species in *Botanische Zeitung* in 1865 after a Mr. Lueddemann, the first person to flower this plant in Europe.

Growth habit: upright, monopodial. **Leaves:** up to 25 cm long by 6 cm wide. **Inflorescences:** erect, bear several flowers at one time, continue to bloom for several months on adult plants. **Flower colour:** base colour white or ivory with magenta bars; labellum narrow, carmine and yellow; blooms up to 4 cm in diameter. **Dorsal sepal:** lanceolate, up to 2 cm long by 6 mm wide. **Petals:** ovate, up to 2 cm long by 7 mm wide. **Lateral sepals:** ovate-elliptic, up to 2 cm long by 1 cm wide. **Labellum:** three-lobed; lateral lobes upright; mid-lobe oblong.

Habitat and Distribution: ***Phalaenopsis lueddemanniana*** is endemic to the Philippines, being found in Apayao, Bataan, Bulacan, Camarines Norte, Kalinga, Laguna, Nueva Vizcaya, Pangasinan, Quezon, Rizal and Sorsogon provinces on Luzon; Palawan and Polillo islands; Leyte island in the Visayas; and Agusan del Norte, Agusan del Sur, Bukidnon, Davao del Norte, Davao Oriental, Davao del Sur, Lanao del Norte and Lanao del Sur provinces on Mindanao. It grows as an epiphyte at elevations to 500 metres.

Photo: David Banks

Phalaenopsis lueddemanniana

Phalaenopsis mariae

Burbidge *ex* Warner and Williams

Section: ***Zebrinae*** Pfitzer

Frederick Burbidge named this plant after his wife when he first collected it on Jolo island in the southern Philippines. Warner and Williams published a valid description of this species in 1883 in *Orchid Album* 2.

Growth habit: upright, monopodial. **Leaves:** bright green, up to 30 cm long by 7 cm wide. **Inflorescences:** semi-pendent, adult plants can have several; flowers about 3 cm in diameter. **Flower colour:** milky white heavily blotched with brown markings; central portion of segments purple; labellum narrow and purple. **Dorsal sepal** and **petals:** narrowly elliptic, up to 1.5 cm long by 4 mm wide. **Lateral sepals:** elliptic-ovate, up to 1.5 cm long by 4 mm wide. **Labellum:** three-lobed; lateral lobes upright; mid-lobe narrowly elliptic.

Photo: Gary Yong Gee

Phalaenopsis mariae

Habitat and Distribution: ***Phalaenopsis mariae*** is widely distributed throughout the provinces of Bukidnon, North and South Cotabato, Davao del Norte, Davao Oriental, Davao del Sur, Lanao del Norte and Davao del Sur on Mindanao; and on the islands of the Sulu Archipelago. It has recently been recorded from Borneo. It normally grows as an epiphyte in quite dense shade at elevations to 1,000 metres.

Photo: David Banks

Phalaenopsis micholitzii

Photo: Jim Cootes

Phalaenopsis pallens

Phalaenopsis micholitzii Rolfe

Section: ***Amboinenses*** Sweet

Robert Rolfe named this species in the *Gardeners' Chronicle* in 1890, in honour of Wilhelm Micholitz, collector of the type specimens.

Growth habit: upright, monopodial. **Leaves:** fleshy, oblong to oblanceolate, up to 15 cm long by 5 cm wide. **Inflorescences:** short, one to several, bear from one to three flowers. **Flower colour:** milky white; labellum has gold markings. **Dorsal sepal:** oblong to elliptic, up to 2.5 cm long by 1.5 cm wide. **Petals:** ovate to elliptic, up to 2.5 cm long by 1.5 cm wide. **Lateral sepals:** oblong to elliptic, up to 2.6 cm long by 1.4 cm wide. **Labellum:** three-lobed, up to 2.2 cm long by 1.6 cm wide; lateral lobes erect; mid-lobe rhomboid with short hairs along the centre.

Habitat and Distribution: *Phalaenopsis micholitzii* is endemic to the Philippines and has been recorded from Camarines Sur province on Luzon and Zamboanga del Norte and Zamboanga del Sur provinces on Mindanao. It grows as an epiphyte at elevations to 400 metres. This species is rare, both in its native habitat and in cultivation.

Phalaenopsis pallens

(Lindley) Reichenbach *f.*

Section: ***Zebrinae*** Pfitzer

This plant has an interesting nomenclatural history. Dr. John Lindley first described it as ***Trichoglottis pallens*** in 1850. Professor Heinrich G. Reichenbach named it ***Stauropsis pallens*** in 1860 and gave it its accepted name in 1864 in *Walpers Annales Botanices*. In 1882 Burbidge named it ***Phalaenopsis lueddemanniana*** var. ***pallens,*** but the name was never accepted. The specific name refers to the flowers' pale colour.

Growth habit: upright, monopodial. **Leaves:** fleshy, oblong to oblanceolate, up to 15 cm long by 5 cm wide. **Inflorescences:** upright and branching; flowers 3 cm in diameter. **Flower colour:** creamy yellow marked with brownish red dashes and lines; labellum white. An albino form of the species has no coloured markings on the segments. **Dorsal** and **lateral sepals:** oblong-elliptic, up to 2 cm long by 6 mm wide. **Petals:** of similar size and shape as other segments. **Labellum:** three-lobed; lateral lobes upright; mid-lobe ovate.

Habitat and Distribution: *Phalaenopsis pallens* is widely distributed throughout the Philippines and has been found in Bulacan, Nueva Vizcaya and Rizal on Luzon; Palawan island; and Bukidnon province on Mindanao. It is epiphytic at elevations to 500 metres.

Phalaenopsis philippinensis

Golamco *ex* Fowlie and C.Z. Tang

Section: ***Phalaenopsis*** Bentham

Andres Golamco Jr. first described this species in the *Philippine Orchid Review* in 1984 but gave no Latin diagnosis. Dr. Jack Fowlie and C.Z. Tang named it correctly in *The Orchid Digest* in 1987. The specific name refers to this plant's country of origin.

Photo: David Banks

Phalaenopsis philippinensis

Growth habit: upright, monopodial. **Leaves:** up to 30 cm long, 7 cm wide, dark green mottled with silvery markings. **Inflorescences:** upright, branch freely on large plants; flowers up to 7 cm in diameter. **Flower colour:** from milky white to very pale pink; lateral lobes of labellum bright yellow with red striping near the centre—a distinguishing feature of this beautiful species. **Dorsal sepal:** elliptic, up to 3.5 cm long, 1.5 cm wide. **Petals:** almost circular, up to 3 cm high, 3 cm wide. **Lateral sepals:** elliptic-ovate, up to 3.5 cm long, 1.5 cm wide. **Labellum:** three-lobed; lateral lobes upright; mid-lobe gradually tapers, two hair-like appendages at the tip.

Habitat and Distribution: ***Phalaenopsis philippinensis*** comes from the mountains of Isabela and Quezon provinces in northeast Luzon. It grows as an epiphyte at elevations of about 1,200 metres, normally close to water, and is endemic to this region.

Notes: This species has been known in collections for a long time and is commonly sold as ***Phalaenopsis* x *leucorrhoda*.** It is a natural hybrid of ***Phalaenopsis aphrodite*** and ***Phalaenopsis schilleriana*. *P. aphrodite*** is found throughout the Philippines. ***P. schilleriana*** is only found in southern Luzon and some of the northern Visayas islands. The natural hybrid ***P.* x *leucorrhoda***, a rare plant in nature, is only found where ***P. schilleriana*** grows, about 400 kilometres south from where ***P. philippinensis*** grows. Its distinguishing feature is the bright yellow lateral lobes of the labellum, which do not occur in ***P.* x *leucorrhoda*.**

Phalaenopsis pulchra

(Reichenbach *f.*) Sweet
Section: ***Zebrinae*** Pfitzer

When this plant was first described in 1875, Professor Heinrich G. Reichenbach named it ***Phalaenopsis leuddemanniana*** var. ***pulchra***. The late Dr. Herman

Photo: David Titmuss

Phalaenopsis pulchra

Sweet gave this plant its currently accepted name in the *American Orchid Society Bulletin* in 1968. The specific name refers to the beauty of the flower.

Growth habit: upright, monopodial, similar to ***Phalaenopsis leuddemanniana***. **Inflorescences:** upright, carry several 3 cm in diameter; star-shaped blooms during the flowering season can last several months. **Flower colour:** solid purple; labellum purple with bright yellow lateral lobes. **Dorsal sepal:** elliptic, up to 1.5 cm long by 5 mm wide. **Petals:** of similar shape and size as the dorsal sepal. **Lateral sepals:** ovate, up to 1.5 cm long by 4 mm wide. **Labellum:** three-lobed; lateral lobes upright; mid-lobe tapered.

Habitat and Distribution: ***Phalaenopsis pulchra*** is endemic to the Philippines where it is found in the provinces of Nueva Ecija and Nueva Vizcaya on Luzon and the island of Leyte in the Visayas. It grows as an epiphyte at elevations to 500 metres.

Phalaenopsis sanderiana

Reichenbach *f.*
Section: ***Phalaenopsis*** Bentham

Professor Heinrich G. Reichenbach named this beautiful species in 1882 in *Flora* 65, in honour of the great British orchid nursery, Sander's and Sons, which employed many collectors in the mid- to late 1800s.

Photo: Jim Cootes

Phalaenopsis sanderiana 'Rosas'

Photo: Jack Jannese

Phalaenopsis schilleriana

Photo: David Banks

Phalaenopsis stuartiana

Growth habit: upright, monopodial. **Leaves:** up to 40 cm long by 10 cm wide with a purplish tinge. **Inflorescences:** upright, up to 60 cm long, branch on older plants; flowers about 6 cm in diameter. **Flower colour:** rose-purple to pale pink. There is an albino form. **Dorsal sepal:** elliptic, up to 3 cm long by 1.5 cm wide. **Petals:** almost circular, up to 3 cm high by 3 cm wide. **Lateral sepals:** ovate, up to 3 cm long by 1.2 cm wide. **Labellum:** three-lobed; lateral lobes upright; mid-lobe cross-shaped with two hair-like appendages at the tip; callus at the junction of side lobe and mid-lobe.

Habitat and Distribution: ***Phalaenopsis sanderiana*** is endemic to South Cotabato, Davao del Sur, Lanao del Sur, Surigao del Norte, Surigao del Sur, Zamboanga del Norte and Zamboanga del Sur provinces on Mindanao island. It is epiphytic at elevations to 500 metres.

Phalaenopsis schilleriana

Reichenbach *f.*
Section: ***Phalaenopsis*** Bentham

Professor Heinrich G. Reichenbach named this pretty species in 1860 in *Hamburger Garten-Blumenzeitung*, after Consul Schiller, the first person in Europe to bloom this floriferous species.

Growth habit: upright, monopodial. **Leaves:** attractively mottled, up to 40 cm long, 8 cm wide. **Inflorescences:** upright, carry many flowers about 5 cm in diameter. **Flower colour:** pink; also an albino form. **Dorsal sepal:** elliptic, up to 2.5 cm long by 1 cm wide. **Petals:** almost circular, up to 2.5 cm high by 2.5 cm wide. **Lateral sepals:** elliptic, up to 2.5 cm long, 1 cm wide. **Labellum:** three-lobed; lateral lobes upright but flaring outward, mid-lobe obovate with two hair-like appendages at the tip.

Habitat and Distribution: ***Phalaenopsis schilleriana*** is endemic to Albay, Laguna, Quezon, Rizal and Sorsogon provinces on Luzon; Marinduque island; and Biliran island in the Visayas. It grows as an epiphyte at elevations to 800 metres.

Phalaenopsis stuartiana

Reichenbach *f.*
Section: ***Phalaenopsis*** Bentham

Professor Heinrich G. Reichenbach named this species in 1881 in the *Gardeners' Chronicle*, after Stuart Low of the firm Hugh Low and Company.

Growth habit: upright, monopodial. **Leaves:** mottled, up to 25 cm long by 7 cm wide. **Inflorescences:** upright, branching; flowers about 4 cm in diameter. **Flower colour:** creamy white; inner half of lateral sepals beautifully spotted with dark cinnamon blotches over a base colour of sulphur yellow. On some clones all the segments are heavily spotted, but these specimens are rare. **Dorsal sepal:** elliptic, up to 2 cm long by 1 cm wide. **Petals:** almost circular, 2.5 cm high by 2.5 cm wide. **Lateral sepals:** elliptic-oblong, up to 2 cm long by 1 cm wide. **Labellum:** three-lobed; lateral lobes upright, flare outward slightly; mid-lobe round with two hair-like appendages at the tip.

Habitat and Distribution: ***Phalaenopsis stuartiana*** is endemic to Agusan del Norte, Agusan del Sur, Bukidnon, Misamis Oriental, Surigao del Norte and Surigao del Sur provinces on Mindanao where it grows as an epiphyte at low elevations, to 300 metres.

Phalaenopsis sumatrana

Phalaenopsis sumatrana

Korthals and Reichenbach *f.*
Section: ***Zebrinae*** Pfitzer

Korthals and Professor Heinrich G. Reichenbach named this variable species in 1860 in *Hamburger Garten-Blumenzeitung.* The specific name refers to the island of Sumatra where the type specimens were collected.

Growth habit: upright, monopodial. **Leaves:** leathery, up to 30 cm long by 8 cm wide. **Inflorescences:** upright, carry up to five blooms about 4 cm in diameter. **Flower colour:** creamy white flowers with brown barring; colour scheme extremely variable. **Dorsal sepal:** oblong-lanceolate, up to 2 cm long by 8 mm wide. **Petals:** lanceolate, up to 2 cm long by 6 mm wide. **Lateral sepals:** oblong-lanceolate, up to 2 cm long by 1 cm wide. **Labellum:** three-lobed; lateral lobes upright; mid-lobe oblong-elliptic.

Habitat and Distribution: ***Phalaenopsis sumatrana*** is widespread, being found in Myanmar, Thailand, Vietnam, Peninsular Malaysia, Sumatra and Borneo. In the Philippines it has been found on Palawan island. It grows as an epiphyte at elevations to 700 metres.

Pholidota Lindley *ex* Hooker

Dr. John Lindley established this genus in 1825. The name was published in Hooker's *Exotic Botany.* The generic name refers to the scales on the bracts on the inflorescence.

The type species for this genus is ***Pholidota imbricata*** Lindl.

There are about 40 species found in mainland Asia, Indonesia, New Guinea, the Philippines and some of the islands of the Pacific. They grow as terrestrials, lithophytes and epiphytes in a wide range of elevations. Three species are found in the Philippines.

Pholidota ventricosa

(Blume) Reichenbach *f.*
Section: ***Chelonanthera*** (Bl.) Butzin

Carl Blume first described this species as ***Chelonanthera ventricosa*** in 1825. Professor Heinrich G. Reichenbach corrected the generic error in 1857 in *Bonplandia.* The specific name refers to a swelling, possibly in the labellum.

Growth habit: upright, sympodial. **Pseudobulbs:** up to 15 cm long by 3 cm wide, tapered. **Leaves:** very leathery, distinct middle vein, up to 50 cm long by 10 cm wide. **Inflorescences:** upright, appear with the new growth, bear up to 60 flowers 1.5 cm in diameter. **Flower colour:** pale yellow. Some clones have bright

Pholidota ventricosa

orange markings on the labellum, others are all the same colour. There is a large bract at the back of each bloom but this does not prevent the flower from fully opening. **Dorsal sepal:** lanceolate, up to 8 mm long by 5 mm wide. **Petals:** linear, up to 1 cm long by 3 mm wide, reflex badly. **Lateral sepals:** ovate, up to 8 mm long by 4 mm wide. **Labellum:** up to 1 cm long by 6 mm wide, divided into two lobes.

Habitat and Distribution: ***Pholidota ventricosa*** is widely distributed throughout Peninsular Malaysia, Vietnam, Indonesia, Borneo and New Guinea. In the Philippines this species has been found in Benguet, Nueva Ecija, Nueva Vizcaya and Pangasinan provinces on Luzon; Leyte island in the Visayas; and Bukidnon, Lanao del Sur and Zamboanga del Norte provinces on Mindanao. I have seen this plant growing both as an epiphyte and a terrestrial in the mountains of central Luzon at an elevation of about 1,200 metres.

Notes: I am not familiar with any other ***Pholidota*** species that grows so large.

Phreatia Lindley

Dr. John Lindley established this genus in 1830 in his *Genera and Species of Orchidaceous Plants.* The generic name refers to the well-like structure formed by the lateral sepals and the labellum.

The type species for this genus is ***Phreatia elegans*** Lindley, from Sri Lanka.

There are approximately 150 species throughout Asia, New Guinea, Australia and the Pacific Islands. The Philippines has approximately 14 members for this genus; 10 of these are endemic.

Photo: David Titmuss

Phreatia secunda

Phreatia secunda (Blume) Lindley

Carl Blume first named this species ***Dendrolirium secundum*** in 1825. Dr. John Lindley transferred the generic name to ***Phreatia*** in 1830 in his *Genera and Species of Orchidaceous Plants.* The specific name refers to the habit of the flowers on the inflorescence, which all face the same direction.

Growth habit: upright, sympodial. **Pseudobulbs:** mere thickenings of leaves around a stem. **Leaves:** leathery, narrowly lanceolate, up to 6.5 cm long by 6.5 mm wide, unevenly notched at the tip. **Inflorescences:** upright, up to 5 cm long, bear up to 30 flowers 3 mm in diameter. **Flower colour:** white to pale green. **Dorsal sepal:** ovate, up to 1 mm long by 0.75 mm wide. **Petals:** lanceolate, up to 0.8 mm long by 0.5 mm wide. **Lateral sepals:** broadly ovate, up to 1.1 mm long by 0.7 mm wide. **Labellum:** circular, about 2 mm in diameter.

Habitat and Distribution: ***Phreatia secunda*** is widespread and has been recorded from Vietnam, Thailand, Malaysia, Indonesia, Sulawesi (Celebes) and Borneo. In the Philippines it has been collected in Nueva Vizcaya, Nueva Ecija and Rizal provinces on Luzon; Leyte island in the Visayas; and Zamboanga del Norte province on Mindanao. It grows as a twig epiphyte in fairly shady localities up to 1,200 metres.

Phreatia sulcata (Blume) J.J. Smith

Carl Blume first described this species as ***Dendrolirium sulcatum*** in 1825. Johannes J. Smith transferred the generic name to ***Phreatia*** in 1905 in *Die Orchideen von Java.* The specific name means furrowed or grooved longitudinally, but I don't know which feature of the plant or flower is thus described.

Growth habit: upright, sympodial. **Pseudobulbs:** ovate, up to 2 cm high by 1.5 cm in diameter. **Leaves:** linear, up to 20 cm long by 2.5 cm wide. **Inflorescence:** up to 15 cm long, bear many small flowers. **Flower colour:** white. **Dorsal sepal:** triangular, 1.5 mm long by 0.75 mm wide. **Petals:** lanceolate, up to 1 mm long by 0.5 mm wide. **Lateral sepals:** triangular, 1.5 mm long by 1 mm wide. **Labellum:** shortly saccate at base, 2 mm long by 2 mm wide, lobe triangular.

Habitat and Distribution: ***Phreatia sulcata*** is widely distributed, being found in Thailand, Indonesia and Borneo. In the Philippines it has been recorded from Albay, Bataan, Ifugao, Mountain Province, Pampanga, Quezon and Rizal provinces on Luzon; Oriental Mindoro province on Mindoro island; Leyte island; Agusan del Norte province on Mindanao; and Basilan island.

Phreatia sulcata

Photo: Dr. Henrik Pedersen

Phreatia vanoverberghii Ames

Professor Oakes Ames named this plant in 1912 in the *Philippine Journal of Science*. The specific name honours Father Vanoverbergh, a Belgian priest, who spent many years in the mountains of central Luzon working with the indigenous tribes.

Growth habit: upright, sympodial. **Pseudobulbs:** reduced to stems, about 5 cm long, hidden by overlapping leaf bases. **Leaves:** linear, up to 30 cm long by 2 cm wide. **Inflorescence:** up to 25 cm long, bear many little flowers about 2 mm in diameter. **Flower colour:** white. **Dorsal sepal:** ovate, 1 mm long by 1 mm wide. **Petals:** triangular at base, round above, 1.5 mm long by 0.75 mm wide. **Lateral sepals:** ovate, 1 mm long, form a small mentum. **Labellum:** clawed, triangular, the basal portion elliptic.

Habitat and Distribution: ***Phreatia vanoverberghii*** is endemic to the mountains of central Luzon and has been recorded from the provinces of Abra, Apayao, Benguet, Ifugao, Kalinga and the Mountain Province. It grows as an epiphyte at elevations between 600 and 2,200 metres.

Phreatia vanoverberghii

Photo: Dr. Henrik Pedersen

Plocoglottis Blume

Carl Blume established this genus in 1825 in *Bijdragen tot de Flora van Nederlandsch Indie.* The generic name refers to the binding of the labellum to the base of the column.

The type species for the genus is ***Plocoglottis javanica*** Bl.

There are about 30 species that are found throughout Asia and New Guinea. The Philippines has at least six species; five are endemic.

Plocoglottis acuminata Blume

Carl Blume named this attractive species in 1849 in *Museum Botanicum Lugdano-Batavum.* The specific name probably refers to the acuminate (pointed) tips of the segments.

Growth habit: upright, sympodial. **Pseudobulbs:** cylindrical, up to 9 cm high by 1 cm in diameter, unifoliate. **Leaves:** elliptic to lanceolate, up to 25 cm long by 10 cm wide; petiole to 10 cm long. **Inflorescences:** upright, usually longer than leaves, bear up to 12 flowers about 3 cm in diameter. **Flower colour:** yellow with reddish spots and blotches on inner half of segments; labellum yellow with two red blotches near the base. **Dorsal sepal:** linear to lanceolate, up to 2.5 cm long by 6 mm wide, curved over column. **Petals:** linear, acuminate, up to 2.5 cm long by 4 mm wide. **Lateral sepals:** lanceolate, turn down, 2.5 cm long by 8 mm wide; apical half spread out. **Labellum:** wedge-shaped, up to 2.3 cm long by 8 mm wide; apex bends down.

Plocoglottis acuminata

Habitat and Distribution: ***Plocoglottis acuminata*** has been found in Peninsular Malaysia, Sumatra, Java and Borneo. In the Philippines it is recorded from Quezon province on Luzon and Leyte island in the Visayas. It grows as a terrestrial on the banks of rivers or ravines, where it is continually damp, often in the shade at elevations between 300 and 600 metres.

Podochilus Blume

Carl Blume established this genus in 1825 in *Bijdragen tot de Flora van Nederlandsch Indie.* The generic name refers to two appendices at the base of the labellum.

The type species for this genus is ***Podochilus lucescens*** Bl.

There are about 75 species in the genus and they have been recorded from Sri Lanka, the Himalayas, Myanmar (Burma), Thailand, Malaysia, Indonesia, Borneo, New Guinea, some Pacific islands and the Philippines, which has approximately 10 species.

Podochilus intricatus Ames

Professor Oakes Ames named this species in the *Philippine Journal of Science* in 1913. The specific name means intricate, but I am not certain which feature of the plant or flower is so described.

Growth habit: semi-pendulous, sympodial. **Pseudobulbs:** very slender, branching, up to 20 cm long. **Leaves:** two-ranked, leathery, up to 1.4 cm long by 2 mm wide. **Inflorescences:** appear from the tip of the

Podochilus intricatus

pseudobulb, up to 1.5 cm long. **Flower colour:** white; labellum white with two purple spots. **Dorsal sepal:** oblong-lanceolate, up to 2 mm long by 1 mm wide. **Petals:** oblong-elliptic, up to 2 mm long by 1 mm wide. **Lateral sepals:** triangular-lanceolate, tapered to a point, up to 3 mm long by 1.5 mm wide. **Labellum:** rhombic, two-forked, 3 mm long by 1.5 mm wide.

Habitat and Distribution: ***Podochilus intricatus*** is endemic to the Philippines and has been recorded from Quezon province in Luzon and Agusan del Norte, Agusan del Sur, Davao del Norte and Davao del Sur provinces on Mindanao; and Basilan island, south of Zamboanga. It is epiphytic in shaded locations at elevations between 300 and 1,100 metres.

Polystachya Hooker

William J. Hooker established this genus in 1825 in his *Exotic Flora*. The generic name refers to the many branches or spikes on the inflorescences of some species in this genus.

The type species for this cosmopolitan genus is ***Polystachya minuta*** (Aubl.) Hook.

Members for this genus are found throughout Africa, Madagascar, Asia and tropical America. There are approximately 150 species, which grow mainly as epiphytes and lithophytes. The Philippines has only one member of this very cosmopolitan genus.

Polystachya concreta

(Jacquin) Garay and Sweet

Jacquin first described this species in 1760 as ***Epidendrum concretum*** from plants collected in the West Indies. Hooker then named this species ***Polystachya concreta*** in 1824 in *Exotic Flora*. In 1974 Garay and Sweet did a study of the distribution and found many of the then-accepted names were only synonyms of this widely distributed orchid. In fact, in *Orquideologia* Dr. Leslie Garay and Dr. Herman Sweet list 54 synonyms for this orchid. The specific name is from the Latin *concresco,* meaning to grow together, and refers to the adnation of the lateral sepals to the full length of the column foot.

Growth habit: upright, sympodial. **Pseudobulbs:** up to 6 cm long by 2 cm in diameter, tapered. **Leaves:** enclose the pseudobulb, up to 20 cm long by 3 cm wide. **Inflorescences:** upright, terminal, up to 30 cm long, branching; flowers small, labellum uppermost. **Flower colour:** sulphur-yellow; labellum a little paler. **Dorsal sepal:** ovate, up to 3 mm long by 2 mm wide. **Petals:** linear, up to 3 mm long by 1 mm wide. **Lateral sepals:** triangular, up to 3 mm long by 3 mm wide. **Labellum:** three-lobed; lateral lobes small; mid-lobe almost circular.

Photo: David Banks

Polystachya concreta

Habitat and Distribution: ***Polystachya concreta*** is the most widely distributed epiphytic orchid known to me. It is found in tropical Africa, Madagascar, Mauritius, Sri Lanka, northeast India, the Nicobar Islands, Thailand, Laos, Vietnam, Peninsular Malaysia, Indonesia, Borneo, Sulawesi (Celebes), Florida, Mexico, Guyana, Surinam, Brazil and the West Indies. In the Philippines it has been found in the provinces of Benguet, Bulacan, Mountain Province, Pangasinan, Quezon, Rizal and Zambales on Luzon; the island of Bohol in the Visayas; and the province of Lanao del Sur on Mindanao. It usually grows as an epiphyte and occasionally a lithophyte at elevations to 1,200 metres.

Pomatocalpa Breda

Jacob van Breda established this genus in 1827 in *Genera et Species Orchidearum et Asclepiadearum.* The generic name refers to shape of the labellum.

The type species for this genus is ***Pomatocalpa spicata*** Breda.

There are approximately 30 species, which are found throughout Asia, New Guinea, Northern Australia and some of the Pacific Islands. The Philippines has five members for this genus.

Pomatocalpa bicolor (Lindley) J.J. Smith

Photo: Jim Cootes

Pomatocalpa bicolor

Francisco Blanco first named this species ***Epidendrum lineare***. Dr. John Lindley then named it ***Cleisostoma bicolor***. Johannes J. Smith changed the generic concept in 1912 in *Natuurkundig Tijdschrift voor Nederlandsch-Indie.* The specific name refers to the bicoloured flowers.

Growth habit: upright, monopodial. **Stems:** up to 30 cm tall by 1 cm in diameter. **Leaves:** strap-shaped, leathery, up to 13 cm long by 3 cm wide. **Inflorescences:** upright, sometimes branch, up to 30 cm tall, bear up to 20 flowers about 1.5 cm in diameter at the tips. **Flower colour:** cream to yellowish with purplish blotches. **Dorsal sepal:** oblanceolate, obtuse, about 1 cm long by 5 mm wide. **Petals:** broadly spathulate, about 8 mm long by 3 mm wide. **Lateral sepals:** elliptical, sickle-shaped, about 1 cm long by 5 mm wide. **Labellum:** three-lobed, 7 mm long by 3 mm wide; mid-lobe circular, heart-shaped; lateral lobes triangular.

Habitat and Distribution: ***Pomatocalpa bicolor*** is endemic to the Philippines and has been recorded from Bataan, Benguet, Ilocos Norte, Laguna, Nueva Vizcaya and Rizal provinces on Luzon; Polillo island, near Quezon province; Bohol, Leyte, Negros (Negros Oriental province) and Romblon islands in the Visayas; Palawan and Busuanga islands; and Agusan Del Norte province on Mindanao. It grows as either an epiphyte or lithophyte at elevations to 500 metres.

Pomatocalpa spicata Breda

Photo: Jim Cootes

Pomatocalpa spicata

Jacob van Breda named this species in 1827 in *Genera et Species Orchidearum et Asclepiaearum.* The generic name refers to its spike-like inflorescences.

Growth habit: upright, monopodial. **Leaves:** very leathery, up to 20 cm long by 5 cm wide, unequally lobed at the tips. **Inflorescences:** pendulous, carry many 1 cm diameter flowers. **Flower colour:** yellow; white labellum. **Dorsal sepal:** elliptic, about 5 mm long by 3 mm wide. **Petals:** lanceolate, up to 4 mm long by 2.5 mm wide. **Lateral sepals:** ovate, about 5 mm long by 3 mm wide, curve around the spur of the labellum. **Labellum:** small, heart-shaped with a spur about 4 mm long by 3 mm wide.

Habitat and Distribution: ***Pomatocalpa spicata*** is a widely distributed species, being found in northeast India, the Andaman Islands, Myanmar (Burma), Laos, Vietnam, Thailand, Malaysia, Indonesia and Borneo. In the Philippines it has been found in Bataan, Bulacan, Cagayan, Laguna and Rizal provinces on Luzon; Balabac, Busuanga and Palawan islands; Leyte island in the Visayas; and Agusan del Norte, Agusan del Sur, Davao del Norte and Davao del Sur provinces on Mindanao. It grows as an epiphyte or lithophyte at elevations to 1,000 metres.

Pteroceras Hasskarl

Justus Hasskarl first described this genus in 1842 in *Flora* 25. The generic name refers to the wing-like appendages at the base of the labellum.

The type species for this genus is ***Pteroceras teres*** (Bl.) Holtt.

There are approximately 20 species, which are found in Thailand, Peninsular Malaysia, Indonesia and the Philippines. All plants in this genus have flowers that may last for a few hours to two days (depending on the species). Some of the blooms are very showy. There are six members of the genus found in the Philippines.

Pteroceras leopardinum

(Parish and Reichenbach *f.*) Seidenfaden and Smitinand
Synonym: ***Thrixspermum leopardinum*** Parish and Rchb. *f.*

Parish and Reichenbach first named this charming species ***Thrixspermum leopardinum*** in 1874. It has several other synonyms. Doctors Seidenfaden and Smitinand transferred the generic name to ***Pteroceras*** in *Orchids of Thailand* Volume 4 in 1963. The specific name refers to the spotted floral segments, which resemble the markings of a leopard.

Growth habit: upright, monopodial. **Stems:** slightly curved, occasionally branching, about 8 mm in diameter. **Leaves:** leathery, linear-lanceolate, up to 14.5 cm long by 2.1 cm wide. **Inflorescences:** several appear simultaneously, up to 6 cm long, bear up to 10 blooms. **Flower colour:** sepals and petals orange to yellow, sometimes white, with brown to violet spots; labellum white with purple markings on the side lobes. **Dorsal sepal:** ovate, up to 5 mm long by 3.2 mm wide. **Petals:** oblong to linear-lanceolate, up to 5 mm long by 1.7 mm wide. **Lateral sepals:** broadly ovate, up to 7 mm long by 4.5 mm wide. **Labellum:** three-lobed, 5 mm long by 3 mm in diameter; side lobes erect.

Habitat and Distribution: ***Pteroceras leopardinum*** is a widely distributed species that has been recorded from India, Myanmar (Burma), Thailand, Vietnam and Borneo. In the Philippines it has been found in Benguet province on Luzon. It grows as an epiphyte at elevations between 450 and 1,700 metres.

Photo: David Banks

Pteroceras leopardinum

Pteroceras longicalcarum

(Ames and Rolfe) Garay

Professor Oakes Ames and Robert Rolfe first named this species ***Sarcochilus longicalcarus*** in 1915. Dr. Leslie Garay transferred the generic name in 1972 in the *Botanical Museum Leaflets of Harvard University*. The specific name refers to the long spur of the mid-lobe of the labellum.

Growth habit: upright, monopodial, becomes semi-pendulous with age. **Leaves:** leathery, linear to oblong, up to 15 cm long by 2.5 cm wide. **Inflorescences:** pendulous, up to a metre long, bear up to 100 blooms. **Flower colour:** creamy yellow, brown blotches and spotting; labellum white with dark brown tip. **Dorsal sepal:** obovate to elliptic, 6 mm long by 3 mm wide. **Petals:** oblanceolate, 6 mm long by 2.5 mm wide. **Lateral sepals:** similar size and shape to dorsal sepal. **Labellum:** three-lobed, 1 cm long by 3 mm wide; lateral lobes erect; mid-lobe elongated, spurred, curved, swollen at the tip.

Photo: David Banks

Pteroceras longicalcarum

Photo: Jim Cootes

Pteroceras unguiculatum

Habitat and Distribution: ***Pteroceras longicalcarum*** is endemic to the Philippines where it is found in Bataan, Bulacan, Camarines Norte, Camarines Sur, Laguna, Mountain Province, Quezon and Sorsogon provinces on Luzon; Oriental Mindoro province on Mindoro island; Leyte island in the Visayas; and Agusan del Norte, Agusan del Sur, Davao del Norte, Davao Oriental and Davao del Sur provinces on Mindanao. It grows as an epiphyte, usually in fairly shaded locations, at elevations to 1,000 metres.

Pteroceras unguiculatum

(Lindley) H.Æ. Pedersen

Dr. John Lindley named this plant ***Sarcochilus unguiculatus*** in 1840. In 1993 Henrik Pedersen transferred the generic name to ***Pteroceras*** in the Danish publication *Opera Botanica* 117. The specific name means contracted at the base into a claw, but I am not certain which floral feature this describes.

Growth habit: semi-pendulous, monopodial. **Leaves:** leathery, up to 25 cm long by 3.5 cm wide, linear to oblong, unequally lobed at the apex. **Inflorescences:** pendulous, up to 30 cm long, carry up to eight blooms at one time; inflorescence bears flowers over a long period; flowers about 4.5 cm in diameter, appear at regular intervals. They last barely six hours but are very sweetly perfumed. **Flower colour:** sepals and petals cream to pale orange; labellum cream to pale orange; side lobes streaked with reddish brown; mid-lobe yellowish spotted with red. **Dorsal sepal:** lanceolate, slightly concave, up to 2.8 cm long by 8 mm wide. **Petals:** lanceolate, slightly concave, up to 2.5 cm long by 7 mm wide. **Lateral sepals:** lanceolate, concave, up to 3 cm long by 8 mm wide. **Labellum:** three-lobed, 1.2 cm long by 8 mm wide; side lobes erect, oblong, rounded at apex, 6 mm long by 3 mm wide; mid-lobe slipper-shaped, prominent callus.

Habitat and Distribution: ***Pteroceras unguiculatum*** is widely distributed throughout the Andaman Islands, Peninsular Malaysia, Indonesia, Borneo and Sulawesi (Celebes). In the Philippines it has been found in Albay, Apayao, Camarines Sur, Cavite, Ilocos Norte, Isabela, Kalinga, Quezon and Sorsogon provinces on Luzon; Palawan island; Oriental Mindoro province on Mindoro island; Leyte island and Negros Oriental province on Negros island in the Visayas; North and South Cotabato and Surigao del Norte provinces on Mindanao; and Basilan island. This is a lowland species and is seldom found above 500 metres. It grows as an epiphyte, often in full sun, on tree trunks and branches and occasionally on rocks.

Renanthera Loureiro

Juan Loureiro named this genus in 1790 in the *Flora Cochinchinensis*. The generic name refers to the kidney-shaped anther of some species.

The type species for this beautiful genus is ***Renanthera coccinea*** Lour.

The genus has about 15 species, which are found in China, India, Thailand, Peninsular Malaysia, Vietnam, Laos, the Philippines, Indonesia, Borneo and New Guinea. Four species are found in the Philippines.

Renanthera monachica Ames

Professor Oakes Ames named this beautiful species in Fascicle V of his *Studies in the Family Orchidaceae* in 1915. The specific name refers to a fanciful resemblance to a cute Spanish dancing girl. H.M. Curran collected the type specimens in Zambales, Luzon. The specimen first flowered under cultivation in February 1908.

Growth habit: upright, monopodial. **Leaves:** stiff and leathery, about 15 cm long, 1 cm wide and 3 mm thick with a purplish tinge. **Inflorescences:** semi-pendulous, branching, carry up to 50 flowers about 4 cm in diameter. **Flower colour:** orange, spotted with red. **Dorsal** and **lateral sepals:** lanceolate, up to 2 cm long by 5 mm wide. **Petals:** lanceolate, up to 1.5 cm long by 3 mm wide. **Labellum:** three-lobed; lateral lobes upright; mid-lobe small, almost circular.

Habitat and Distribution: ***Renanthera monachica*** is endemic to the Philippines where it is found in Quezon, Rizal and Zambales provinces on Luzon and Surigao del Norte province on Mindanao. It usually grows as an epiphyte at elevations to 500 metres.

Renanthera monachica

Renanthera philippinensis

Ames and Quisumbing

Professor Oakes Ames and Dr. Eduardo Quisumbing named this glorious species in the *Philippine Orchid Review* in 1950. The specific name refers to this plant's country of origin.

Renanthera philippinensis

Growth habit: upright to semi-pendulous, monopodial. **Leaves:** leathery, up to 10 cm long by 4 cm wide, unequally lobed at the apex. **Inflorescences:** branching, bear many flowers up to 2.5 cm across the lateral sepals. **Flower colour:** bright red to orange; labellum red, white and yellow; column red; anther cap red with a white vertical stripe. **Dorsal sepal:** narrowly oblanceolate, up to 2 cm long by 5.5 mm wide. **Petals:** narrowly oblanceolate, up to 1.8 cm long by 5 mm wide. **Lateral sepals:** oblanceolate, up to 2.9 cm long by 1.1 cm wide. **Labellum:** three-lobed; lateral lobes squarish, upright; mid-lobe ovate, small spur.

Habitat and Distribution: ***Renanthera philippinensis*** is endemic to the Philippines and has been found in the provinces of Bataan, Cagayan, Camarines, Mountain Province, Quezon and Rizal on Luzon; the island of Palawan; Leyte and Samar islands in the Visayas; the island of Dinagat; and the provinces of Agusan del Norte, Agusan del Sur, Surigao del Norte and Surigao del Sur on Mindanao. It normally grows as an epiphyte in mangrove swamps, which makes the collection from the Mountain Province a little doubtful, as this province is totally land-locked.

Renanthera storiei Reichenbach *f.*

Professor Heinrich G. Reichenbach named this plant in 1880 in the *Gardeners' Chronicle* after the plant's collector James Storie.

Growth habit: upright, monopodial, up to 4 metres long. **Leaves:** leathery, about 12 cm long by 3 cm wide. **Inflorescences:** semi-pendulous, branch freely, carry up to 100 very showy blooms about 6 cm in diameter. **Flower colour:** bright red. **Dorsal sepal:** linear, up to 2.5 cm high by 5 mm wide. **Petals:** of similar size and shape as the dorsal sepal. **Lateral sepals:** spathulate, up to 3.5 cm long by 1.2 cm wide. **Labellum:** three-lobed; lateral lobes upright; mid-lobe small, oblong.

Photo: Jim Cootes

Renanthera storiei

Habitat and Distribution: ***Renanthera storiei*** is endemic to the Philippines where it is found in Bataan, Camarines Norte, Camarines Sur, Mountain Province, Rizal, Sorsogon and Zambales provinces on Luzon; Palawan island; Leyte and Samar islands in the Visayas; Dinagat island; and Agusan del Norte, Agusan del Sur, Surigao del Norte and Surigao del Sur provinces on Mindanao. It grows in bright light, usually as an epiphyte at elevations to 1,000 metres.

Rhynchostylis Blume

Carl Blume named this genus in 1825 in *Bijdragen tot de Flora van Nederlandsch Indie.* The generic name refers to the beak-like column of the flower.

The type species for this small genus is ***Rhynchostylis retusa*** (L.) Blume.

There are four species in the genus and they are found throughout Southeast Asia. There are two species found in the Philippines; one is endemic.

Rhynchostylis praemorsa

(Willdenow) Blume

This species first became known in 1805 when Willdenow named it ***Aerides praemorsa***. Carl Blume gave this plant its currently accepted name in 1825, in *Bijdragen tot de Flora van Nederlandsch Indie.* The specific name refers to the leaf tips, which are jagged.

Growth habit: upright, monopodial. **Leaves:** very leathery, arch gracefully, up to 30 cm long by 3 cm wide, lightly striped with green lengthwise. **Inflorescences:** pendulous, carry many flowers 1.5 cm in diameter. **Flower colour:** white, spotted with purple; labellum tipped with white at the apex, pink at the base. **Dorsal sepal:** elliptic, up to 7 mm long by 4 mm wide. **Petals:** oblong-lanceolate, up to 7 mm long by 3 mm wide. **Lateral sepals:** ovate, up to 7 mm long by 6 mm wide. **Labellum:** strap-shaped, saccate.

Photo: David Titmuss

Rhynchostylis praemorsa

Habitat and Distribution: ***Rhynchostylis praemorsa*** is endemic to the Philippines and has been recorded from Bataan, Bulacan, Cavite and Rizal on Luzon. It grows as an epiphyte at elevations to 500 metres.

Robiquetia Gaudichaud

Charles Gaudichaud established this genus in 1826 in *Botanique, In Freycinet, L. de, Voyage atour du Monde.* The generic name was not generally accepted until 1912 when J.J. Smith did a revision of the ***Aeridinae*** (syn. ***Sarcanthinae***). Since that time there have been a number of transfers to and from this genus by many notable botanists and taxonomists. The generic name honours a French chemist, Pierre Robiquet.

The type species for this genus is ***Robiquetia ascendens*** Gaud.

Current estimates have about 40 species in this genus. Plants have been found in India, Myanmar (Burma), Laos, Cambodia, Vietnam, China, Malaysia, Sumatra, Java, Borneo, New Guinea, northern Australia and the islands of the Pacific. Five species are found in the Philippines.

Photo: David Titmuss

Robiquetia cerina

Photo: David Banks

Robiquetia compressa

Robiquetia cerina (Reichenbach *f.*) Garay

Professor Heinrich G. Reichenbach was the first to name this plant. Dr. Leslie Garay gave it this generic name in 1972 in the *Botanical Museum Leaflets of Harvard University*. The specific name refers to the waxy nature of the blooms.

Growth habit: semi-pendulous, monopodial. **Leaves:** leathery, about 20 cm long by 4 cm wide; foliage has a greyish tinge. **Inflorescences:** pendulous, bearing many flowers about 8 mm in diameter. The habit of the raceme reminds me of a bunch of grapes. **Flower colour:** purplish. **Dorsal** and **lateral sepals:** broadly ovate, concave, up to 5 mm long by 4 mm wide. **Petals:** almost round, up to 5 mm high by 5 mm wide. **Labellum:** three-lobed; lateral lobes very small; tip of mid-lobe triangular.

Habitat and Distribution: *Robiquetia cerina* is endemic to the Philippines where it is found in Bataan, Benguet, Camarines Norte, Camarines Sur, Laguna, Mountain Province and Rizal provinces on Luzon; Negros Occidental province on Negros island in the Visayas; and Agusan del Norte and Bukidnon provinces on Mindanao. It grows as an epiphyte, usually on tree trunks at elevations over 350 metres.

Robiquetia compressa

(Lindley) Schlechter

Dr. John Lindley first described this species as ***Saccolabium compressum***. Schlechter changed the generic concept to ***Robiquetia*** in 1913 in *Fedde, Repertorium Specierum Novarum, Beihefte.* The specific name refers to the flattened stem of this lovely species.

Growth habit: upright to semi-pendulous, monopodial. **Stems:** up to 60 cm long by 5 mm wide, flat toward the apex. **Leaves:** leathery, linear, unequally bilobed at the tip, up to 15 cm long by 3 cm wide. **Inflorescences:**

appear opposite the leaf, sometimes branching, up to 20 cm long, bearing many small flowers. **Flower colour:** pure white; side lobes of labellum yellow and dark brown. **Dorsal sepal:** ovate, concave, up to 4.5 mm long by 2 mm wide. **Petals:** broadly ovate, 3.5 mm long by 2.5 mm wide. **Lateral sepals:** obovate, wedge-shaped, tip broad, 4 mm long by 3 mm wide. **Labellum:** three-lobed; side lobes erect, squarish; mid-lobe fleshy; spur curves forward, about 1 cm long.

Habitat and Distribution: *Robiquetia compressa* is endemic to the Philippines and has been recorded from Aurora, Bataan, Camarines Sur, Catanduanes, Quezon, Rizal and Sorsogon provinces on Luzon; Oriental Mindoro province on Mindoro; Leyte and Panay islands in the Visayas; and Davao del Norte, Davao Oriental, Davao del Sur, Misamis Occidental and Zamboanga del Norte provinces on Mindanao. It grows as an epiphyte, usually on tree trunks, in bright light, at elevations between 550 and 2,900 metres.

Sarcophyton Garay

Dr. Leslie Garay established this genus in 1972 in the *Botanical Museum Leaflets of Harvard University*. The genus is only known from the Philippines where there are two species. The generic name refers to the fleshy nature of the plants.

The type species for this small genus is ***Sarcophyton crassifolium*** (Lindl.) Garay.

Photo: David Banks

Sarcophyton pachyphyllus

Sarcophyton pachyphyllus

(Ames) Garay

Professor Oakes Ames named this plant ***Sarcanthus pachyphyllus*** in 1915. Dr. Leslie Garay transferred the generic name in 1972. The specific name refers to the very thick leaves. W. Schultze collected the type specimens on the slopes of Mayon Volcano in Albay province on Luzon. The plants flowered in Manila in May 1912.

Growth habit: upright, monopodial. **Leaves:** up to 40 cm long by 3 cm wide by up to 1 cm thick. **Inflorescences:** upright, branch readily, carry up to 300 flowers about 7 mm in diameter. **Flower colour:** sepals and petals white to cream; labellum pale purple. **Dorsal sepal:** elliptic, up to 3.5 mm long by 2 mm wide. **Petals:** oblong-elliptic, up to 3.5 mm long by 2 mm wide. **Lateral sepals:** ovate, 3.5 mm long by 2 mm wide. **Labellum:** three-lobed; lateral lobes small; mid-lobe squarish.

Habitat and Distribution: *Sarcophyton pachyphyllus* is endemic to the Philippines where it is found in the provinces of Albay, Ilocos Norte and Rizal on Luzon. It grows as an epiphyte at elevations to 500 metres.

Spathoglottis Blume

Carl Blume established this genus in 1825 in *Bijdragen tot de Flora van Nederlandsch Indie.* The generic name refers to the shape of the labellum.

The type species for this extremely showy genus is ***Spathoglottis plicata*** Bl.

Spathoglottis is a genus of terrestrial orchids found in mainland Asia, Indonesia, New Guinea, northern Australia and the Pacific islands. The Philippines has six representatives.

Photo: Jim Cootes

Spathoglottis chrysantha

Spathoglottis chrysantha Ames

Professor Oakes Ames named this lovely species in 1908 in Fascicle II of his *Studies in the Family Orchidaceae.* The specific name refers to the colour of the blooms. A.D.E. Merrill collected the type specimens in Benguet province, Luzon, in February 1904.

Growth habit: upright, sympodial. **Pseudobulbs:** about 1.5 cm in diameter, carry up to four leaves. **Leaves:** very thin, textured, up to 60 cm long by 2 cm wide. **Inflorescences:** upright, above the foliage, bear up to 10 flowers about 4 cm in diameter. **Flower colour:** deep yellow; labellum yellow with rusty-red lateral lobes. **Dorsal** and **lateral sepals:** elliptic-oblong, up to 2.5 cm long by 9 mm wide. **Petals:** elliptic-oblong, up to 2.2 cm long by 1 cm wide. **Labellum:** narrow, tapering toward the base, about 2 cm long; lateral lobes narrow, linear, upright.

Habitat and Distribution: *Spathoglottis chrysantha* is endemic to Luzon island where it is found in Albay, Benguet, Ifugao, Mountain Province, Nueva Ecija, Nueva Vizcaya and Pangasinan provinces. It grows terrestrially at elevations over 1,400 metres.

Spathoglottis kimballiana var. *angustifolia* Ames

Professor Oakes Ames named this variety in 1915 in his *Studies in the Family Orchidaceae*, Fascicle V. The varietal name refers to the narrow leaves, the only distinguishing feature from the species.

Growth habit: upright, sympodial. **Pseudobulbs:** about 2 cm in diameter, ovoid, bear up to four leaves. **Leaves:** very thinly textured, up to 65 cm long by 4 cm wide, purple-tinged at base. **Inflorescences:** upright, well above the foliage, bear many flowers up to 8 cm in diameter (only three or four are open at any one time). **Flower colour:** bright yellow; base of labellum spotted with reddish purple. **Dorsal** and **lateral sepals:** elliptic, up to 4 cm long by 2 cm wide. **Petals:** elliptic, up to 4 cm long by 2 cm wide. **Labellum:** three-lobed; mid-lobe narrow, widens to a fan-like tip, base has two small, triangular lobes; side lobes sickle-shaped, broad.

Habitat and Distribution: *Spathoglottis kimballiana* var. ***angustifolia*** has been recorded from Borneo. In the Philippines it has been found in Albay, Ilocos Norte and Isabela provinces on Luzon and Panay (province of Capiz) and Sibuyan islands in the Visayas. It grows as a terrestrial, usually in full sun at elevations to 300 metres.

Notes: This magnificent species is popular in the Philippines and is frequently seen in gardens. Six pots with four or five inflorescences per pot is a very spectacular sight.

Photo: Jim Comber

Spathoglottis kimballiana var. ***angustifolia***

Spathoglottis plicata Blume

Carl Blume named this plant in 1825 in *Bijdragen tot de Flora van Nederlandsch Indie.* The specific name refers to the pleated habit of the leaves.

Growth habit: upright, sympodial. **Pseudobulbs:** about 2 cm in diameter, carry up to four leaves. **Leaves:** very thin textured, up to 40 cm long by 10 cm wide, appear pleated. **Inflorescences:** upright, above the foliage, bear up to 20 flowers about 4 cm in diameter. **Flower colour:** extremely variable ranging from rose-purple to pink to white and every shade in between; labellum fan-shaped and a darker colour than the other segments of the flower. **Dorsal** and **lateral sepals:** ovate-lanceolate, up to 2 cm long by 8 mm wide. **Petals:** ovate, up to 2 cm long by 1 cm wide. **Labellum:** three-lobed; lateral lobes upright; mid-lobe thin, widens to a fan-like tip.

Habitat and Distribution: ***Spathoglottis plicata*** is the most widely spread member of this lovely genus. It has been recorded from India, the Ryukyu Islands, Taiwan, Peninsular Malaysia, Thailand, Myanmar (Burma), Vietnam, Indonesia, New Guinea, north-eastern Australia, the Solomon Islands, New Caledonia, Fiji, Tonga and Samoa. In the Philippines it is found in almost every province. It grows as a terrestrial at elevations to 1,500 metres, usually in full sun.

Notes: This beautiful species is a common garden plant in the Philippines. There are many colour forms.

Photo: Jim Cootes

Spathoglottis plicata

Photo: Jim Cootes

Spathoglottis plicata var. ***alba***

Spathoglottis vanoverberghii

Ames

Professor Oakes Ames named this species in 1920 in his *Studies in the Family Orchidaceae*, Fascicle VI. The plant was named after Father Vanoverbergh, a Belgian priest who worked with the native people of central Luzon for a number of years. He also collected the type specimens, in the Mountain Province, in December 1910.

Growth habit: upright, sympodial. **Pseudobulbs:** about 4 cm long by 2.5 cm wide. **Leaves:** up to 40 cm long by 5 cm wide. The plant has no leaves when in flower. **Inflorescences:** upright, about 25 cm long, carry up to a dozen blooms 4 cm in diameter (only one or two are open at a time). **Flower colour:** bright canary yellow. **Dorsal** and **lateral sepals:** elliptic-lanceolate, up to 2 cm long by 8 mm wide. **Petals:** ovate, up to 2 cm long by 1 cm wide. **Labellum:** three-lobed; lateral lobes upright; mid-lobe thin but widens to a fan-like tip.

Photo: Jim Cootes

Spathoglottis vanoverberghii

Habitat and Distribution: ***Spathoglottis vanoverberghii*** is endemic to the mountains of the provinces of Benguet and Mountain Province in central Luzon. It grows as a terrestrial at elevations over 1,000 metres.

Staurochilus Ridley *ex* Pfitzer

Henry Ridley established this genus in 1896 in his *Material for a Flora of the Malayan Peninsular*, but it was Ernst Pfitzer who wrote the generic description in 1900. ***Staurochilus*** is closely related to ***Trichoglottis***. The major difference between the two genera is that the inflorescence in ***Staurochilus*** is always erect and occasionally branched.

The type species for this showy genus is ***Staurochilus fasciatus*** Rchb. *f.*

There are approximately 14 species in the genus and their centre of distribution is in the Philippines, although they are found throughout Asia and some of the islands of the Pacific. The generic name refers to the shape of the labellum in some species. Approximately nine species are found in the Philippines.

Photo: Jim Cootes

Staurochilus fasciatus

Photo: David Banks

Staurochilus ionosma

Staurochilus fasciatus

Reichenbach *f.*
Synonym: ***Trichoglottis fasciatus*** Rchb. *f.*

Professor Heinrich G. Reichenbach named this species in 1872 in the *Gardeners' Chronicle*. The specific name refers to the broad, parallel stripes of colour on the segments of the flower.

Growth habit: upright, monopodial, up to 75 cm long. **Leaves:** leathery, up to 12 cm long by 3 cm wide. **Inflorescences:** upright, carry up to eight blooms 4 cm in diameter; flowers pleasantly fragrant. **Flower colour:** yellow, barred with brown; labellum white, cross-shaped. **Dorsal** and **lateral sepals:** oblong-elliptic, up to 2 cm long by 5 mm wide. **Petals:** lanceolate, up to 2 cm long by 5 mm wide. **Labellum:** three-lobed; mid-lobe cross-shaped.

Habitat and Distribution: ***Staurochilus fasciatus*** is the most widely distributed member of the genus, being found in Peninsular Malaysia, Thailand, Laos, Vietnam, Indonesia and Borneo. In the Philippines it has been found in Aurora, Quezon, Rizal and Sorsogon provinces on Luzon; Leyte island in the Visayas; and Agusan del Norte province on Mindanao. It grows as an epiphyte at elevations to 800 metres.

Staurochilus ionosma

(Lindley) Schlechter
Synonym: ***Trichoglottis ionosma*** (Lindl.) J.J. Sm.

Dr. John Lindley first named this species ***Cleisostoma ionosmum*** in 1847, in the *Botanical Register*. Dr. Rudolf Schlechter transferred the generic name to ***Staurochilus***. The specific name refers to the violet-like perfume of the flowers.

Growth habit: upright, monopodial, up to 1 metre tall. **Leaves:** very leathery, up to 30 cm long by 2.5 cm wide, arch gracefully, tips unequally bilobed. **Inflores-**

cences: upright, branching, carry up to 25 sweetly perfumed blooms 3 cm in diameter. **Flower colour:** yellow with pale brown blotches; labellum white with a few red spots. **Dorsal** and **lateral sepals:** oblanceolate, up to 1.5 cm long by 5 mm wide. **Petals:** similar but marginally smaller than the sepals. **Labellum:** three-lobed; lateral lobes small; mid-lobe heart-shaped.

Habitat and Distribution: ***Staurochilus ionosma*** is found in the Ryukyu Islands and Taiwan. In the Philippines it has been found in Bataan, Camarines Norte, Camarines Sur, Quezon and Rizal provinces on Luzon and Agusan del Norte province on Mindanao. It grows as an epiphyte at elevations between 300 and 1,300 metres.

Staurochilus loheriana

(Kränzlin) E.A. Christenson
Synonym: ***Trichoglottis loheriana*** (Krzl.) L.O. Wms.

Dr. Fritz Kränzlin originally named this species ***Acampe loheriana*** but Eric Christenson transferred it to ***Staurochilus*** in 1993 in *The Proceedings of the 14th World Orchid Conference.* The specific name honours the collector Augustus Loher.

Growth habit: upright, monopodial. **Leaves:** very leathery, up to 30 cm long by 2 cm wide by 5 mm thick. **Inflorescences:** upright, carry up to 12 flowers 2 cm in diameter. **Flower colour:** green and brown with white on the back; labellum white with orange and red markings. **Dorsal** and **lateral sepals:** ovate-elliptic, up to 1 cm long by 4 mm wide. **Petals:** of similar shape to the sepals but marginally smaller. **Labellum:** three-lobed; lateral lobes small; mid-lobe obovate. All the floral segments are very fleshy.

Habitat and Distribution: ***Staurochilus loheriana*** is endemic to Quezon and Rizal provinces on Luzon. It grows as an epiphyte at elevations to 900 metres.

Staurochilus luzonensis Ames

Synonym: ***Trichoglottis luzonensis*** Ames

Professor Oakes Ames named this lovely species in 1915 in the fifth Fascicle of *Studies in the Family Orchidaceae*, after the island of Luzon in the Philippines where the original specimen was collected. L. Escritor collected the type specimen in Quezon province, Luzon, in June 1913.

Growth habit: upright, monopodial, up to 60 cm tall. **Leaves:** leathery, up to 20 cm long by 5 cm wide. **Inflorescences:** upright, up to 45 cm long, will branch; blooms about 5 cm in diameter. **Flower colour:** creamy yellow with brownish blotches; labellum cream. **Dorsal** and **lateral sepals:** spathulate, up to 2.5 cm long by 5 mm wide. **Petals:** spathulate and of similar size to the other segments. **Labellum:** three-lobed; lateral lobes upright; mid-lobe narrowly elliptic, covered with small hairs.

Habitat and Distribution: ***Staurochilus luzonensis*** is endemic to the Philippines where it is found in Aurora, Camarines Sur and Quezon provinces on Luzon and Agusan del Norte province on Mindanao. It grows as an epiphyte at elevations to 500 metres.

Staurochilus loheriana

Staurochilus luzonensis

Thecostele Reichenbach *f.*

Professor Heinrich G. Reichenbach established this monotypic genus in 1857 in *Orchideae Zollingeriana itineris primi recensentur. - Bonplandia* 5. The generic name refers to the box-like base of the column.

The single species is widely distributed throughout Asia.

Thecostele alata

(Roxburgh) Parish and Reichenbach *f.*

William Roxburgh first named this plant ***Cymbidium alatum*** in 1832. The Reverend Charles Parish and Professor Heinrich G. Reichenbach transferred the generic name to ***Thecostele*** in 1874 in the *Transactions of the Linnaean Society.* The specific name refers to the wing-like arms on the column.

Growth habit: upright, sympodial. **Pseudobulbs:** clustered, cylindrical, up to 6 cm long by 2.5 cm in diameter, usually grooved, unifoliate. **Leaves:** leathery, up to 15 cm long by 4 cm wide, oblong-lanceolate, prominent middle rib. **Inflorescences:** pendulous, slender, up to 50 cm long, bear many blooms (only a few open at a time) up to 1.5 cm in diameter. **Flower colour:** base colour creamy white blotched and dotted with reddish purple. **Dorsal sepal:** oblong, up to 1 cm long by 4 mm wide. **Petals:** linear, up to 8 mm long by 2 mm wide. **Lateral sepals:** ovate, up to 1 cm long by 5 mm wide. **Labellum:** three-lobed; lateral lobes small, upright; mid-lobe large, spreading, short hairs on the upper surface.

Habitat and Distribution: ***Thecostele alata*** is widely distributed throughout Asia, being found in Northeast India, Myanmar (Burma), Thailand, Laos, Vietnam, Malaysia, Sumatra, Java and Borneo. In the Philippines it has been found on the islands of Palawan and Leyte in the Visayas. It grows as an epiphyte at altitudes of up to 1,000 metres.

Photo: David Banks

Thecostele alata

Thrixspermum Loureiro

Juan Loureiro named this genus of monopodial orchids in 1790 in his *Flora Cochinchinensis.* The generic name refers to the hair-like seeds.

The type species for this genus is ***Thrixspermum centipeda*** Lour.

There are approximately 150 species found throughout Asia, the Pacific Islands and Australia. This is the largest genus of monopodial orchids. The Philippines has at least 22 species.

Thrixspermum robinsonii Ames

Professor Oakes Ames named this species in 1915 in *Studies in the Family Orchidaceae* Fascicle V. C.B. Robinson collected the type specimen on Mount Makiling, Laguna province, Luzon, in December 1912 and was honoured by having this species named after him.

Growth habit: upright, monopodial. **Stems:** short, only 2 to 3 cm tall. **Leaves:** leathery, oblong-lanceolate, up to 4 cm long by 6 mm wide. **Inflorescences:** upright, longer than leaves, bear many flowers in succession (only one bloom open at a time). **Flower colour:** white, tinged with pale purple. **Dorsal sepal:** elliptic-lanceolate, 6 mm long by 4 mm wide. **Petals:** elliptic-lanceolate, 5.5 mm long by 3 mm wide. **Lateral sepals:** unequally ovate, 5.5 mm long by 4 mm wide. **Labellum:** three-lobed, slightly saccate; lobes round; mid-lobe hairy.

Habitat and Distribution: ***Thrixspermum robinsonii*** is endemic to the Philippines and has been recorded from Bataan, Bicol, Laguna, Quezon and Sorsogon provinces on Luzon; Leyte and Negros Occidental and Negros Oriental provinces on Negros island in the Visayas; Agusan del Norte and Agusan del Sur provinces on Mindanao; and Camiguin island north of Mindanao. It grows as an epiphyte on mossy tree branches at elevations from 600 to 2,150 metres.

Thrixspermum robinsonii

Thrixspermum subulatum

(Blume) Reichenbach *f.*

Carl Blume first named this plant ***Dendrocolla subulata*** in the early 1800s. Professor Reichenbach changed the generic name in *Xenia Orchidaceae* in 1867. The specific name possibly refers to the tips of the leaves, which taper to a point.

Growth habit: pendulous, monopodial. **Stems:** up to 40 cm long, flattened, branch frequently, covered with persistent sheaths. **Leaves:** leathery, deeply channelled, up to 10 cm long by 1 cm wide; tips pointed. **Inflorescences:** short, bear up to four flowers. **Flower colour:** cream; labellum cream with orange markings. **Dorsal sepal:** oblong-lanceolate, concave, 8.5 mm long by 5 mm wide. **Petals:** oblong-lanceolate, 7 mm long by 3 mm wide. **Lateral sepals:** ovate-lanceolate, 8 mm long by 4 mm wide. **Labellum:** three-lobed, saccate; lateral lobes broadly sickle-shaped.

Habitat and Distribution: ***Thrixspermum subulatum*** is quite widely spread and has been found in Ambon, Java and Sumatra. In the Philippines it has been recorded from the province of Ilocos Norte and Mountain Province on Luzon. It grows as an epiphyte at elevations of about 1,000 metres.

Thrixspermum subulatum

Notes: The root tips of this plant are orange.

Trichoglottis Blume

Carl Blume established this genus in 1825 in *Bijdragen tot de Flora van Nederlandsch Indie.* The generic name refers to the hairy labellum of some of the species.

The type species for this genus is ***Trichoglottis retusa*** Bl.

There are approximately 70 species, which are found in the Himalayas, Myanmar (Burma), Thailand, Peninsular Malaysia, Indochina (Laos, Cambodia, Vietnam), Indonesia, Borneo, Sulawesi (Celebes), New Guinea, the north of Australia and the islands of the Pacific. The Philippines has about 11 representatives.

Trichoglottis amesiana

L.O. Williams

Louis O. Williams named this species in 1933 in the *Philippine Journal of Science*, after Professor Oakes Ames who named many Philippine orchids.

Growth habit: pendulous, monopodial, reaches over 60 cm in length. **Leaves:** leathery, up to 14 cm long, 1.5 cm wide. **Inflorescences:** short, appear opposite leaves, bear up to 10 flowers about 1.5 cm in diameter. **Flower colour:** brown, segments edged with green; labellum white with a purple blotch. **Dorsal sepal:** elliptic, up to 7 mm long, 3 mm wide. **Petals:** oblanceolate, up to 7 mm long, 3 mm wide. **Lateral sepals:** elliptic, up to 6.5 mm long, 3 mm wide. **Labellum:** three-lobed; lateral lobes upright; mid-lobe ovate.

Habitat and Distribution: ***Trichoglottis amesiana*** is endemic to the mountains of Cavite and Nueva Ecija and Nueva Vizcaya in central Luzon. It grows as an epiphyte in quite dense shade at elevations of about 1,200 metres.

Photo: David Titmuss

Trichoglottis amesiana

Photo: Jim Cootes

Trichoglottis atropurpurea

Trichoglottis atropurpurea

Reichenbach *f.*

Professor Heinrich G. Reichenbach named this magnificent species in 1876 in *Linnaea*. The specific name refers to the dark-coloured flowers.

Growth habit: upright, monopodial, reaches 60 cm in height. **Leaves:** up to 8 cm long, 4 cm wide. **Inflorescences:** short, single-flowered, appear opposite the leaves; blooms up to 6 cm in diameter. **Flower colour:** very dark reddish purple, almost black; labellum bright pink, quite hairy along the central ridge; throat golden yellow. **Dorsal sepal:** ovate, up to 2.1 cm long, 1.2 cm wide. **Petals:** oblanceolate, up to 2.1 cm long, 8 mm wide. **Lateral sepals:** ovate-lanceolate, up to 2.2 cm long, 1.5 cm wide. **Labellum:** five-lobed, 2.4 cm long, 1.4 cm wide; central ridge hairy; distal three lobes form a cross.

Habitat and Distribution: ***Trichoglottis atropurpurea*** is endemic to the Philippines where it is found in the province of Quezon on Luzon; the islands of Catanduanes and Palawan; Leyte island in the Visayas; and the provinces of Agusan del Norte, Agusan del Sur, Davao del Norte, Davao Oriental, Davao del Sur, Surigao del Norte and Surigao del Sur on Mindanao. It grows as an epiphyte at elevations of up to 300 metres, and has been found in mangrove swamps.

Notes: This species is better known as ***Trichoglottis brachiata***, but Reichenbach's name has priority by 46 years. A study of the microfiche of the type specimen of Trichoglottis atropurpurea leaves no doubt as to the synonymy of ***Trichoglottis brachiata***. My thanks to Paul Ormerod who drew this species to my attention.

I had previously wondered why this spectacular species had escaped description until 1922, when it was so widely distributed throughout the Philippines.

Photo: Jim Cootes

Trichoglottis bataanensis

Photo: David Titmuss

Trichoglottis geminata

Trichoglottis bataanensis Ames

Professor Oakes Ames named this species in 1905 in Fascicle I of *Studies in the Family Orchidaceae*, after the Philippine province of Bataan where the type specimens were collected in May 1904 by T.E. Borden.

Growth habit: pendulous, monopodial, reaches lengths of 40 cm. **Leaves:** leathery, up to 10 cm long by 1.5 cm wide. **Inflorescences:** short, carry up to six flowers 1 cm in diameter which appear opposite the leaves. **Flower colour:** pale yellow spotted with pale brown, whitish in the centre; labellum pale purple. **Dorsal sepal:** elliptic, up to 5 mm long by 2 mm wide. **Petals:** oblong, up to 5 mm long by 2 mm wide. **Lateral sepals:** oblong, up to 5 mm long by 3 mm wide. **Labellum:** three-lobed; lateral lobes small; mid-lobe circular.

Habitat and Distribution: ***Trichoglottis bataanensis*** is endemic to the Philippines and has been found in Bataan, Laguna, Nueva Ecija, Nueva Vizcaya, Pangasinan, Quezon, Rizal and Zambales provinces on Luzon; Oriental Mindoro province on Mindoro; Palawan and Polillo islands; Agusan del Norte, Agusan del Sur, Surigao del Norte and Surigao del Sur provinces on Mindanao; and the Sulu Archipelago. It grows as an epiphyte at elevations to 200 metres.

Trichoglottis geminata J.J. Smith

Synonym: ***Trichoglottis wenzelii*** Ames

Johannes J. Smith named this species in 1905 in *Die Orchideen von Ambon.* The specific name refers to the two-flowered inflorescence.

Growth habit: upright to semi-pendulous, monopodial, 40 cm or more tall. **Leaves:** leathery, up to 12 cm long by 2 cm wide. **Inflorescences:** short, carry up to two blooms, appear opposite the leaves; flowers 1.5 cm in diameter. **Flower colour:** olive green with maroon bars around floral segments; labellum white and pink, pointed. **Dorsal sepal:** oblong-ovate, up to 1 cm long by 5 mm wide. **Petals:** oblanceolate, up to 8 mm long by 4 mm wide. **Lateral sepals:** broadly obovate, up to 1 cm long by 7 mm wide. **Labellum:** three-lobed; lateral lobes small and upright; mid-lobe oblanceolate, distal end narrows to a tail-like tip.

Habitat and Distribution: ***Trichoglottis geminata*** is widely distributed, being found throughout Borneo, Sulawesi (Celebes) and possibly Sumatra. In the Philippines it has been found in Camarines Sur, Laguna and Quezon provinces on Luzon; Polillo island; Leyte, Negros Occidental and Negros Oriental provinces on Negros island and Samar island in the Visayas; Agusan del Norte, Agusan del Sur, North and South Cotabato, Davao del Norte, Davao Oriental, Davao del Sur, Surigao del Norte, Surigao del Sur, Zamboanga del Norte and Zamboanga del Sur provinces on Mindanao; and Basilan and Tawitawi islands in the Sulu Archipelago. It grows as an epiphyte at low elevations up to 200 metres.

Trichoglottis latisepala Ames

Professor Oakes Ames named this species in 1909 in the *Philippine Journal of Science*. The specific name refers to the broad lateral sepals of the flowers.

Photo: David Titmuss

Trichoglottis latisepala

Growth habit: pendulous, monopodial, reaches lengths of 1 metre. **Leaves:** leathery, up to 12 cm long by 1.5 cm wide, distinct purple colour when grown in bright light. **Inflorescences:** short, appear opposite the leaves, bear clusters of up to six blooms about 1 cm in diameter. **Flower colour:** pale purple. **Dorsal sepal:** oblong, up to 5 mm long by 3 mm wide. **Petals:** oblanceolate, up to 5 mm long by 2.5 mm wide. **Lateral sepals:** shaped like a bird's wings (in flight), up to 6 mm long by 5 mm wide. **Labellum:** three-lobed; lateral lobes small; mid-lobe round.

Habitat and Distribution: *Trichoglottis latisepala* is endemic to the Philippines where it has been found in Aurora, Camarines Sur, Laguna, Nueva Ecija, Nueva Vizcaya, Quezon and Rizal on Luzon; on Bancalan, Catanduanes, Lumbucan and Palawan islands; in Negros Occidental and Negros Oriental on Negros island and on Samar island in the Visayas; in Agusan del Norte, Bukidnon, Davao del Norte, Zamboanga del Norte and Zamboanga del Sur on Mindanao; and on Tawitawi in the Sulu Archipelago. It grows as an epiphyte at elevations to 1,000 metres.

Trichoglottis philippinensis Lindley

Photo: Jack Jannese

Trichoglottis philippinensis (dark)

Photo: David Banks

Trichoglottis philippinensis (pale, spotted)

Dr. John Lindley named this plant in 1845 in the *Annals and Magazine of Natural History*. ***Trichoglottis philippinensis*** is closely related to ***Trichoglottis brachiata***, and some botanists consider them to be varieties. The specific name refers to this plant's country of origin: the Philippines.

Growth habit: upright, monopodial, reaches a height of 60 cm. **Leaves:** up to 7 cm long by 3.5 cm wide. **Inflorescences:** short, single-flowered, appear opposite the leaf; star-shaped flowers up to 4 cm in diameter. The variation in colour is remarkable. **Flower colour:** base colour greenish yellow. There is brown spotting and blotching on the sepals and petals and this is where the variation occurs; some clones are almost totally brown and others have almost no brown markings at all. Every conceivable colour combination in between occurs. Labellum is white and cross-shaped. **Dorsal sepal:** lanceolate, up to 2 cm long by 8 mm wide. **Petals:** lanceolate, up to 2 cm long by 5 mm wide. **Lateral sepals:** ovate, up to 2 cm long by 1.5 cm wide. **Labellum:** five-lobed; distal three lobes form a cross-like structure.

Habitat and Distribution: ***Trichoglottis philippinensis*** has recently been found in Borneo. In the Philippines it has been found in the provinces of Bulacan, Camarines Norte, Camarines Sur, Cavite and Rizal on Luzon; the islands of Catanduanes and Palawan; Cebu and the provinces of Negros Occidental and Negros Oriental on the island of Negros in the Visayas; the provinces of Davao del Norte, Davao Oriental, Davao del Sur, Misamis Occidental, Surigao del Norte, Surigao del Sur, Zamboanga del Norte and Zamboanga del Sur on Mindanao; and Tawitawi in the Sulu Archipelago. It grows as an epiphyte at elevations to 300 metres.

Trichotosia Blume

Carl Blume named this genus in 1825 in *Bijdragen tot de Flora van Nederlandsch Indie.* The generic name refers to the hairy labellum of some of the species.

The type species for this genus of interesting plants is ***Trichotosia pauciflora*** Bl.

Trichotosia is an Asian genus of unusual orchids that are readily distinguishable from the closely related ***Eria*** by the hairy flower bracts, the usually hairy leaves and always hairy leaf sheaths. There are approximately 50 species known to science; seven are found in the Philippines.

Trichotosia fastigiatifolia

(Ames) Kränzlin

Professor Oakes Ames first named this plant ***Eria fastigatifolia*** in 1908. Dr. Fritz Kränzlin transferred the generic concept to ***Trichotosia*** after a study of the genus ***Eria*** in 1911. The specific name refers to the clustered leaves, which are close together on the upper third of the pseudobulbs.

Growth habit: upright to semi-pendulous, sympodial. **Pseudobulbs:** stem-like, up to 80 cm long by 1 cm wide. **Leaves:** grass-like, up to 20 cm long by 8 mm wide. **Inflorescences:** about 4 cm long, appear from opposite the leaves, bear up to 14 flowers about 1 cm in diameter. **Flower colour:** white and pink; labellum white and pink. **Dorsal sepal:** oblong-elliptic, up to 5 mm long by 3 mm wide. **Petals:** oblong-spathulate, 4 mm long by about 2 mm wide. **Lateral sepals:** ovate-oblong, 5 mm long by 3 mm wide. **Labellum:** three-lobed; side lobes triangular; mid-lobe bilobed, margin irregularly toothed.

Habitat and Distribution: ***Trichotosia fastigiatifolia*** is endemic to the Philippines and has been recorded from Benguet, Ifugao, Laguna, Mountain Province, Quezon, Rizal and Sorsogon provinces on Luzon; Leyte and Negros (Negros Oriental province) islands in the Visayas; and Lanao del Norte, Lanao del Sur Surigao del Norte and Surigao del Sur provinces on Mindanao. It grows as an epiphyte at elevations between 300 and 1,300 metres.

Photo: Dr. Henrik Pedersen

Trichotosia fastigiatifolia

Trichotosia vulpina

(Reichenbach *f.*) Kränzlin

Professor Heinrich G. Reichenbach first named this species ***Eria vulpina*** in 1855. In 1911 Dr. Fritz Kränzlin studied the genus ***Eria*** and decided to split the then section ***Trichotosia*** in to its own genus. The specific name means fox-like, possibly alluding to the colour of the hairs on the leaves of the plant.

Growth habit: semi-pendulous, sympodial, large-growing, stems up to 1 metre long. **Leaves:** up to 10 cm long, 2 cm wide; covered with reddish brown hairs, as are leaf sheaths. **Inflorescences:** pendulous, hairy, appear from near the stem ends, up to 10 cm long;

Photo: David Titmuss

Trichotosia vulpina

bear approximately eight blooms 1 cm in diameter. **Flower colour:** dark red spur; white labellum; cream petals; back of dorsal and lateral sepals covered with short brown hairs. **Dorsal sepal:** elliptic, up to 5 mm long, 3 mm wide. **Petals:** oblong, up to 6 mm long, 3 mm wide. **Lateral sepals:** triangular, up to 8 mm high, 5 mm wide; join to form a short spur. **Labellum:** oblong, tapers to an angled tip.

Habitat and Distribution: ***Trichotosia vulpina*** is endemic to the Philippines where it is found in Laguna, Mountain Province, Pampanga, Quezon, Rizal and Sorsogon on Luzon; Oriental Mindoro on Mindoro; Bohol, Leyte, Negros (Negros Oriental province) and Sibuyan islands in the Visayas; Camiguin island, north of Mindanao; Zamboanga del Norte and Zamboanga del Sur on Mindanao; and Basilan island. It grows as an epiphyte at elevations to 1,600 metres.

Tuberolabium Yamamoto

Yoshimatsu Yamamoto proposed this genus in 1924 in the *Botanical Magazine, Tokyo.* The name alludes to the shape of the labellum, which resembles a tuber.

The type species for this genus is ***Tuberolabium kotoense*** Yamamoto.

Tuberolabiums have been recorded from Assam, Thailand, Malaysia, Taiwan, the Philippines, Indonesia, Borneo, New Guinea, Australia, the Solomon Islands, New Caledonia, Fiji, Vanuatu and Tahiti. There are three species found in the Philippines.

Tuberolabium quisumbingii

(L.O. Williams) E.A. Christenson
Synonym: ***Saccolabium quisumbingii*** L.O. Wms.

Louis Williams first named this species in 1938 and it then went through a few name changes until Eric Christenson tidied up the names in *Lindleyana* in 1992. The specific name honours Dr. Eduardo Quisumbing, a Filipino botanist who did much research on Philippine orchids, naming many new species with Professor Oakes Ames of the United States.

Growth habit: upright, monopodial. **Leaves:** leathery, up to 10 cm long by 2.5 cm wide, unequally lobed at the tips. **Inflorescences:** as long or longer than the leaves, bear about 20 blooms up to 1.5 cm in diameter. **Flower colour:** petals and sepals pale green; labellum white, pale purple blotches, side lobes edged with reddish brown. **Dorsal sepal:** elliptic, 6 mm long by 4 mm wide. **Petals:** narrowly obovate, 5.5 mm long by 3 mm wide. **Lateral sepals:** ovate, 6 mm long by 3 mm wide. **Labellum:** fleshy, three-lobed; lateral lobes squarish; mid-lobe short and thick.

Photo: David Banks

Tuberolabium quisumbingii

Habitat and Distribution: ***Tuberolabium quisumbingii*** is endemic to the Philippines and has been found in the provinces of Laguna and Rizal on Luzon. It grows as an epiphyte at elevations to 300 metres.

Tuberolabium rhopalorrachis

(Rchb. *f.*) J.J. Wood

Professor Reichenbach first named this species ***Dendrocolla rhopalorrachis***, in 1856. Dr. Leslie Garay transferred the generic name to ***Trachoma*** in 1972 in *Botanical Museum Leaflets of Harvard University*. The generic name was transferred to ***Tuberolabium*** by Jeffery Wood in 1992, in the *Nordic Journal of Botany*. The specific name refers to the club-shaped inflorescence.

Growth habit: upright, monopodial. **Leaves:** leathery, linear, up to 9.3 cm long, 1.7 cm wide. **Inflorescences:** thick, fleshy, upright, bearing up to six blooms. **Flower colour:** white; base of sepals and petals striped with dark purple; side lobes of labellum yellow. **Dorsal sepal:** elliptic to ovate, 6 mm long by 3.5 mm wide. **Petals:** oblanceolate, 5.5 mm long by 2 mm wide. **Lateral sepals:** elliptic-ovate, 6 mm long by 3.5 mm wide. **Labellum:** fleshy, three-lobed, side lobes upright, mid-lobe triangular.

Photo: David Banks

Tuberolabium rhopalorrachis

Habitat and Distribution: ***Tuberolabium rhopalorrachis*** is a widely distributed species that has been found in Thailand, Malaysia, Sumatra, Java, Ambon and Borneo. In the Philippines it has been recorded from the provinces of Nueva Ecija, Nueva Vizcaya and Rizal in Luzon. It grows as an epiphyte, usually close to the ground at elevations of between 500 and 1,200 metres.

Vanda Jones *ex* R. Brown

Sir W. Jones established this genus in 1795 in *Asiatic Researches* but did not designate a type species. R. Brown in 1820 established ***Vanda roxburghii*** as the type species for this genus, but this plant is now considered to be a synonym of ***Vanda tessellata*** (Roxb.) G. Don. The generic name comes from the Sanskrit word for ***Vanda tessellata***.

The type species for this popular genus is ***Vanda tessellata*** (Roxb.) G. Don.

Vandas are widely spread throughout Asia, being found in India, China, Myanmar (Burma), Thailand, Peninsular Malaysia, Vietnam, Indonesia, Borneo, Sulawesi Celebes), New Guinea, the north of Australia and the Philippines. Current estimates have about 40 species in this showy genus; nine are found in the Philippines.

Vanda javierae Tiu *ex* Fessel and Lückel

Section: ***Hastifera*** E.A. Christenson

Danny Tiu first named this plant in the *Philippine Orchid Review* in 1984 but did not give a Latin diagnosis. Hans Fessel and Emil Lückel formally named ***Vanda javierae*** in 1990 in *Die Orchidee*. The orchid was named after Alicia Javier, the owner of an orchid nursery in the Philippines.

Growth habit: upright, monopodial, up to 35 cm tall. **Leaves:** leathery, up to 20 cm long, 2.5 cm wide.

Vanda javierae

Vanda lamellata

Vanda lamellata var. ***boxalli***

Inflorescences: upright, carry up to six flowers of good shape about 6 cm in diameter. **Flower colour:** pure white; labellum has pale pinkish brown markings in the base. **Dorsal sepal:** spathulate, up to 3 cm long, 1 cm wide. **Petals:** spathulate, up to 3 cm long, 2 cm wide. **Lateral sepals:** spathulate, up to 3 cm long, 1.5 cm wide. **Labellum:** fan-shaped, large for a *Vanda.*

Habitat and Distribution: ***Vanda javierae*** is endemic to a very small mountainous part of Pangasinan province in central Luzon. It is epiphytic at elevations of about 1,200 metres, usually close to water.

Vanda lamellata Lindley

Section: ***Vanda*** Jones *ex* R. Brown

Dr. John Lindley named this species in 1838 in the *Botanical Register.* The specific name refers to the thin ridges on the labellum.

Growth habit: upright, monopodial, up to 40 cm tall. **Leaves:** leathery, thin, up to 20 cm long by 2 cm wide. **Inflorescences:** upright, carry up to 20 flowers about 3 cm in diameter. **Flower colour:** pale yellow with brown markings; labellum pink; all segments reflex a little. **Dorsal sepal:** spathulate, up to 1.5 cm long by 1 cm wide. **Petals:** spathulate, up to 1.5 cm long by 5 mm wide. **Lateral sepals:** obovate, up to 1.5 cm long by 5 mm wide. **Labellum:** three-lobed; lateral lobes upright; mid-lobe obovate.

Habitat and Distribution: ***Vanda lamellata*** is found in Taiwan and Borneo. In the Philippines it is recorded from Abra, Bataan, Batangas, Benguet, Cavite, Ifugao, Laguna, Quezon, Rizal and Zambales provinces on Luzon; Culion island; Oriental Mindoro province on Mindoro island; Palawan; Panay island in the Visayas; Davao del Sur on Mindanao; and the Sulu Archipelago. It is epiphytic in bright light at elevations to 400 metres.

All the several varieties of this species resemble the type species for shape and size of floral segments.

1) ***Vanda lamellata*** var. ***boxalli*** Rchb. *f.* has more flowers on the inflorescence (up to 40 blooms), the segments are cream with the inner half of the lateral sepals dark brown and the labellum is rose-purple. It is found in Zambales province on Luzon and in southern Taiwan. It is epiphytic at elevations to 1,000 metres.

Photo: Danny Tiu

Vanda lamellata var. ***flava***

Photo: David Titmuss

Vanda lamellata var. ***remediosae***

2) ***Vanda lamellata*** var. ***flava*** Valmayor and Tiu. The flowers are all yellow-green with no brown markings. This is a rare variety and its locality is unknown.

3) ***Vanda lamellata*** var. ***remediosae*** Ames and Quisumbing has almost white segments; the inner half of the lateral sepals is greenish; the labellum is pale purple. It has been recorded from the Sulu Archipelago in the southern Philippines. It grows as an epiphyte at elevations up to 300 metres.

Vanda luzonica Loher *ex* Rolfe

Section: ***Deltoglossa*** E.A. Christenson

Augustus Loher first described this plant in 1915, but rather briefly. Robert Rolfe gave a better description in *The Orchid Review* the same year. The specific name refers to Luzon island, where this orchid occurs.

Growth habit: upright, monopodial, old plants become semi-pendulous. The plant can grow very tall, reaching heights of over 1.5 metres; specimens this size can have many side growths. **Leaves:** leathery, up to 45 cm long, arch gracefully. **Inflorescences:** upright, bear up to 12 flowers about 4 cm in diameter. **Flower colour:** sepals and petals white with purple markings; labellum solid dark pink. **Dorsal sepal:** oblanceolate, up to 2 cm long by 7 mm wide. **Petals:** spathulate, up to 2 cm long by 1 cm wide. **Lateral sepals:** spathulate, up to 2 cm long by 1 cm wide. **Labellum:** three-lobed; lateral lobes small and upright; mid-lobe round.

Habitat and Distribution: ***Vanda luzonica*** is endemic to the Philippines and has been found in Bulacan, Rizal and Zambales provinces on Luzon. It grows as an epiphyte at an elevation of around 500 metres.

Notes: This species was once quite common on Mount Pinatubo, but the volcanic eruptions during 1991 probably destroyed many of the plants.

Vanda merrillii Ames and Quisumbing

Section: ***Deltoglossa*** E.A. Christenson

Professor Oakes Ames and Dr. Eduardo Quisumbing named this species in 1932 in the *Philippine Journal of Science*. They named the orchid after Dr. Elmer Merrill, who did a lot of work on Philippine plants.

Growth habit: upright, monopodial; old plants become semi-pendulous; grows into a large plant reaching 1.5 metres in height. **Leaves:** leathery, up to 30 cm long by 3 cm wide. **Inflorescences:** upright, carry up to 10 flowers about 4 cm in diameter, which in some clones are very sweetly perfumed. **Flower colour:** yellow,

Photo: David Banks

Vanda luzonica

Photo: Jim Cootes

Vanda merrillii 'Peter'

heavily marked with red, appearing as if varnished; labellum brown with reddish stripes. **Dorsal sepal:** spathulate, up to 2 cm long by 1 cm wide. **Petals:** spathulate, up to 2 cm long by 1.3 cm wide. **Lateral sepals:** spathulate, up to 2 cm long by 1 cm wide. **Labellum:** three-lobed; lateral lobes small and upright; mid-lobe round.

Habitat and Distribution: ***Vanda merrillii*** is endemic to the Philippines and has been found in the provinces of Aurora, Nueva Ecija and Quezon on Luzon and Negros Occidental and Negros Oriental on Negros in the Visayas. It is epiphytic at elevations of about 500 metres.

Vanda roeblingiana Rolfe

Section: ***Hastifera*** E.A. Christenson

Robert Rolfe named this species in 1894 in the *Kew Bulletin* after Charles Roebling of the state of New Jersey in the USA.

Growth habit: upright, monopodial, up to 40 cm tall. **Leaves:** leathery, up to 20 cm long by 2 cm wide. **Inflorescences:** upright, carrying up to 15 blooms 4 cm in diameter. **Flower colour:** yellow, heavily marked with red; labellum brown. **Dorsal sepal:** oblanceolate, up to 2 cm long by 8 mm wide. **Petals:** spathulate, up to 2 cm long by 9 mm wide. **Lateral sepals:** obovate, up to 2 cm long by 1 cm wide. **Labellum:** three-lobed; lateral lobes small and upright; mid-lobe fan-shaped, fringed, large for a *Vanda*.

Photo: David Banks

Vanda roeblingiana 'Jannine'

Habitat and Distribution: ***Vanda roeblingiana*** is endemic to the mountains of Benguet and the Mountain Province in central Luzon at about 1,600 metres. It is epiphytic, sometimes in quite heavy shade.

Photo: Elias Javier

Vanda scandens

Photo: David Banks

Vanda ustii

Vanda scandens Holttum

Section: ***Hastifera*** E.A. Christenson

Professor R. Eric Holttum named this species in 1950 in the *Sarawak Museum Journal*. The specific name refers to the plants' climbing nature.

Growth habit: upright, monopodial, semi-pendulous in mature specimens, up to about 1.5 metres tall. **Leaves:** leathery, up to 50 cm long, 2.5 cm wide, arch gracefully. **Inflorescences:** upright, carry up to six flowers 3 cm in diameter. **Flower colour:** base colour yellow, finely spotted with brown markings. **Dorsal sepal:** spathulate, up to 2 cm long, 8 mm wide. **Petals:** spathulate, up to 2 cm long, 8 mm wide. **Lateral sepals:** obovate, up to 2 cm long, 1 cm wide. **Labellum:** three-lobed; lateral lobes small and upright; two small 'arms' on the mid-lobe are a distinguishing characteristic for this infrequently seen species.

Habitat and Distribution: ***Vanda scandens*** occurs in Borneo. It has been recorded from Mindanao and Palawan island without the exact localities having been noted. It grows as an epiphyte up to 500 metres.

Vanda ustii Golamco, Claustro and de Mesa

Section: ***Deltoglossa*** E.A. Christenson

Andres Golamco Jr., Alicia de Leon Claustro and Fr. Pompeyo de Mesa named this species in the *Waling-Waling Review* (Official Publication of the Philippine Orchid Society) in February 2000. The specific name, an acronym for University of Santo Tomas, in Manila, is pronounced you-ess-tee-ee-eye.

Growth habit: upright, monopodial, becomes semi-pendulous in mature specimens, up to 1 metre tall with thick roots of up to 3 metres long. **Leaves:** alternate, curve gracefully, up to 35 cm long by 2 cm wide, unequally bilobed at the tip. **Inflorescences:** horizontal to pendulous, bear up to 12 blooms about 5 cm in diameter. **Flower colour:** sepals and petals creamy yellow; on some clones there are tan markings near the apex of the segments; labellum magenta, white at the base. **Dorsal sepal:** spathulate, up to 2.5 cm long by 1.7 cm wide. **Petals:** spathulate, up to 2.3 cm long by 1.7 cm wide. **Lateral sepals:** spathulate, up to 2.5 cm long by 1.7 cm wide. **Labellum:** three lobed, up to 1.8 cm long by 1.3 cm wide; mid-lobe heart-shaped; side lobes squarish.

Habitat and Distribution: ***Vanda ustii*** is endemic to the Philippines and is only known from Nueva Ecija, Nueva Vizcaya and Ifugao on Luzon. It grows as an epiphyte at elevations of about 1,250 metres.

Notes: This species has been exported from the Philippines as ***Vanda luzonica*** var. ***immaculata***.

Zeuxine Lindley

Dr. John Lindley established this genus of terrestrial orchids in 1826 in *Collectanea Botanica Appendix.* The generic name refers to the partial union of the labellum and the column.

The type species for this genus is ***Zeuxine sulcata*** (Roxburgh) Lindley.

The genus is widespread and has been found in Africa, Madagascar and throughout tropical Asia to the Fiji islands in the Pacific. Of about 75 species, nine occur in the Philippines.

Zeuxine flava

(Wallich *ex* Lindley) Bentham *ex* Hooker *f.*

Wallich first described this species in Lindley's *Genera and Species of Orchidaceous Plants* as ***Monochilus flavum*** in 1840. Bentham transferred the generic name in Hooker's *Icones Plantarum* in 1890. The specific name refers to the yellow flowers.

Growth habit: upright, sympodial. **Pseudobulbs:** none, the plant grows from a rhizome to a height of about 25 cm. **Leaves:** ovate, acute, up to 5 cm long by 2.5 cm wide. **Inflorescence:** terminal, about 10 cm high, bears many small blooms. **Flower colour:** yellow. **Dorsal sepal:** ovate, acute, 4 mm long by 2 mm wide. **Petals:** falcate, 2.5 mm long by 1.5 mm wide. **Lateral sepals:** falcate, 2.5 mm long by 1.5 mm wide. **Labellum:** 2.5 mm long by 3.5 mm wide, bilobed at the apex; side lobes enclose the column.

Habitat and Distribution: ***Zeuxine flava*** is a widely spread species and has been found throughout the Himalayas, in Myanmar (Burma) and Thailand. In the Philippines it has recently (1994) been found on Mount Makiling in the province of Laguna on Luzon. It grows as a terrestrial at elevations above 400 metres.

Notes: This is the first record of this species from the Philippines and specimens are held at the Philippine National Herbarium under the number Pedersen 645.

My sincere thanks go to Dr. Henrik Pedersen for bringing this species to my attention.

Photo: Dr. Henrik Pedersen

Zeuxine flava

CULTIVATION

When offered orchid species from the Philippines, many orchid growers, particularly those living in more temperate climates, will immediately ask: "Do I need a heated greenhouse in order to grow these plants?"

The answer to that question, for many of the species, is "yes", especially if the plants come from the lowlands where the species enjoy high evening temperatures (seldom less than 15 degrees Celsius), high humidity and bright light—conditions that, for the average hobbyist, are expensive to reproduce. The sad fact is that the majority of the showy, horticulturally attractive species come from the lowlands and require a minimum winter temperature of at least 15 degrees Celsius, even at night, for the plants to grow and flower to their full potential.

Please do not despair, there are many orchid species from the Philippines that can be successfully cultivated without the need for winter warmth, particularly plants of high elevations where the major problem may be how to keep them cool in hot weather. The flower size of these species may be smaller than their lowland cousins, but they compensate by having more brightly coloured blooms. These are often produced in great numbers and frequently last a long time.

Cultivation notes are given here for orchids at three distinct elevations. The description of each species in this book includes the elevation where it grows, whether it is a lowland, mid-elevation or high elevation species. General requirements for each elevation are outlined briefly here.

Lowland Species (sea level up to 500 metres)
If these plants are to be cultivated successfully in more temperate climates they need to be cultivated in a heated enclosure with a winter minimum temperature, at night, of at least 15 degrees Celsius.

Mid-elevation Species (500 metres up to 1,300 metres)
These species need the same conditions as the lowland species but with a minimum night temperature in the winter of about 10 degrees Celsius.

Highland Species (1,300 metres up to 2,500 metres)
Plants that grow at high elevations in tropical environments are usually plants that enjoy bright light, high humidity and cooler temperatures. The main problem is keeping them cool in warm weather. This can be achieved by using misting systems with sensors that detect a drop in humidity. The misters are then activated for a period of time until the humidity has returned to an acceptable level. More shade cloth and continual air circulation should also be used to assist in the cooling process.

The successful cultivation of any orchid needs to take into account the following factors:
1) pot or mount,
2) light or shade,
3) humidity and air movement,
4) water,
5) fertiliser and
6) temperature.

These variables listed above are discussed next: they are not listed in any order of importance.

Orchid plants always need to have continual air movement, high humidity and usually a minimum of 50 percent shade. The correct temperature and the right amount of water to ensure a healthy root system are other requirements. In addition, regular applications (at least once a week) of both organic and inorganic fertiliser are extremely beneficial to the growth and flowering of the plants.

POT OR MOUNT?

The methods of growing orchids are as many and as varied as the plants themselves. They can be grown either in pots or on mounts in a wide range of microclimates that can be created in growing areas. Most species are adaptable as to how they will grow, but some prefer to grow on a mount.

Pots

There are two choices of pots in which to grow orchids, terracotta or plastic. Both have their merits and disadvantages.

Terracotta Pots

Advantages:
- Plants will dry out more quickly.
- Pots are much heavier, so plants that are top heavy are less likely to fall over.

Disadvantages:
- If the pot is dropped it will usually break (even worse if it lands on your foot).
- Terracotta pots are much more expensive than plastic pots.
- Roots usually adhere to the pot, which is a disadvantage when repotting, as the roots will be damaged no matter how careful you are. It is best to allow the plants to dry out completely before repotting so damage to roots will be minimised.

Plastic Pots

Advantages:
- Plants will stay damp for a longer period, therefore will require less watering.
- They are cheaper to buy.
- The drainage is exceptionally good.
- They can be disposed of after use to lessen the risk of spreading viruses.

Disadvantages:
- They can become brittle with age.
- Black plastic pots can generate a lot of heat if subjected to direct summer sunlight.

Potting Mixes

I like to keep my potting mixes as simple as possible. The best bark to use is the one of your own choice, but it is better to add about 25 percent pebbles of the same size as the bark to keep the mixture open. This allows air to get to the roots of the potted plant. The extra weight of the pebbles helps to keep those top-heavy plants upright. If a plant likes to stay damp, it is also wise to add a small percentage (10 percent) of perlite so dampness will be maintained for a longer period.

I have recently grown selected species in plain sphagnum moss with great success. All the plants tried in 'spag' have progressed almost beyond expectations. The new root growth is remarkable, to say the least, and if a plant has a good root system it won't be long before everything else associated with the plant will also improve. It has been suggested that chopped polystyrene foam be mixed 50-50 with the 'spag' to keep it more open and allow more air to get to the roots.

Mounts

The most frequently used mounts are either natural cork bark or tree fern fibre. Occasionally weathered hardwood planks, such as fence palings, are used. Some growers use the branches of the *Melaleuca* (paper-bark) or *Callistemmon* (bottlebrush) species with some success, but these seem to rot fairly quickly and sometimes get wood borers in them.

Natural Cork

Advantages:
- Plants dry out almost immediately after watering.
- It is lightweight.
- It does not seem to deteriorate.

Disadvantages:
- Plants dry out almost immediately after watering; species that prefer to be damp at all times may require watering more often.
- It is becoming expensive.
- It is not readily available.
- It can get borers in the bark.

Tree Fern

There are many different species of tree ferns available and some are better for mounting plants on than others. The two southern Australian species, *Cyathea cooperi* and *Cyathea australis*, have close-knit fibres that retain moisture well but seem to break down fairly quickly (within a couple of years). There are two species from far-northern Queensland, Australia, that have fibres that are more open and harder, and any plant mounted on these species seems to enjoy having its roots roaming among the fibres.

Advantages:
- Plants can be kept damp for a longer period.

Disadvantages:
- Mounts (particularly the southern Australian species) eventually go sour. This can cause the death of the mounted plant if it is not removed in time.
- It is hard to get legally.
- It is not readily available.
- It is expensive when it can be obtained.

Attaching an Orchid to a Mount

The best method of attaching a plant to a mount is by tying it on with plastic-coated copper wire. If using fishing line I would suggest a breaking strain of at least 6.5 kilos (20 lbs), because this is less likely to cut the rhizome of the plant (and your fingers). Old panty hose are also successful, but they seem to rot fairly quickly. Always place a little sphagnum **on top** of the rhizome where you are tying on the plant so there will be a little extra moisture while the plant is getting established.

The natural wood mounts are not recommended.

A Note on Microclimates

Microclimates are created when there are localities within a growing area that receive more light or shade, more or less air movement. Some plants will grow much better when placed near the ground; others prefer bright light so should be grown near the roof. Frequently a big difference can be seen in the growth of a plant by moving it just a metre—up or down, left or right, across to another bench; the choices are innumerable. If a plant does not seem to be growing as well as you think it should, ***move it****; it may make all the difference. How many times have we given a plant that has not flowered to another grower and within weeks it blooms?*

LIGHT OR SHADE?

The majority of the plants described in these pages are epiphytes, growing on the branches or trunks of trees. Several of the species have adapted to a life on rocks, and others spend their time on the forest floor trying to keep above the continually falling leaf litter. The species that grow on the outer branches of trees always get the most light and air movement during the day. They usually spend the day in full sunlight. The plants that grow on the inner branches, closer to the trunk, and on the trunk itself usually spend a fair percentage of the day in shade, but this is variable depending on the season. This factor will vary if the tree is deciduous.

The terrestrial species do not get much direct sun during the day. They normally have large, thinly textured leaves to try to capture as much light as possible so the plant can grow and flower. Usually terrestrial species require 80 percent shade. Species that have smaller, more leathery foliage usually are plants that grow in about 50 percent shade. Those species which have terete or small, very leathery foliage are normally plants that grow in bright light. These enjoy a minimum of 30 percent shade.

HUMIDITY AND AIR MOVEMENT

I do not know of any orchid species that does not enjoy high humidity and continual air movement. High humidity is easily provided in the greenhouse by keeping the pathway and the area under the growing benches always damp. Continual air movement is provided by electric fans, which are kept going 24 hours a day.

I have been in the forests of the Philippines many times and have noticed that wherever one is in the forest there is always air movement, no matter how little. The plants that grow on the outer limbs of the trees enjoy the most air movement. Those species that grow on the floor of the forest receive the least amount.

WATER

More orchids are killed from over-watering than any other cause. How much to water is one of the major challenges for any novice grower. In general plants growing in pots require less water than those growing on mounts. I am of the opinion that you cannot over-water a plant grown on a mount.

Potted plants have their roots growing in a bark medium; this should drain freely to prevent the roots from rotting. Orchids will not prosper unless they have a healthy root system. It is always difficult to know when to water. One good method is to check the weight of a pot. A pot in need of water feels considerably lighter than a container containing damp potting mix.

FERTILISER

Use fertilisers often, but in small amounts; usually **half** of what the manufacturer recommends is good. If the container suggests 10 millimetres per 10 litres of water, make the mixture 10 millilitres per 20 litres of water. It is also beneficial to use several different brands of fertiliser because no two fertilisers have the same trace elements. These are most important for the plant to grow to its greatest potential. It is very important to flush out pots occasionally with liberal amounts of pure water to remove the unused fertiliser salts.

TEMPERATURE

To grow exotic tropical plants in more temperate climates some form of artificial heat has to be provided during the colder months of the year. How that heat is provided is an individual choice. I have found electric fan heaters to be most successful, but they seem a little expensive when the electricity bill for the winter quarter arrives. There really is no cheap method to heat a growing area.

There is a great deal of pleasure to be derived from successfully growing and flowering plants that are not native to the grower's country. The rewards are even greater when the plants are from a tropical country and have been cultivated under artificial conditions.

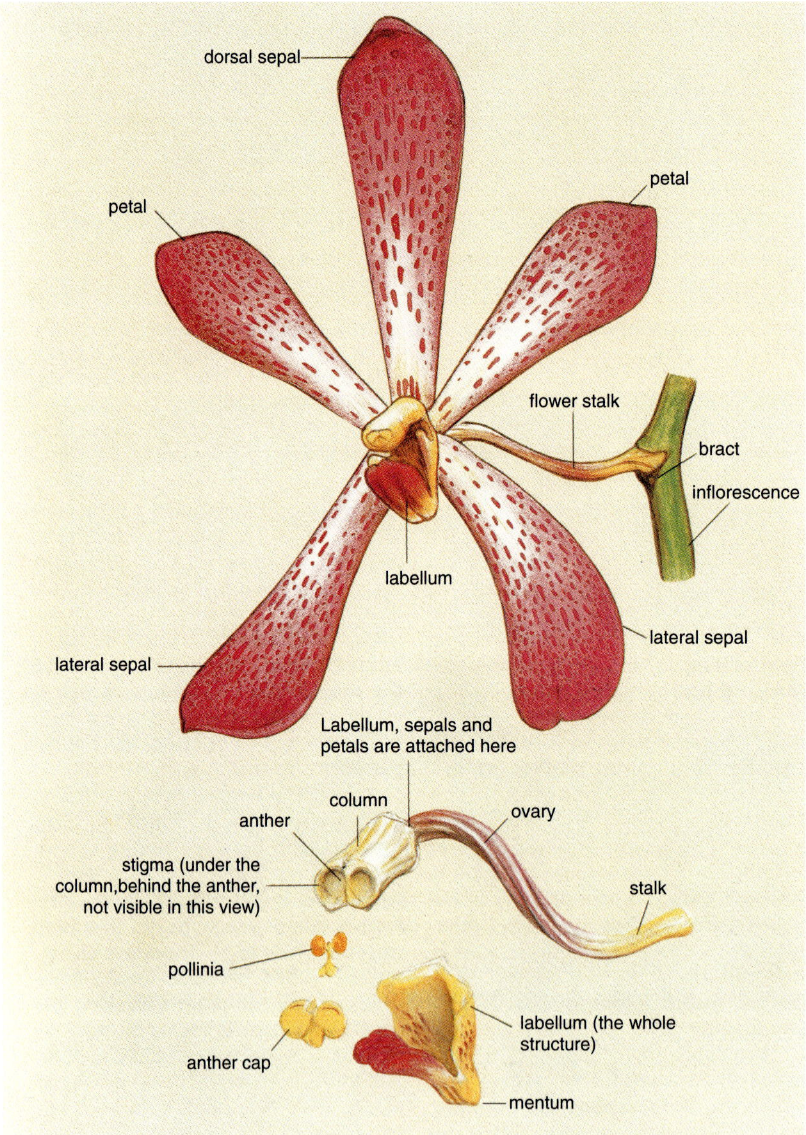

GLOSSARY

acuminate — gradually tapering to a point

adnate — attached by the whole length, usually along the column foot

arcuate — arched or curved

auricle(s) — a lobe, especially when at the base of a leaf or petal

bilabiate — with two lips

carinae — ridges, usually in the labellum

clavate — club-shaped

connate — joined or united

coriaceous — leathery

distichous — leaves or blooms arranged in two opposite rows along a stem

dorsal sepal — the uppermost of the outer three segments

endemism — only found in one area or country; unique to that region

epichele — the apical portion of a labellum

epiphyte — growing on another plant but not as a parasite

falcate — sickle-shaped

filiform — thread-like

fimbriate — fringed

fusiform — spindle-shaped, broadest in the middle, tapered at both ends

heteranthous — an apical inflorescence produced on a separate shoot that does not develop a pseudobulb or leaves

inflorescence — flowering portion of the orchid

keels — ridges, usually in the labellum

labellum — the third petal, usually modified and different to the other petals

lanceolate — narrow and tapered at both ends

lateral sepal — the lowermost two segments of the outer three segments

lithophyte — growing on rocks

mentum — spur-like extension, combining the column foot and the bases of the lateral sepals

monopodial — growing in only one direction

nectary — similar to a spur, but may contain some reward for the pollinator

nonresupinate — labellum uppermost

oblanceolate — tapering more toward the base than the apex

obovate — inversely ovate

obtuse — blunt or rounded at the apex

paleae — small flag-like bracts on the tips and edges of the dorsal sepal or petals in some species of the Bulbophyllinae

papillous — a surface covered with small nipple-like protruberances

pedicel — the single stalk of a flower

petiole — the stalk at the base of a leaf

porrect — directed outward and forward

pseudobulb — thickened, bulb-like internode

pseudoumbellate — falsely umbellate; see umbellate

rhizome — stem from which roots grow and the tip produces leaves, stems or pseudobulbs

saccate — shaped like a sac

scape — a leafless stalk or stem

sepals — the outer floral segments

spathulate — like a spatula or spoon, narrowed at the base

spur — a hollow, slender projection from the disc of the labellum

staminode — an abortive stamen without an anther as in *Paphiopedilum*

stelidia — pointed, finger-like projections from the tip of the column, usually pointing forward or downward (see the picture of ***Bulbophyllum mearnsii***, page 42).

subumbellate — not quite in the form of an umbel

synsepalum — two sepals joined as in *Paphiopedilum*

sympodial — growth in which each new shoot could terminate in an inflorescence

terete — circular in cross-section, cylindrical and usually tapering

terrestrial — growing in or on the ground

type species — the actual herbarium specimen after which a new species or genus has been named

umbel — an inflorescence in which the clustered pedicels arise from a single point (see the picture of ***Cirrhopetalum cumingii***, page 54)

umbellate — in the form of an umbel

unifoliate — bearing a single leaf

REFERENCES

American Orchid Society. 1974. *An Orchidist's Glossary.*

Ames O. 1905 to 1922. *Studies in the Family Orchidaceae* Fascicles I to VII. (Reprint) Bishen Singh Mahendra Pal Singh, India.

Ames O. 1922 to 1930. *Schedulae Orchidianae* Volumes 1 to 10. (Reprint) Bishen Singh Mahendra Pal Singh, India.

Banks D.P. 1999. *Tropical Orchids of Southeast Asia.* Periplus Editions (HK) Ltd.

Banks D.P. and Cootes J.E. 1995. 'The Genus *Dendrochilum*—A Guide to the Species in Cultivation (Part 1)'. *Orchids Australia* 7(3): 48–55.

Banks D.P. and Cootes J.E. 1995. 'The Genus *Dendrochilum*—A Guide to the Species in Cultivation (Part 2)'. *Orchids Australia* 7(4): 57–59.

Banks D.P. and Cootes J.E. 1995. 'The Genus *Dendrochilum*—A Guide to the Species in Cultivation (Part 3)'. *Orchids Australia* 7(5): 40–49.

Braem G.J. 1988. *Paphiopedilum.* Brucke-Verlag Kurt Schmersow D-3200 Hildesheim, Germany.

Chan C., Lamb A., Shim P., Wood J. 1994. *Orchids of Borneo* 1. The Sabah Society, Kota Kinabalu, Sabah, Malaysia.

Christenson E.A. 1992. 'Notes on Asiatic Orchids'. *Lindleyana* 7(2): 88–94.

Christenson E.A. 1993. 'Taxonomy of the Aeridinae with an Infrageneric Classification of *Vanda* Jones ex R. Br.' *Proceedings of the 14th World Orchid Conference.* J. Thompson Colour Printers, Glasgow.

Christenson E.A. 1994. 'Significant collections of Orchidaceae conserved in Herbarium Hamburgense (HBG)'. *Brittonia* 46(4): 346.

Clements M.A. 1999. 'Two New Species of Orchidaceae from the Philippines'. *Australian Orchid Review* 64(6): 6–10.

Clements M.A. and Jones D.L. 1997. 'A preliminary Taxonomic Review of *Grastidium* Bl. and *Eriopexis* (Schltr.) Brieger (Orchidaceae)'. *Lasianthera* Vol. 1, issue 2: 52–128.

Comber J.B. 1990. *Orchids of Java.* Royal Botanic Gardens, Kew.

Cootes J.E. 1996. 'The Genus *Vanda*—A Selection'. *Orchids Australia* 8(3): 51–58.

Cootes J.E. 1998. '*Dendrobium* Section *Formosae* as represented in the Philippines'. *Orchids Australia* 10(4): 34–38.

Cootes J.E. and Banks D.P. 1998. '*Amesiella monticola,* A New Species of Orchidaceae from the Philippines'. *Orchids Australia* 10(5): 26–28.

Cootes J.E. and Banks D.P. 1997. 'Philippine Dendrochilums'. *Orchids,* August 1997: 800–807.

Cootes J.E. and Christenson E.A. 1995. 'The Orange Flowered *Ascocentrum* species—let's get their names right!' *Orchids Australia* 7(6): 18–19.

Cribb P.J. 1983. 'A Revision of *Dendrobium* Sect. *Latouria* (Orchidaceae)'. *Kew Bulletin* 38(2).

Cribb P.J. 1986. 'A Revision of *Dendrobium* Sect. *Spatulata* (Orchidaceae)'. *Kew Bulletin* 41(3).

Cribb P.J. 1987. *The Genus Paphiopedilum.* Kew: Royal Botanic Gardens; Portland, Oregon: Timber Press.

Davis R.S. and Steiner M.L. 1952. *Philippine Orchids.* The William Frederick Press, New York.

de Vogel E. 1988. *Orchid Monographs* Vol. 3. Revisions in Coelogyninae. (Orchidaceae) III The Genus *Pholidota.* E.J. Brill, Leiden, The Netherlands.

de Vogel E. 1992. *Orchid Monographs* Vol. 6. Revisions in Coelogyninae. Rijksherbarium/Hortus Botanicus, Leiden University.

Dolera N.A. 1976–78. *The Malayan Orchid Review* Vol. 13.

Du Puy D. and Cribb P. 1988. *The Genus Cymbidium.* Timber Press, Portland, Oregon.

Fessel H.H. and Peter Balzer. 1999. *A Selection of Native Philippine Orchids.* Times Editions, Singapore.

Garay L.A. et al. 1994. 'The genus *Cirrhopetalum* and the genera of the *Bulbophyllum* alliance'. *Nordic Journal of Botany* 14(6): 609–646.

Golamco A. 1991. *Philippines' Book on Orchids.* jemma, Inc. Publishing Group, Cainta, Rizal, Philippines.

Hickey M. and King C. 2000. *The Cambridge Illustrated Glossary of Botanical Terms.* University Press, Cambridge, United Kingdom.

Höbel R. 1979. *The Philippines.* Robert Rovera, Ltd, Hong Kong.

Jones D.L. 1988. *Native Orchids of Australia.* Reed Books, Frenchs Forest, N.S.W., Australia.

Leavitt R.G. 1909. 'The Genus *Eria* in the Philippines'. Botany Volume 4, *Philippine Journal of Science,* C Number 3: 201–245.

Lewis B. and Cribb P. 1989. *Orchids of Vanuatu.* Royal Botanic Gardens, Kew.

Lewis B. and Cribb P. 1991. *Orchids of the Solomon Islands and Bougainville.* Royal Botanic Gardens, Kew.

Lin Tsan-Piao. 1975. *Native Orchids of Taiwan* Vol. 1. Southern Materials Center, Taipei, R.O.C.

Lin Tsan-Piao. 1987. *Native Orchids of Taiwan* Vol. 2. Southern Materials Center, Taipei, R.O.C.

Lin Tsan-Piao. 1988. *Native Orchids of Taiwan* Vol. 3. Southern Materials Center, Taipei, R.O.C.

Madulid D.A. 1995. *A Pictorial Cyclopedia of Philippine Ornamental Plants.* Bookmark Inc., Makati, Metro Manila.

Mayr H. 1998. *Orchid Names and Their Meanings.* English Translation by M. Schmucker. A.R.G. Gantner Verlag K.-G.

Minderhoud M.E. and de Vogel E.F. 1986. 'A Taxonomic Revision of the Genus *Acriopsis* Reinw. ex Blume'. *Orchid Monographs* Volume 1. E.J. Brill/Rijksherbarium, Leiden.

O'Byrne P. 1994. *Lowland Orchids of Papua New Guinea.* SNP Publishers, Singapore.

Pedersen H.Æ. 1993. 'The Genus *Pteroceras*—a taxonomic revision'. *Opera Botanica* 117.

Pedersen H.Æ. 1997. 'The genus *Dendrochilum* (Orchidaceae) in the Philippines—a taxonomic revision'. *Opera Botanica* 131.

Pedersen H.Æ, Wood J.J. and Comber J.B. 1997. 'A revised subdivision and bibliographical survey of *Dendrochilum* (Orchidaceae)'. *Opera Botanica* 130.

Raulerson L. and Rinehart A. 1992. *Ferns and Orchids of the Mariana Islands.*

Schlechter R. 1982. *The Orchidaceae of German New Guinea* (English Translation). The Australian Orchid Foundation, Melbourne.

Schultes R.E. and Pease A.P. 1963. *Generic Names of Orchids. Their Origin and Meaning.* Academic Press, New York and London.

Science Education Center. 1971. *Plants of the Philippines.* Diliman, Quezon City: U.P. Press.

Seidenfaden G. 1973. 'Notes on *Cirrhopetalum* Lindl'. *Dansk Botanisk Arkiv* 29(1).

Seidenfaden G. 1977. 'A Note on *Dendrobium serpens* (Hk. *f.*) Hk. *f.*' *Gardeners' Bulletin, Singapore.* Vol. 30, 269–274.

Seidenfaden G. 1985. 'Orchid Genera in Thailand XII. *Dendrobium* Sw.'. *Opera Botanica* 83.

Seidenfaden G. 1986. 'Orchid Genera in Thailand XIII. Thirty-Three Epidendroid Genera'. *Opera Botanica* 89.

Seidenfaden G. 1988. Orchid Genera in Thailand XIV. Fifty-Nine Vandoid Genera'. *Opera Botanica* 95.

Seidenfaden G. and Wood J.J. 1992. *The Orchids of Peninsular Malaysia and Singapore.* Olsen and Olsen, Fredensborg, Denmark.

Siegerist E.S. 1997. 'Novelties in the *Bulbophyllum* Alliance'. *Orchids* Vol. 66, No. 6: 574–581. The American Orchid Society, Inc.

Smith J.J. 1984. (Reprint) *Die Orchideen von Java.* Figuren Atlas. Bishen Singh Mahendra Pal Singh, India.

Stearn W.T. 1986. *Botanical Latin.* David and Charles, London.

Stearn W.T. 1990. *Orchid Research Newsletter* 15: 9.

Stearn W.T. 1992. *Stearn's Dictionary of Plant Names for Gardeners.* Cassell Publishers Ltd, London.

Tiu D.A. 1997. 'A Quandary: *Dendrobium sanderae* var. *luzonicum*, a Natural Hybrid?' *Waling-waling* Vol. 5 No. 2: 30–31.

Valmayor H.L. 1984. *Orchidiana Philippiniana.* Eugenio Lopez Foundation, Inc., Manila, Philippines.

Valmayor H.L. (Ed.) 1981. *The Complete Writings of Dr. Eduardo A. Quisumbing on Philippine Orchids.* Eugenio Lopez Foundation, Inc., Manila, Philippines.

Vermeulen J.J. 1991. *Orchids of Borneo 2, Bulbophyllum.* Royal Botanic Gardens, Kew and Toihaan Publishing Co., Kota Kinabalu, Sabah, Malaysia.

Wood J.J., Beaman R.S. and Beaman J.H. 1993. *The Plants of Mount Kinabalu* 2. *Orchids.* Royal Botanic Gardens, Kew.

Wood J.J. and Cribb P. 1994. *A Checklist of the Orchids of Borneo.* Royal Botanic Gardens, Kew.

APPENDIX

The ***Bulbophyllinae*** and the ***Dendrobiinae*** are the two groups of orchids which have the largest number of species in the Philippines.

These large groups are arranged into sections of similar and related species. This makes the identification of the species somewhat easier, particularly if you know to which section the species belongs.

My sincere thanks go to Jaap Vermeulen (Netherlands) for originally arranging the Philippine ***Bulbophyllum*** species into sections. I added section ***Lepidorhiza*** because the flowers are readily separable from the related section ***Intervallatae***. After I received Vermeulen's arrangement of the bulbophyllums, Doctors Garay, Hamer and Siegerist published their findings of a study of the species that were previously considered to be from the section ***Cirrhopetalum***. The section ***Cirrhopetalum*** was reinstated, correctly I believe, to generic status along with an arrangement for similar species, i.e., those plants with umbellate or capitate inflorescences. A number of new genera and sections were created to accommodate species previously considered to belong to section ***Cirrhopetalum*** as stated in the previous sentence. I have attempted to merge the findings of both groups of taxonomists. If my arrangement is successful only time will tell. My sincere thanks to Dr. Garay who provided much assistance with this fascinating group of orchids.

Also, my thanks are due to Dr. Howard Wood (USA) and Danny Tiu (Philippines) for their assistance with the ***Dendrobiinae***.

Bulbophyllinae

Bulbophyllum sections in alphabetical order

§ *Altisceptrum* J.J. Sm.

Many-flowered, elongate (not subumbellate) inflorescence; petals two-thirds length of dorsal sepal or longer, widest close to their base, either gradually narrowing toward their tip, or somewhat constricted halfway and with a thickened top part; smaller pseudobulbs and flowers never marked with purple.

Type species for this section is ***Bulbophyllum elongatum*** (Bl.) Hassk.

Philippine species from this section: ***Bulbophyllum elongatum*** (Bl.) Hassk., ***Bulbophyllum nemorale*** L.O. Wms. and ***Bulbophyllum penduliscapum*** J.J. Sm.

§ *Aphanobulbon* Schltr.

Very small pseudobulbs compared to the size of the plants; white, yellow, orange or even brick red flowers, hardly ever stained with purple.

Type species for this section is ***Bulbophyllum odoratum*** (Bl.) Lindl.

Philippine species from this section: ***Bulbophyllum calophyllum*** L.O. Wms., ***Bulbophyllum curranii*** Ames, ***Bulbophyllum dasypetalum*** Rolfe *ex* Ames, ***Bulbophyllum escritorii*** Ames, ***Bulbophyllum flavescens*** (Bl.) Lindl., ***Bulbophyllum invisum*** Ames, ***Bulbophyllum lancifolium*** Ames, ***Bulbophyllum lepantense*** Ames, ***Bulbophyllum leytense*** Ames, ***Bulbophyllum longipetiolatum*** Ames, ***Bulbophyllum marivelense*** Ames, ***Bulbophyllum masaganapense*** Ames, ***Bulbophyllum mutabile*** (Bl.) Lindl., ***Bulbophyllum odoratum*** (Bl.) Lindl., ***Bulbophyllum reilloi*** Ames, ***Bulbophyllum rhizomatosum*** Ames and Schweinf., ***Bulbophyllum sempiternum*** Ames, ***Bulbophyllum subaequale*** Ames, ***Bulbophyllum vagans*** Ames, ***Bulbophyllum vermiculare*** Hook. *f.* and ***Bulbophyllum zambalense*** Ames.

§ *Biflorae* Garay, Hamer and Siegerist

Caudate, fleshy, clavate-tipped lateral sepals and edentate column. In fully matured flowers the lateral sepals separate soon after the flowers open fully.

Type species for this section is ***Bulbophyllum biflorum*** T. and B.

Philippine species from this section: ***Bulbophyllum biflorum*** T. and B.

§ *Careyana* Pfitz.

Pseudobulbs with one, coriaceous, very persistent leaf; flowers with the lateral sepals adherent along their lower margins; margins of petals entire to erose, rarely serrate; labellum with occasionally indistinct, more or less forward pointing, auricles near its base.

Type species for this section is ***Bulbophyllum careyanum*** Spreng.

Philippine species from this section: ***Bulbophyllum cupreum*** Lindl.

§ *Cirrhopetaloides*

Garay, Hamer and Siegerist

Plants rhizomatous; pseudobulbs distant or more or less approximate; inflorescences rather slender, pseudoumbellate; in mature flowers lateral sepals basally connate for a short distance; outer margins incurved and often seemingly connivent, at least temporally, along median area, apical part free and often setaceous; petals without a mucro, stelidia present.

Type species for this section is ***Bulbophyllum longissimum*** (Ridl.) Ridl.

Philippine species from this section: ***Bulbophyllum surigaense*** Ames and Quis.

§ *Corymbosia* (Bl.) Pfeiffer

Plants rhizomatous; pseudobulbs variously distant, well developed; inflorescence rather slender, subumbellate; flowers rather small on long filiform pedicels; sepals elongate, narrow with more or less incurved or inrolled margins.

Type species for this section is ***Bulbophyllum laxiflorum*** (Bl.) Lind.

Philippine species from this section: ***Bulbophyllum laxiflorum*** (Bl.) Lindl.

§ *Desmosanthes* Bl.

Pseudobulbs distinct; inflorescence with two or more flowers; rachis very short; inflorescence subumbellate; flowers white, yellow, orange or bright red, no purple; sepals and petals glabrous or papillose, not ciliate.

Type species for this section is ***Bulbophyllum angustifolium*** (Bl.) Lindl.

Philippine species from this section: ***Bulbophyllum arrectum*** Krzl., ***Bulbophyllum cephalophorum*** Garay et al., ***Bulbophyllum chrysendetum*** Ames, ***Bulbophyllum lancilabium*** Ames and ***Bulbophyllum stellatum*** Ames.

§ *Ephippium* Schltr.

Petals less than half the length of dorsal sepal; dorsal sepal different in shape and/or texture when compared to lateral sepals.

Type species for this section is ***Bulbophyllum ciliatum*** Bl.

Philippine species from this section: ***Bulbophyllum maxillare*** (Lindl.) Rchb. *f.*

§ *Globiceps* Schltr.

Flowers arranged in a very dense raceme, usually in a regular pattern; lateral sepals with their upper margins recurved together form a saucer-like structure around the lip; petals not truly ciliate; column with distinct teeth along the lower and upper margins.

Type species for this section is ***Bulbophyllum globiceps*** Schltr.

Philippine species from this section: ***Bulbophyllum coniferum*** Ridl., ***Bulbophyllum cuneatum*** Rolfe *ex* Ames, ***Bulbophyllum doryphoroide*** Ames, ***Bulbophyllum ebracteolatum*** Krzl., ***Bulbophyllum merrittii*** Ames and ***Bulbophyllum saurocephalum*** Rchb. *f.*

§ *Hirtula* Ridley

Usually elongate but not subumbellate inflorescence, with two flowers or more; sepals of approximately equal length and shape; distinctly ciliate petals.

Type species for this section is ***Bulbophyllum hirtulum*** Ridl.

Philippine species from this section: ***Bulbophyllum lasioglossum*** Rolfe *ex* Ames and ***Bulbophyllum negrosianum*** Ames.

§ *Hyalosema* Schltr.

Inflorescence one-flowered; flower large with large dorsal sepal. Petals acuminate or with a thickened tip.

Type species for this section is ***Bulbophyllum grandiflorum*** Bl.

Philippine species from this section: ***Bulbophyllum antenniferum*** (Lindl.) Rchb. *f.*

§ *Hybochilus* Schltr.

Plants small; inflorescence single-flowered; lateral sepals free; labellum usually with a median crest on upper side, a partly papillose surface and often with a ciliate margin.

Type species for this section is ***Bulbophyllum acutum*** J.J. Sm.

Philippine species from this section: ***Bulbophyllum colubrimodum*** Ames, ***Bulbophyllum elmeri*** Ames, ***Bulbophyllum erratum*** Ames, ***Bulbophyllum exile*** Ames, ***Bulbophyllum glandulosum*** Ames, ***Bulbophyllum gnomoniferum*** Ames, ***Bulbophyllum lipense*** Ames, ***Bulbophyllum melanoglossum*** Krzl., ***Bulbophyllum pampangense*** Ames, ***Bulbophyllum sensile*** Ames, ***Bulbophyllum toppingii*** Ames and ***Bulbophyllum williamsii*** Hawkes.

§ *Intervallatae* Ridley

Many-flowered inflorescence(s) in which only one large flower opens at a time; dorsal sepal 2 cm or more long.

Type species for this section is ***Bulbophyllum tardaelorens*** Ridl.

Philippine species from this section: ***Bulbophyllum cleistogamum*** Ridl.

§ *Lepidorhiza* Schltr.

More or less papillose roots; long and slightly twisted lateral sepals; prominent petals; flowers do **not** face downward; fleshy labellum usually has two upright arms at the basal end; extended multi-flowered inflorescence with dorsal sepal at least 2 cm long.

Type species for this section is ***Bulbophyllum klabatense*** Schltr.

Philippine species from this section: ***Bulbophyllum basisetum*** J.J. Sm., ***Bulbophyllum carunculatum*** Garay et al., ***Bulbophyllum cootesii*** M.A. Clem., ***Bulbophyllum cryptophoranthus*** Garay, ***Bulbophyllum levanae*** Ames, ***Bulbophyllum mearnsii*** Ames, ***Bulbophyllum nasseri*** Garay, ***Bulbophyllum nymphopolitanum*** Krzl., ***Bulbophyllum orthoglossum*** Wendl. and Krzl., ***Bulbophyllum papulosum*** Garay and ***Bulbophyllum recurvilabre*** Garay.

§ *Leptopus* Schltr.

One-flowered inflorescence; lip thick and very soft and of spongy texture; column with distinct stelidia; distinctly widened and flattened column foot.

Type species for this section is ***Bulbophyllum leptopus*** Schltr.

Philippine species from this section: ***Bulbophyllum aestivale*** Ames, ***Bulbophyllum bolsteri*** Ames, ***Bulbophyllum catenulatum*** Krzl., ***Bulbophyllum chryseum*** (Krzl.) Ames, ***Bulbophyllum cubicum*** Ames, ***Bulbophyllum dissolutum*** Ames, ***Bulbophyllum elassoglossum*** Siegerist, ***Bulbophyllum fenixii*** Ames, ***Bulbophyllum graciliscapum*** Ames and Rolfe, ***Bulbophyllum leptocaulon*** Krzl., ***Bulbophyllum maquilingense*** Ames and Quis., ***Bulbophyllum pardalotum*** Garay et al., ***Bulbophyllum sapphirinum*** Ames and ***Bulbophyllum woelfliae*** Garay et al.

§ *Medusa* Pfitz.

Plants rhizomatous, repent; pseudobulbs remote; inflorescence densely capitate; flowers all-directional; lateral sepals divergent at base, hence flowers never bilabiate, often much elongated, even caudate.

Type species for this section is ***Bulbophyllum medusae*** (Lindl.) Rchb. *f.*

Philippine species from this section: ***Bulbophyllum chrysendetum*** Ames, ***Bulbophyllum stellatum*** Ames and ***Bulbophyllum vaginatum*** (Lindl.) Rchb. *f.*

§ *Monilibulbus* J.J. Sm.

Creeping rhizome; distinct pseudobulbs which are close together, prostrate and covering or even partly enveloping the rhizome; one-flowered inflorescence.

Type species for this section is ***Bulbophyllum tenellum*** (Bl.) Lindl.

Philippine species from this section: ***Bulbophyllum alboroseum*** Ames, ***Bulbophyllum exquisitum*** Ames, ***Bulbophyllum mindorense*** Ames, ***Bulbophyllum pleurothalloides*** Ames and ***Bulbophyllum schefferi*** (Ktze.) Schltr.

§ *Nematorhizis* Ridley

Plants small; inflorescence one-flowered; lateral sepals usually free; lip without a median crest on upper side; surface and margins glabrous.

Type species for this section is not yet designated.

Philippine species from this section: ***Bulbophyllum alagense*** Ames, ***Bulbophyllum bontocense*** Ames, ***Bulbophyllum canlaonense*** Ames and ***Bulbophyllum halconense*** Ames.

§ *Oxysepalum* (Wight) Rchb. *f.*

Rhizome patent or pendulous. Inflorescence one-flowered, numerous flowers along rhizome. Lip without ridges, glabrous.

Type species for this section is ***Bulbophyllum cryptanthum*** Cogn.

Philippine species from this section: ***Bulbophyllum profusum*** Ames.

§ *Pahudiella* Garay, Hamer and Siegerist

Plants rhizomatous; pseudobulbs distant, cylindric; inflorescence erect, subumbellate, few to several flowered; flowers variable in size on slender, but never filiform, pedicels; sepals broad without inrolled margins.

Type species for this section is ***Bulbophyllum subumbellatum*** Ridl.

Philippine species from this section: ***Bulbophyllum aeolium*** Ames, ***Bulbophyllum monstrabile*** Ames and ***Bulbophyllum uniflorum*** (Bl.) Hassk.

§ *Pelma* (Finet) Schltr.

Rhizome pendulous; petals distinct (not very small); lip has concavity near its base where a knob on the column foot fits.

Type species for this section is ***Bulbophyllum absconditum*** J.J. Sm.

Philippine species from this section: ***Bulbophyllum savaiense*** subsp. ***subcubicum*** (J.J. Sm.) J.J. Verm.

§ *Sestochilus* (Breda) Benth. and Hook. *f.*

Large flowers, solitary or in a subumbellate inflorescence; petals more than half as long as the dorsal sepal.

Type species for this section is ***Bulbophyllum lobbii*** Lindl.

Philippine species from this section: ***Bulbophyllum alsiosum*** Ames, ***Bulbophyllum cheiri*** Lindl., ***Bulbophyllum cornutum*** (Bl.) Rchb. *f.*, ***Bulbophyllum dearei*** Rchb. *f.*, ***Bulbophyllum emiliorum*** Ames and Quis., ***Bulbophyllum facetum*** Garay et al., ***Bulbophyllum lobbii*** Lindl., ***Bulbophyllum macranthum*** Lindl., ***Bulbophyllum membranifolium*** Hook. *f.*, ***Bulbophyllum piestoglossum*** J.J. Verm. and ***Bulbophyllum santosii*** Ames.

Species of Uncertain Position

Bulbophyllum lasiopetalum Krzl., ***Bulbophyllum papillipetalum*** Ames and ***Bulbophyllum peramoenum*** Ames.

Excluded Species

Bulbophyllum balaeniceps Rchb. *f.* (a species from South America).

Cirrhopetalum Lindley

§ *Cirrhopetalum* Lindley

Dorsal sepal and petals hirsute or glabrous without paleae. In fully mature flowers the lateral sepals porrect from a divergent base twisted once so as to form a convex blade united through conjoined outer margins, and thus forming a tunnel-like passage around the lip.

Type species for this genus and section is ***Cirrhopetalum umbellatum*** (Forst. *f.*) Hook. and Arn.

Philippine species from this section: ***Cirrhopetalum baucoense*** (Ames) Garay et al., ***Cirrhopetalum brevibrachiatum*** Schltr., ***Cirrhopetalum cumingii*** Lindl., ***Cirrhopetalum cyclosepalon*** (Carr) Garay et al., ***Cirrhopetalum dolichoblepharon*** Schltr., ***Cirrhopetalum gusdorfii*** (J.J. Sm.) Garay et al., ***Cirrhopetalum lepidum*** (Bl.) Schltr., ***Cirrhopetalum loherianum*** Krzl., ***Cirrhopetalum makoyanum*** Rchb. *f.*, ***Cirrhopetalum nutans*** Lindl., ***Cirrhopetalum trisetum*** (Ames) Garay et al., ***Cirrhopetalum umbellatum*** (Forst. *f.*) Hook. and Arn. and ***Cirrhopetalum weberi*** (Ames) Seng.

§ *Recurvae* Garay, Hamer and Siegerist

In every respect the flowers are similar to those in section ***Cirrhopetalum*** with the exception that they always exhibit a spatial separation which is occupied by the petals; lateral sepals arcuately recurved.

Type species for this section is ***Cirrhopetalum curtisii*** Hook. *f.*

Philippine species from this section: ***Cirrhopetalum sibuyanense*** (Ames) Garay et al. and ***Cirrhopetalum zamboangense*** (Ames) Garay et al.

§ *Wallichii* Garay, Hamer and Siegerist

In fully mature flowers the lateral sepals form a subparallel base gradually folding inward to form a convex blade united through connate outer margins, and thus forming an opening around the lip without a true tunnel-like passage; at a later stage, near post floration, the lateral sepals may split into free segments.

Type species for this section is ***Cirrhopetalum wallichii*** Lindl.

Philippine species from this section: ***Cirrhopetalum puguahaanense*** (Ames) Garay et al.

Mastigion Garay, Hamer and Siegerist

Lateral sepals more or less free at base then involute so as to become partially connivent along the outer margins, with free, long-caudate apices, whip-like.

Type species for this genus is ***Mastigion appendiculatum*** (Rolfe) Garay, Hamer and Siegerist.

Philippine species from this genus: ***Mastigion putidum*** (T. and B.) Garay et al.

Osyricera Blume

Inflorescence racemose with a more or less fusiform rachis; anther bullate.

Type species for this genus is ***Osyricera crassifolia*** Bl.

Philippine species from this genus: ***Osyricera erosipetala*** (C. Schweinf.) Garay et al.

Rhytionanthos

Garay, Hamer and Siegerist

Lateral sepals involute, firmly united along both margins, horn- or pouch-like.

Type species for this genus is ***Rhytionanthos cornutum*** (Lindl.) Garay, Hamer and Siegerist.

Philippine species from this genus: ***Rhytionanthus plumatum*** (Ames) Garay et al.

Synarmosepalum

Garay, Hamer and Siegerist

Inflorescence densely racemose; lateral to pseudobulb; sepals connate two-thirds basally; petals linear.

Type species for this genus is ***Synarmosepalum kittredgei*** Garay, Hamer and Siegerist.

Philippine species from this genus: ***Synarmosepalum kittredgei*** Garay et al.

Dendrobiinae

Sections of Philippine ***Dendrobium*** species

The genus ***Dendrobium*** has been a constant source of problems for taxonomists for many years. The most commonly used arrangement was that devised by Rudolf Schlechter, which was created in the early 1900s. It is now somewhat out-of-date, because of modern technology. The subgeneric concepts of Schlechter are not used here. The sections are placed in alphabetical order.

§ *Aporum* (Bl.) Lindl.

Stems wiry; no fleshy, swollen internodes; leaves laterally compressed, broadest at base. No conical wart beneath apex of labellum.

Type species for this section is ***Dendrobium lobatum*** Bl.

Philippine species from this section: ***Dendrobium albayense*** Ames, ***Dendrobium aloifolium*** (Bl.) Rchb. *f.*, ***Dendrobium basilanense*** Ames, ***Dendrobium distichum*** (Presl) Rchb. *f.*, ***Dendrobium escritorii*** Ames, ***Dendrobium lunatum*** Lindl., ***Dendrobium merrillii*** Ames, ***Dendrobium quisumbingii*** Hawkes and Heller, ***Dendrobium sinuosum*** Ames and ***Dendrobium wenzelii*** Ames.

§ *Bolbidium* Lindl.

Small tufted plants; close stems of a few nodes, the uppermost internodes swollen, carrying two leaves between which appear the one-flowered inflorescences successively or two to three at the same time.

Type species for this section is ***Dendrobium pachyphyllum*** (Kze.) Bakh. *f.*

Philippine species from this section: ***Dendrobium cabadbarense*** Ames and ***Dendrobium hymenanthum*** Rchb. *f.*

§ *Breviflores* Hk. *f.*

Stems quite long and leafy when young, only the older (often leafless) ones bearing several shortly pedunculate multi-flowered inflorescences along their length. Flowers are quite small and have a large saccate mentum, the labellum is three-lobed, concave and hairy either on the inside or along its margins.

Type species for this section is ***Dendrobium bicameratum*** Lindl.

Philippine species from this section: ***Dendrobium hercoglossum*** Rchb. *f.* and ***Dendrobium pristinum*** Ames.

§ *Calcarifera* J.J. Sm.

Fairly large plants with stems thickened along their length. Plants are leafy from base to apex when young but lose their leaves later, the basal ones dropping first. Stems may be either leafy or bare when in flower. Inflorescences lateral, of several to many flowers, often pendulous but sometimes the blooms are placed on the lower side of horizontal stems. Flowers with a long mentum, which points away from the pedicel and ovary; labellum unlobed, usually broadened suddenly above a narrow claw, often with a projection of some sort on the upper surface of this claw.

Type species for this section is ***Dendrobium sanguinolentum*** Lindl.

This section can be split into two groups because of the growth habit of the plants.

1. Plants that branch from along the pseudobulb and are pendulous growers

Philippine species from this section: ***Dendrobium ceraula*** Rchb. *f.*, ***Dendrobium chameleon*** Ames and its varieties, ***Dendrobium serratilabium*** L.O. Wms., ***Dendrobium victoriae-reginae*** Loher and its variety and ***Dendrobium yeageri*** Ames and Quis.

2. Plants that don't branch, usually with an upright or semi-pendulous growth habit

Philippine species from this section: ***Dendrobium amethystoglossum*** Rchb. *f.*, ***Dendrobium bunuanense*** Ames, ***Dendrobium cerinum*** Rchb. *f.*,

Dendrobium crassimarginatum L.O. Wms., ***Dendrobium diffusum*** L.O. Wms., ***Dendrobium fairchildiae*** Ames and Quis., ***Dendrobium guerreroi*** Ames and Quis., ***Dendrobium ionopus*** Rchb. *f.*, ***Dendrobium obrienianum*** Krzl., ***Dendrobium phillipsii*** Ames and Quis., ***Dendrobium profusum*** Rchb. *f.*, and ***Dendrobium zamboangense*** Ames.

§ *Conostalix* Krzl.

Leaf sheaths hairy; base of labellum connate with sides of column foot, forming a spur.

Type species for this section is ***Dendrobium lobbii*** Tjsm. and Binn.

Philippine species from this section: ***Dendrobium nemorale*** L.O. Wms.

§ *Crumenata* Pfitz.

Only a few of the internodes, near the base, swollen and fleshy, the upper part of the stem is thin and wiry. Leaves sometimes terete.

This section can be divided into four groups because of the structure of the leaves of the plant.

Type species for this section is ***Dendrobium crumenatum*** Sw.

Philippine species from this section:

Leaves conduplicate: ***Dendrobium blumei*** Lindl., ***Dendrobium bukidnonense*** Ames and Quis., ***Dendrobium crumenatum*** Sw. and ***Dendrobium planibulbe*** Lindl.

Leaves laterally flattened: ***Dendrobium equitans*** Krzl. and ***Dendrobium ventricosum*** Krzl.

Numerous terete leaves: ***Dendrobium carinatum*** (Linn.) Willd., ***Dendrobium gerlandianum*** Krzl., ***Dendrobium junceum*** Lindl., ***Dendrobium modestum*** Rchb. *f.*, ***Dendrobium multiramosum*** Ames, ***Dendrobium philippinense*** Ames and ***Dendrobium polytrichum*** Ames.

Single terete leaf: ***Dendrobium usteri*** Schltr. and ***Dendrobium usterioides*** Ames.

§ *Dendrobium* Swartz

Flowers medium to large in few to many flowered inflorescences. Labellum without side lobes, upper surface usually papillous to pubescent, without ornaments except for a callus at base.

Type species for this section is ***Dendrobium moniliforme*** (Linn.) Sw.

Philippine species from this section: ***Dendrobium aclinia*** Rchb. *f.*, ***Dendrobium anosmum*** Lindl. and its varieties, ***Dendrobium aphyllum*** (Roxb.) Fischer, ***Dendrobium heterocarpum*** Lindl. and ***Dendrobium tetrachromum*** Rchb. *f.*

§ *Distichophyllum* Hk. *f.*

Stems with rather close distichous leaves often of thick texture, persistent when flowering; inflorescences with very short scape of one (to two) flowers. Flowers rather small; labellum mostly with distinct side lobes, fleshy, stiff, often connate with column foot at base; epichile with keels or warts; mentum about the same length as the dorsal sepal.

Type species for this section is ***Dendrobium uniflorum*** Griff.

Philippine species from this section: ***Dendrobium quadrisulcatum*** J.J. Sm. and ***Dendrobium uniflorum*** Griff. and its variety.

§ *Dolichocentrum* Schltr.

The distinguishing features of this section are the thick roots; thin stems; grass-like leaves; usually single, large, white (sometimes pale pink) flowers; labellum with claw and expanded, undivided front lobe of the labellum, large spur with the distally fused lateral sepals.

Type species for this section is ***Dendrobium furcatum*** Schltr.

Philippine species from this section: ***Dendrobium auriculatum*** Ames and Quis., ***Dendrobium miyasakii*** Ames and Quis. and ***Dendrobium papilio*** Loher.

§ *Formosae* (Benth. and Hk. *f.*) Hk. *f.*

Leaf sheaths and also sometimes other parts of the plant have dark brown or blackish hairs; flowers usually near apex of stem, large, white, with coloured markings on the labellum.

Type species for this section is ***Dendrobium formosum*** Lindl.

Philippine species from this section: ***Dendrobium dearei*** Rchb. *f.*, ***Dendrobium sanderae*** Rolfe and its varieties and ***Dendrobium schuetzei*** Rolfe.

§ *Fugacia* J.J. Sm.

Leaves without sheaths, placed near the top of fairly long stems that consist of several internodes, the uppermost swollen. Normally this section bears single-flowered inflorescences, which usually last only one day.

Type species for this section is ***Dendrobium spurium*** (Bl.) J.J. Sm.

Philippine species from this section: ***Dendrobium balzerianum*** Fessel and Lückel, ***Dendrobium orbilobulatum*** Fessel and Lückel and ***Dendrobium spurium*** (Bl.) J.J. Sm.

§ *Latouria* (Bl.) Miq.

Stems placed close together on the shortly creeping rhizome, usually swollen over much of their length. Leaves without basal sheaths, two or three placed near the stem apex; inflorescences many-flowered usually with a stout peduncle, often arising from nodes very close to the stem apices. Flowers rather large, opening widely; labellum is three-lobed.

Type species for this section is ***Dendrobium spectabile*** (Bl.) Miq.

Philippine species from this section are ***Dendrobium macrophyllum*** A. Rich. and ***Dendrobium setigerum*** Ames *ex* M.A. Clements.

§ *Oxyglossum* Schltr.

Small tufted plants; flowers fairly large; ovaries winged or distinctly ribbed.

Type species for this section is ***Dendrobium cyanocentrum*** Schltr.

Philippine species from this section: ***Dendrobium pentapterum*** Schltr.

§ *Oxystophyllum* (Bl.) Miq.

Stems wiry; no fleshy, swollen internodes; leaves laterally compressed, broadest at base; small, conical wart beneath apex of labellum.

Type species for this section is ***Dendrobium concinnum*** (Bl.) Miq.

Philippine species from this section: ***Dendrobium cultratum*** Schltr. and ***Dendrobium elmeri*** Ames.

§ *Pedilonum* (Bl.) Lindl.

Flowers of medium size in short inflorescences, usually with a long narrow mentum; lateral sepals often connate on under-side; labellum with a narrow claw, often with varying ornaments.

Type species for this section is ***Dendrobium secundum*** (Bl.) Lindl.

Philippine species from this section: ***Dendrobium bullenianum*** Rchb. *f.*, ***Dendrobium goldschmidtianum*** Krzl., ***Dendrobium ramosii*** Ames and ***Dendrobium secundum*** (Bl.) Lindl.

§ *Platycaulon* Schltr.

Pseudobulbs distinctly flattened. The plants of this section could not be confused with members of any other ***Dendrobium*** section.

Type species for this section is ***Dendrobium lamellatum*** Lindl.

Philippine species from this section: ***Dendrobium milaniae*** Fessel and Lückel and ***Dendrobium platycaulon*** Rolfe.

§ *Spatulata* Lindl.

Large epiphytes, usually with long, thick, upright stems which are leafy throughout at the time of flowering. Leaves have stem-enveloping sheaths at the base; apices normally unequal. One or more inflorescences emerge laterally from the upper half of the stem; they are typically upright, many-flowered and have a stiff peduncle. Flowers open widely; petals usually longer than the sepals; labellum three-lobed with keels or ridges along most of its length.

Type species for this section is ***Dendrobium lineale*** Rolfe.

Philippine species from this section: ***Dendrobium bicaudatum*** Reinw. *ex* Lindl., ***Dendrobium busuangense*** Ames, ***Dendrobium conanthum*** Schltr. and ***Dendrobium taurinum*** Lindl.

§ *Strongyle* Lindl.

Stems wiry; no fleshy, swollen internodes; leaves terete.

Type species for this section is ***Dendrobium subulatum*** (Bl.) Lindl.

Philippine species from this section: ***Dendrobium marivelense*** Ames, ***Dendrobium mindanaense*** Ames and ***Dendrobium pseudoequitans*** Fessel and Lückel.

Genera in *Dendrobiinae* (excluding *Dendrobium*)

Cadetia Gaud.

Cadetia microphyton (L.O. Wms.) E.A. Christenson

Diplocaulobium (Rchb. f.) Krzl.

Diplocaulobium clemensiae (Ames) A.D. Hawkes

Epigeneium Gagnep.

§ *Racemosum* Garay and Romero

Epigeneium acuminatum (Rolfe) Summerh.
Epigeneium stella-silvae (Loher and Krzl.) Summerh.
Epigeneium treacherianum (Rchb. *f.* *ex* Hook. *f.*) Summerh.
Epigeneium treacherianum fma. ***album*** Hort.

Flickingeria Hawkes

§ *Flickingeria* Seidenf.

Flickingeria fimbriatolabellum (Hay.) Hawkes
Flickingeria scopa (Lindl.) Brieger

§ Bilobulatae Seidenf.

Flickingeria bancana (J.J. Sm.) Hawkes
Flickingeria candoonense (Ames) Seid.
Flickingeria chrysographota (Ames) Hawkes
Flickingeria eurora (Ames) Hawkes
Flickingeria interjecta (Ames) Hawkes
Flickingeria junctiloba Fessel and Lückel
Flickingeria pseudoconvexa (Ames) Hawkes
Flickingeria unicornis (Ames) Hawkes
Flickingeria xantholeuca (Rchb. *f.*) Hawkes

§ Plicatiles Seidenf.

Flickingeria fimbriata (Bl.) Hawkes
Flickingeria loheriana (Krzl.)
Flickingeria luxurians (J.J. Sm.) Hawkes
Flickingeria purpureostelidia (Ames) Hawkes

Grastidium Bl.

Grastidium acuminatissimum Bl.
Grastidium agusanense (Ames) M.A. Clem. and D.L. Jones
Grastidium alagense (Ames) M.A. Clem. and D.L. Jones
Grastidium aliciae (Ames and Quis.) M.A. Clem. and D.L. Jones
Grastidium leytense (Ames) M.A. Clem. and D.L. Jones
Grastidium luzonense (Lindl.) M.A. Clem. and D.L. Jones
Grastidium ornithoflorum (Ames) M.A. Clem. and D.L. Jones
Grastidium pergracile (Ames) M.A. Clem. and D.L. Jones
Grastidium verruculosum (Ames) M.A. Clem. and D.L. Jones

A CHECKLIST OF PHILIPPINE ORCHID SPECIES

Species marked with an asterisk (*) are illustrated and described in this book.

GENERA AND SPECIES

Abdominea J.J. Sm.
- ***Abdominea minimiflora*** J.J. Sm

Acanthephippium Bl.
- ****Acanthephippium mantinianum*** Linden and Cogn.
- ***Acanthephippium*** sp. (solid red flowers)
- ***Acanthephippium sylhetense*** Lindl.

Acriopsis Reinw. *ex* Bl.
- ***Acriopsis indica*** Wight
- ****Acriopsis javanica*** Reinw. *ex* Blume
- ***Acriopsis javanica*** var. ***floribunda*** Ames

Aerides Lour.
- ***Aerides augustiana*** Rolfe
- ***Aerides burbidgei*** Rchb. *f.*
- ****Aerides lawrenciae*** Rchb. *f.*
- ***Aerides lawrenciae*** var. ***amesiana*** Sander *ex* Krzl.
- ****Aerides lawrenciae*** var. ***fortichii*** Ames and Quis.
- ***Aerides lawrenciae*** var. ***punctata*** Ames and Quis.
- ***Aerides lawrenciae*** var. ***sanderiana*** Sander *ex* Krzl.
- ****Aerides leeana*** Rchb. *f.*
- ***Aerides leeana*** var. ***alba*** Val. and Tiu
- ***Aerides leeana*** var. ***smithii*** Ames and Quis.
- ***Aerides odorata*** fma. ***alba*** Hort.
- ****Aerides odorata*** Lour.
- ****Aerides quinquevulnera*** Lindl.
- ****Aerides quinquevulnera*** var. ***farmeri*** (Boxall) Stein
- ****Aerides quinquevulnera*** var. ***flava*** Val. and Tiu
- ***Aerides quinquevulnera*** var. ***marginata*** (Rchb. *f.*) Stein
- ***Aerides quinquevulnera*** var. ***punctata*** Val. and Tiu
- ****Aerides quinquevulnera*** var. ***purpurata*** Rchb. *f.*
- ***Aerides quinquevulnera*** var. ***schadenbergiana*** Stein
- ***Aerides savageana*** Sander *ex* Veitch

Aglossorrhyncha Schltr.
- ***Aglossorrhyncha*** sp.

Agrostophyllum Bl.
- ***Agrostophyllum elmeri*** Ames
- ***Agrostophyllum elongatum*** (Ridl.) Schuitemann
- ****Agrostophyllum inocephalum*** (Schauer) Ames
- ***Agrostophyllum leytense*** Ames
- ***Agrostophyllum longivaginatum*** Ames
- ***Agrostophyllum luzonense*** Ames and Quis.
- ***Agrostophyllum malindangense*** Ames
- ***Agrostophyllum mearnsii*** Ames
- ***Agrostophyllum merrillii*** Ames
- ***Agrostophyllum mindanaense*** Ames
- ***Agrostophyllum philippinense*** Ames
- ***Agrostophyllum saccatilabium*** Ames and Quis.
- ***Agrostophyllum stipulatum*** ssp. ***stipulatum***
- ***Agrostophyllum wenzelii*** Ames

Amesiella Schltr. *ex* Garay
- ***Amesiella minor*** Senghas
- ****Amesiella monticola*** J.E. Cootes and D.P. Banks
- ****Amesiella philippinensis*** (Ames) Garay

Anoectochilus Bl.
- ***Anoectochilus*** sp.

Aphyllorchis Bl.
- ***Aphyllorchis halconensis*** Ames
- ***Aphyllorchis montana*** Rchb. *f.*
- ***Aphyllorchis pallida*** Bl.

Apostasia Bl.
- ***Apostasia wallichii*** R. Br. *ex* Wall.

Appendicula Bl.
- ***Appendicula anceps*** Bl.
- ****Appendicula buxifolia*** Bl.
- ***Appendicula clemensiae*** Ames
- ***Appendicula cornuta*** Bl.
- ***Appendicula crotalina*** (Ames) Schltr.
- ***Appendicula cuneata*** Ames
- ***Appendicula elmeri*** Ames

Appendicula fenixii (Ames) Schltr.
Appendicula irigensis Ames
Appendicula leytensis Ames
Appendicula lucbanensis Ames
Appendicula luzonensis Ames
Appendicula malindangensis (Ames) Schltr.
Appendicula maquilingensis Ames
Appendicula merrillii Ames
****Appendicula micrantha*** Lindl.
Appendicula negrosiana Ames
Appendicula pendula Bl.
Appendicula perplexa Ames
Appendicula polyantha Ames
Appendicula reflexa Bl.
Appendicula tagalensium Krzl.
Appendicula undulata var. ***longicalcarata*** (Rolfe) Ames
Appendicula weberi Ames
Appendicula xytriophora Rchb. *f.*

Arachnis Bl.
Arachnis flosaeris (Linn.) Rchb. *f.*
Arachnis flosaeris var. ***insignis***
****Arachnis longicaulis*** (Schltr.) L.O. Wms.
Arachnis longicaulis var. ***flavescens*** Val. and Tiu

Ascocentrum Schltr.
****Ascocentrum aurantiacum*** ssp. ***philippinense*** E.A. Christenson
Ascocentrum aurantiacum ssp. ***philippinense*** var. ***citrinum*** Rchb. *f.*

Ascochilus Ridley
Ascochilus calceolaris (Teij. and Binn.) Garay
Ascochilus leytensis (Ames) Garay
Ascochilus siamensis Ridl.

Ascoglossum Schltr.
****Ascoglossum calopterum*** (Rchb. *f.*) Schltr.

Bogoria J.J. Sm.
Bogoria merrillii (Ames) Garay

Brachypeza Garay
Brachypeza zamboangensis (Ames) Garay

Bromheadia Lindl.
Bromheadia alticola Ridl.
Bromheadia devogelii J. Kruizinga
Bromheadia finlaysoniana (Lindl.) Rchb. *f.*

Bulbophyllum Thouars
Bulbophyllum aeolium Ames
Bulbophyllum aestivale Ames
****Bulbophyllum alagense*** Ames
Bulbophyllum alboroseum Ames
Bulbophyllum alsiosum Ames
****Bulbophyllum antenniferum*** (Lindl.) Rchb. *f.*
Bulbophyllum apodum Hook. *f.*
Bulbophyllum arrectum Krzl.
Bulbophyllum basisetum J.J. Sm.
****Bulbophyllum biflorum*** Teij. and Binn.
Bulbophyllum bolsteri Ames
Bulbophyllum bontocense Ames
Bulbophyllum calophyllum L.O. Wms.
Bulbophyllum canlaonense Ames
****Bulbophyllum carunculatum*** Garay et al.
****Bulbophyllum catenulatum*** Krzl.
****Bulbophyllum cephalophorum*** Garay et al.
Bulbophyllum cheiri Lindl.
Bulbophyllum chrysendetum Ames
Bulbophyllum chryseum (Krzl.) Ames
Bulbophyllum clandestinum Lindl.
****Bulbophyllum cleistogamum*** Ridl.
Bulbophyllum colubrimodum Ames
Bulbophyllum coniferum Ridl.
****Bulbophyllum cootesii*** M.A. Clem.
Bulbophyllum cornutum (Bl.) Rchb. *f.*
Bulbophyllum cryptophoranthus Garay
Bulbophyllum cubicum Ames
Bulbophyllum cuneatum Rolfe *ex* Ames
****Bulbophyllum cupreum*** Lindl.
Bulbophyllum curranii Ames
Bulbophyllum dasypetalum Rolfe *ex* Ames
****Bulbophyllum dearei*** Rchb. *f.*
Bulbophyllum dissolutum Ames
Bulbophyllum doryphoroide Ames
Bulbophyllum ebracteolatum Krzl.
****Bulbophyllum elassoglossum*** Siegerist
Bulbophyllum elmeri Ames
Bulbophyllum elongatum (Bl.) Hassk.
****Bulbophyllum emiliorum*** Ames and Quis.
Bulbophyllum erratum Ames
Bulbophyllum escritorii Ames
Bulbophyllum exile Ames
Bulbophyllum exquisitum Ames
****Bulbophyllum facetum*** Garay et al.
Bulbophyllum fenixii Ames
Bulbophyllum flavescens (Bl.) Lindl.
Bulbophyllum glandulosum Ames
Bulbophyllum gnomoniferum Ames
Bulbophyllum graciliscapum Ames and Rolfe
****Bulbophyllum halconense*** Ames
Bulbophyllum invisum Ames
Bulbophyllum lancifolium Ames
Bulbophyllum lancilabium Ames
Bulbophyllum lancipetalum Ames

Bulbophyllum lasioglossum Rolfe *ex* Ames
Bulbophyllum lasiopetalum Krzl.
Bulbophyllum laxiflorum (Bl.) Lindl.
Bulbophyllum lepantense Ames
Bulbophyllum leptocaulon Krzl.
Bulbophyllum levanae Ames
Bulbophyllum leytense Ames
Bulbophyllum lipense Ames
Bulbophyllum lobbii Lindl.
Bulbophyllum longipetiolatum Ames
Bulbophyllum macranthum Lindl.
Bulbophyllum maquilingense Ames and Quis.
Bulbophyllum marivelense Ames
Bulbophyllum masaganapense Ames
Bulbophyllum maxillare (Lindl.) Rchb. *f.*
Bulbophyllum mearnsii Ames
Bulbophyllum melanoglossum Krzl.
Bulbophyllum membranifolium Hook. *f.*
Bulbophyllum merrittii Ames
Bulbophyllum mindorense Ames
Bulbophyllum monstrabile Ames
Bulbophyllum mutabile (Bl.) Lindl.
Bulbophyllum nasseri Garay
Bulbophyllum negrosianum Ames
Bulbophyllum nemorale L.O. Wms.
Bulbophyllum nymphopolitanum Krzl.
Bulbophyllum odoratum (Bl.) Lindl.
Bulbophyllum orthoglossum Wendl. and Krzl.
Bulbophyllum pampangense Ames
Bulbophyllum papillipetalum Ames
Bulbophyllum papulosum Garay
Bulbophyllum pardalotum Garay et al.
Bulbophyllum penduliscapum J.J. Sm.
Bulbophyllum peramoenum Ames
Bulbophyllum piestoglossum Verm.
Bulbophyllum pleurothalloides Ames
Bulbophyllum profusum Ames
Bulbophyllum recurvilabre Garay
Bulbophyllum reilloi Ames
Bulbophyllum rhizomatosum Ames and Schweinf.
Bulbophyllum santosii Ames
Bulbophyllum sapphirinum Ames
Bulbophyllum saurocephalum Rchb. *f.*
Bulbophyllum savaiensis ssp. ***subcubicum*** (J.J. Sm.) J.J. Verm.
Bulbophyllum schefferi (Ktze) Schltr.
Bulbophyllum sempiternum Ames
Bulbophyllum sensile Ames
Bulbophyllum sp. aff. ***cheiri*** Lindl.
Bulbophyllum sp. aff. ***fenixii*** Ames
Bulbophyllum stellatum Ames
Bulbophyllum subaequale Ames
Bulbophyllum surigaense Ames and Quis.
Bulbophyllum toppingii Ames
Bulbophyllum tortuosum (Bl.) Lindl.
Bulbophyllum trigonosepalum Krzl.
Bulbophyllum unguiculatum Rchb. *f.*
Bulbophyllum uniflorum (Bl.) Hassk.
Bulbophyllum vagans Ames
Bulbophyllum vaginatum (Lindl.) Rchb. *f.*
Bulbophyllum vermiculare Hook. *f.*
Bulbophyllum williamsii Hawkes
Bulbophyllum woelfliae Garay et al.
Bulbophyllum zambalense Ames
Cadetia Gaud.
Cadetia microphyton (L.O. Wms.) E.A. Christenson
Cadetia sp.
Calanthe R. Brown
Calanthe angustifolia (Bl.) Lindl.
Calanthe conspicua Lindl.
Calanthe davaensis Ames
Calanthe halconensis Ames
Calanthe hennisii Loher
Calanthe lacerata Ames
Calanthe lyroglossa Rchb. *f.*
Calanthe maquilingensis Ames
Calanthe mcgregorii Ames
Calanthe mindorensis Ames
Calanthe pulchra (Bl.) Lindl.
Calanthe rubens Ridley
Calanthe triplicata (Willem.) Ames
Calanthe vestita Lindl.
Cephalantheropsis Guill.
Cephalantheropsis halconensis (Ames) Liu and Siu
Cephalantheropsis longipes (Hk. *f.*) Ormd.
Cephalantheropsis obcordata (Lindl.) Ormd.
Ceratocentron Senghas
Ceratocentron fesselii Senghas
Ceratostylis Bl.
Ceratostylis caespitosa L.O. Wms.
Ceratostylis dataensis Ames
Ceratostylis elmeri Ames
Ceratostylis incognita J.T. Atwood and J. Beckner
Ceratostylis latipetala Ames
Ceratostylis loheri L.O. Wms.
Ceratostylis mindanaensis Ames

Ceratostylis philippinensis Rolfe *ex* Ames
Ceratostylis ramosa Ames and Rolfe
****Ceratostylis retisquama*** Rchb. *f.*
Ceratostylis rubra Ames
****Ceratostylis senilis*** Rchb. *f.*
****Ceratostylis subulata*** Bl.
Ceratostylis wenzelii Ames
Chamaeanthus Schltr.
Chamaeanthus wenzelii Ames
Cheirostylis Bl.
Cheirostylis chinensis Rolfe
Cheirostylis octodactyla Ames
Chelonistele Pfitzer
****Chelonistele sulphurea*** (Bl.) Pfitz.
Chrysoglossum Bl.
Chrysoglossum ornatum Bl.
Cirrhopetalum Lindley
Cirrhopetalum auratum Lindl.
Cirrhopetalum baucoense (Ames) Garay et al.
****Cirrhopetalum brevibrachiatum*** Schltr.
****Cirrhopetalum cumingii*** Lindl.
Cirrhopetalum cyclosepalon (Carr) Garay et al.
Cirrhopetalum dolichoblepharon Schltr.
Cirrhopetalum gusdorfii (J.J. Sm.) Garay et al.
Cirrhopetalum lepidum (Bl.) Schltr.
****Cirrhopetalum loherianum*** Krzl.
****Cirrhopetalum makoyanum*** Rchb. *f.*
Cirrhopetalum merrillii Garay
****Cirrhopetalum nutans*** Lindl.
Cirrhopetalum puguahaanense (Ames) Garay et al.
Cirrhopetalum sibuyanense (Ames) Garay et al.
****Cirrhopetalum trisetum*** (Ames) Garay et al.
Cirrhopetalum umbellatum (Forst. *f.*) Hook. and Arn.
****Cirrhopetalum weberi*** (Ames) Senghas
Cirrhopetalum zamboangense (Ames) Garay et al.
Cleisostoma Bl.
Cleisostoma dealbatum (Lindl.) Garay
Cleisostoma pugioniforme (Klotzsch) Garay
Cleisostoma sagittatum Bl.
Cleisostoma striolatum (Rchb. *f.*) Garay
****Cleisostoma subulatum*** Bl.
Cleisostoma uraiensis (Hayata) Garay and Sweet
****Cleisostoma weberi*** (Ames) Garay
Coelogyne Lindl.
****Coelogyne asperata*** Lindl.
****Coelogyne bilamellata*** Lindl.
Coelogyne candoonense Ames
****Coelogyne chloroptera*** Rchb. *f.*
Coelogyne confusa Ames
Coelogyne elmeri Ames
Coelogyne integerrima Ames
Coelogyne loheri Rolfe
Coelogyne longirachis Ames
****Coelogyne marmorata*** Rchb. *f.*
****Coelogyne merrillii*** Ames
Coelogyne minutissima Krzl.
Coelogyne palawanense Ames
Coelogyne pandurata Lindl.
****Coelogyne quinquelamellata*** Ames
Coelogyne ramosii Ames
Coelogyne remediosae Ames and Quis.
****Coelogyne rochussenii*** de Vriese
Coelogyne sp. Section ***Speciosae***
Coelogyne sparsa Rchb. *f.*
Coelogyne swaniana Rolfe
Coelogyne vanoverberghii Ames
Collabium Bl.
Collabium formosanum Hayata
Cordiglottis J.J. Sm.
Cordiglottis filiformis (J.D. Hook.) Garay
Corybas Salisbury
Corybas laceratus L.O. Wms.
Corybas merrillii Ames
Corybas ramosianus J. Drans.
Corymborkis Thouars
Corymborkis veratrifolia (Reinw.) Bl.
Crepidium Bl.
Crepidium alagensis (Ames) M.A. Clem. and D.L. Jones
Crepidium balabacense (Ames) Szlach.
Crepidium bancanoides (Ames) Szlach.
Crepidium bataanense (Ames) M.A. Clem. and D.L. Jones
Crepidium bilobum (Lindl.) Szlach.
Crepidium binabayense (Ames) Szlach.
Crepidium bracteosum (Ames) Szlach.
Crepidium copelandii (Ames) Szlach.
Crepidium davaensis (Ames) Szlach.
Crepidium dentatum (Ames) Szlach.
Crepidium elmeri (Ames) Szlach.
Crepidium epidendrum (Ames) M.A. Clem. and D.L. Jones
Crepidium hutchinsonianum (Ames) Szlach.
Crepidium lilacina (Ames) M.A. Clem. and D.L. Jones
Crepidium macgregorii (Ames) M.A. Clem. and D.L. Jones

Crepidium merrillii (Ames) Szlach.
Crepidium mindaenaense (Ames) M.A. Clem. and D.L. Jones
Crepidium mindorense (Ames) M.A. Clem. and D.L. Jones
Crepidium negrosianum (Ames) M.A. Clem. and D.L. Jones
Crepidium propinquum (Ames) Szlach.
Crepidium purpureiflorum (Ames) M.A. Clem. and D.L. Jones
Crepidium quadridentatum (Ames) Szlach.
Crepidium quadrilobum (Ames) Szlach.
Crepidium ramosii (Ames) Szlach.
Crepidium rizalense (Ames) M.A. Clem. and D.L. Jones
Crepidium taylorii (Ames) Szlach.
Crepidium uncatum (Ames) M.A. Clem. and D.L. Jones
Crepidium wenzelii (Ames) Szlach.
Crepidium williamsii (Ames) Szlach.
Cryptostylis R. Brown
Cryptostylis acutata J.J. Sm.
Cryptostylis arachnites (Bl.) Hassk.
· **Cryptostylis taiwaniana** Masam.
Cymbidium Swartz
***Cymbidium aliciae** Quis.
***Cymbidium atropurpureum** (Lindl.) Rolfe
***Cymbidium dayanum** Rchb. *f.*
Cymbidium dayanum var. **austrojaponicum** Tuyama
Cymbidium ensifolium var. **misericors** (Hay.) T.P. Lin
Cymbidium finlaysonianum fma. **album** Hort.
***Cymbidium finlaysonianum** Lindl.
Cymbidium gonzalesii Quis.
***Cymbidium pubescens** Lindl.
Cystopus Bl.
Cystopus mindanaensis Ames
Cystopus philippinensis Ames
Cystorchis Bl.
Cystorchis aphylla Ridl.
Cystorchis javanica Bl.
***Cystorchis luzonensis** Ames
Dendrobium Swartz
Dendrobium aclinia Rchb. *f.*
Dendrobium aff. **kiauense**
Dendrobium albayense Ames
***Dendrobium aloifolium** (Bl.) Rchb. *f.*
***Dendrobium amethystoglossum** Rchb. *f.*
***Dendrobium anosmum** Lindl.
***Dendrobium anosmum** var. **dearei** (Rolfe) Ames and Quis.
Dendrobium anosmum var. **giganteum** (Rchb. *f.*) Val. and Tiu
***Dendrobium anosmum** var. **huttonii** (Rchb. *f.*) Ames and Quis.
Dendrobium aphyllum (Roxb.) Fischer
***Dendrobium auriculatum** Ames and Quis.
***Dendrobium balzerianum** Fessel and Lückel
Dendrobium basilanense Ames
***Dendrobium bicaudatum** Reinw. *ex* Lindl.
***Dendrobium blumei** Lindl.
Dendrobium bukidnonense Ames and Quis.
***Dendrobium bullenianum** Rchb. *f.*
Dendrobium bunuanense Ames
Dendrobium busuangense Ames
Dendrobium cabadbarense Ames
***Dendrobium carinatum** (Linn.) Willd.
***Dendrobium ceraula** Rchb. *f.*
***Dendrobium cerinum** fma. **album** Hort.
***Dendrobium cerinum** Rchb. *f.*
***Dendrobium chameleon** Ames
Dendrobium chameleon fma. **album** Hort.
***Dendrobium chameleon** var. **ovatilabium** Ames and Quis.
Dendrobium chameleon var. **ovatilabium** fma. **album** Ames and Quis.
Dendrobium compactum Rolfe
***Dendrobium conanthum** Schltr.
Dendrobium crassimarginatum L.O. Wms.
***Dendrobium crumenatum** Sw.
***Dendrobium cultratum** Schltr.
***Dendrobium dearei** Rchb. *f.*
Dendrobium diffusum L.O. Wms.
***Dendrobium distichum** fma. **album** Hort.
***Dendrobium distichum** (Presl) Rchb. *f.*
Dendrobium elmeri Ames
***Dendrobium equitans** Krzl.
Dendrobium escritorii Ames
***Dendrobium fairchildiae** Ames and Quis.
***Dendrobium fairchildiae** fma. **album** Hort.
***Dendrobium gerlandianum** Krzl.
***Dendrobium goldschmidtianum** Krzl.
***Dendrobium guerreroi** Ames and Quis.
Dendrobium hercoglossum Rchb. *f.*
***Dendrobium heterocarpum** Lindl.
***Dendrobium hymenanthum** Rchb. *f.*
Dendrobium indivisum (Bl.) Miq.
***Dendrobium ionopus** Rchb. *f.*
***Dendrobium junceum** Lindl.

Dendrobium lunatum Lindl.
Dendrobium macrophyllum A. Rich.
Dendrobium marivelense Ames
****Dendrobium merrillii*** Ames
****Dendrobium milaniae*** Fessel and Lückel
****Dendrobium mindanaense*** Ames
Dendrobium miyasakii Ames and Quis.
****Dendrobium modestum*** Rchb. *f.*
Dendrobium multiramosum Ames
****Dendrobium nemorale*** L.O. Wms.
Dendrobium obrienianum Krzl.
****Dendrobium orbilobulatum*** Fessel and Lückel
****Dendrobium papilio*** Loher
****Dendrobium pentapterum*** Schltr.
****Dendrobium philippinense*** Ames
****Dendrobium phillipsii*** Ames and Quis.
Dendrobium planibulbe Lindl.
****Dendrobium platycaulon*** Rolfe
****Dendrobium polytrichum*** Ames
****Dendrobium pristinum*** Ames
****Dendrobium profusum*** Rchb. *f.*
Dendrobium pseudoequitans Fessel and Lückel
Dendrobium quadrisulcatum J.J. Sm.
Dendrobium quisumbingii Hawkes and Heller
****Dendrobium ramosii*** Ames
****Dendrobium sanderae*** Rolfe
****Dendrobium sanderae*** var. ***luzonicum*** Tiu
****Dendrobium sanderae*** var. ***major*** Hort.
Dendrobium sanderae var. ***millerii*** Quis.
****Dendrobium sanderae*** var. ***parviflorum*** Anschutz *ex* Quis.
****Dendrobium sanderae*** var. ***surigaense*** Quis.
****Dendrobium schuetzei*** Rolfe
****Dendrobium secundum*** (Bl.) Lindl.
****Dendrobium secundum*** var. ***niveum*** Rchb. *f.*
****Dendrobium serratilabium*** fma. ***album*** Hort.
****Dendrobium serratilabium*** L.O. Wms.
****Dendrobium setigerum*** Ames *ex* M.A. Clem.
Dendrobium sinuosum Ames
Dendrobium sp. section ***Rhopalanthe***
****Dendrobium spurium*** (Bl.) J.J. Sm.
****Dendrobium taurinum*** fma. ***album*** Val. and Tiu
****Dendrobium taurinum*** Lindl.
Dendrobium tetrachromum Rchb. *f.*
****Dendrobium uniflorum*** Griff.
Dendrobium uniflorum var. ***angustum*** Quis.
Dendrobium x***usitae*** Yukawa
Dendrobium usteri Schltr.
Dendrobium usterioides Ames
****Dendrobium ventricosum*** Krzl.
Dendrobium victoriae-reginae fma. ***album*** Val. and Tiu
****Dendrobium victoriae-reginae*** Loher
****Dendrobium wenzelii*** Ames
****Dendrobium yeageri*** Ames and Quis.
Dendrobium zamboangense Ames

Dendrochilum Bl.
Dendrochilum abortum (Ames) L.O. Wms.
Dendrochilum affine Ames
Dendrochilum amesianum Pedersen
Dendrochilum anfractum var. ***anfractoides*** Ames) L.O. Wms.
Dendrochilum anfractum var. ***anfractum*** (Ames) Pfitz.
Dendrochilum apiculatum (Ames) L.O. Wms.
Dendrochilum apoense T. Hashim.
****Dendrochilum arachnites*** Rchb. *f.*
Dendrochilum asperum L.O. Wms.
****Dendrochilum auriculare*** Ames
Dendrochilum binuangense Ames
Dendrochilum cinnabarinum var. ***cinnabarinum*** Pfitz.
Dendrochilum cinnabarinum var. ***sanguineum*** (Ames) Pedersen
****Dendrochilum cobbianum*** Rchb. *f.*
****Dendrochilum convallariaeforme*** Schauer
Dendrochilum convallariaeforme var. ***minor*** (Ames) Pedersen
****Dendrochilum cootesii*** H.Æ. Pedersen
Dendrochilum copelandii Ames
****Dendrochilum curranii*** Ames
Dendrochilum curranii var. ***serratoi*** (Ames) L.O. Wms.
****Dendrochilum cymbiforme*** Ames
Dendrochilum ecallosum Ames
Dendrochilum edanoi Quis.
Dendrochilum elmeri Ames
Dendrochilum exiguum (Ames) Pedersen
Dendrochilum exile Ames
Dendrochilum eximium (Ames) L.O. Wms.
****Dendrochilum filiforme*** Lindl.
Dendrochilum flexuosum Pedersen
Dendrochilum foxworthyi Ames
****Dendrochilum glumaceum*** Lindl.
Dendrochilum graciliscapum (Ames) Pfitz.
Dendrochilum graminifolium (Ames) Pfitz.
Dendrochilum hastatum Ames
Dendrochilum hutchinsonianum Ames

Dendrochilum irigense Ames
****Dendrochilum javieri*** Magrath et al.
Dendrochilum kingii Hook. *f.*
Dendrochilum latifolium Lindl.
****Dendrochilum latifolium*** var. ***macranthum*** (Schltr.) H.Æ. Pedersen
Dendrochilum loheri Ames
****Dendrochilum longibulbum*** Ames
****Dendrochilum longifolium*** Rchb. *f.*
Dendrochilum longilabre (Ames) Pfitz.
Dendrochilum louisianum Pedersen
Dendrochilum luzonense Ames
Dendrochilum magnum Rchb. *f.*
Dendrochilum maleolens Krzl.
Dendrochilum malindangense Ames
Dendrochilum marginatum (Ames) L.O. Wms.
Dendrochilum mcgregorii Ames
Dendrochilum mearnsii Ames
Dendrochilum merrillii (Ames) Pfitz.
****Dendrochilum microchilum*** (Schltr.) Ames
Dendrochilum mindanaense (Ames) L.O. Wms.
Dendrochilum mindorense Ames
****Dendrochilum niveum*** Ames
Dendrochilum ocellatum (Ames) Pfitz.
Dendrochilum oliganthum (Ames) Pfitz.
Dendrochilum oreophilum (Ames) L.O. Wms.
****Dendrochilum pallideflavens*** Bl.
Dendrochilum pangasinanense Ames
Dendrochilum parvipapillatum Pedersen
****Dendrochilum parvulum*** (Ames) Pfitz.
Dendrochilum parvulum var. ***strictiforme*** (Ames) Pedersen
Dendrochilum perplexum var. ***montanum*** (Ames) Pedersen
Dendrochilum perplexum var. ***perplexum*** (Ames) L.O. Wms.
Dendrochilum philippinense var. ***philippinense*** (Ames) Pfitz.
Dendrochilum philippinense var. ***purpureum*** (Ames) Pedersen
Dendrochilum plocoglottoides Pedersen
Dendrochilum prodigiosum Ames
Dendrochilum propinquum Ames
****Dendrochilum pulcherrimum*** (Ames) L.O. Wms.
Dendrochilum pulogense Ames
Dendrochilum pumilum var. ***pumilum*** Rchb. *f.*
Dendrochilum pumilum var. ***recurvum*** (Ames) Pedersen
****Dendrochilum quadrilobum*** Ames
Dendrochilum quisumbingianum Pedersen
Dendrochilum reniforme Ames
****Dendrochilum rhombophorum*** (Rchb. *f.*) Ames
Dendrochilum rotundilabium L.O. Wms.
****Dendrochilum saccolabium*** Krzl.
Dendrochilum schweinfurthianum (Ames) L.O. Wms.
Dendrochilum simulacrum Ames
****Dendrochilum smithianum*** (Ames) L.O. Wms.
****Dendrochilum stenophyllum*** L.O. Wms.
****Dendrochilum tenellum*** (Nees and Meyen) Ames
****Dendrochilum tenuibulbum*** (Ames) L.O. Wms.
Dendrochilum tenuifolium (Ames) Pfitz.
Dendrochilum tetradactyliferum Pedersen
Dendrochilum turpe (Ames) Pfitz.
****Dendrochilum uncatum*** Rchb. *f.*
Dendrochilum uncatum var. ***longispicatum*** (Ames) Pedersen
Dendrochilum unicallosum L.O. Wms.
Dendrochilum unicorne (Ames) L.O. Wms.
****Dendrochilum vanoverberghii*** Ames
****Dendrochilum wenzelii*** Ames
****Dendrochilum williamsii*** (Ames) Pfitz.
Dendrochilum woodianum Ames
****Dendrochilum yuccaefolium*** L.O. Wms.

Dicerostylis Bl.
Dicerostylis lanceolata Bl.
Dicerostylis rubra (Ames) Schltr.

Didymoplexis Griff.
Didymoplexis pallens Griff.
Didymoplexis philippinensis Ames

Dienia Lind.
Dienia benguetensis (Ames) M.A. Clem. and D.L. Jones
Dienia curranii (Ames) M.A. Clem. and D.L. Jones
Dienia latifolia (Smith) M.A. Clem. and D.L. Jones

Diglyphosa Bl.
Diglyphosa elmeri Ames

Dilochia Lindl.
Dilochia elmeri Ames

Diplocaulobium (Rchb. *f.*) Krzl.
Diplocaulobium clemensiae (Ames) A.D. Hawkes

Dipodium R. Brown.
****Dipodium paludosum*** (Griff.) Rchb. *f.*

Dispersis Swartz
Dispersis philippinensis Schltr.

Doritis Lindl.
Doritis pulcherrima Lindl.

Epiblastus Schltr.

Epiblastus merrillii L.O. Wms.

Epicrianthes Bl.

****Epicrianthes flavofimbriata*** (J.J. Sm.) Garay and Kitt.

Epigeneium Gagnep.

****Epigeneium acuminatum*** (Rolfe) Summerh.

Epigeneium stella-silvae (Loher and Krzl.) Summerh.

Epigeneium treacherianum fma. ***album*** Hort.

****Epigeneium treacherianum*** (Rchb. *f. ex* Hook. *f.*) Summerh.

Epipogium Gmelin

Epipogium roseum (D. Don.) Lindl.

Eria Lind.

Eria albolutea Rolfe

Eria aliciae Quis

Eria anceps Leavitt

****Eria aporoides*** Lindl.

Eria barbifrons Krzl.

****Eria brachystachya*** Rchb. *f.*

****Eria bractescens*** Lindl.

Eria candoonensis Ames

Eria carnicolor Ames

Eria clemensiae Leavitt

Eria compacta Ames

Eria copelandii Leavitt

Eria copelandii var. ***fusiformis*** Leavitt

Eria curranii Leavitt

Eria cylindrostachya Ames

Eria cymbiformis J.J. Sm.

Eria dagamensis Ames

Eria davaensis Ames

****Eria floribunda*** Lindl.

****Eria gigantea*** Ames

Eria hirsutipetala Ames

Eria hutchinsoniana Ames

Eria jarensis Ames

****Eria javanica*** (Sw.) Bl.

Eria lamellata Ames

Eria leavittii Krzl.

Eria loheriana Krzl.

Eria longibracteata Leavitt

Eria longicruris Leavitt

Eria longilabris Lindl.

****Eria longissima*** Ames and Quis.

Eria lyonii Leavitt

Eria macera Ames

Eria major Ridl. *ex* Stapf.

Eria maquilingensis Ames

Eria mcgregorii Ames

Eria mearnsii Leavitt

Eria merrittii Ames

Eria microchila Ames

Eria mindanaensis Ames

****Eria ornata*** (Bl.) Lindl.

****Eria ovata*** Lindl.

****Eria ovata*** var. ***retroflexa*** (Lindl.) Garay and Sweet

Eria palawanensis Ames

Eria perspicabilis Ames

Eria philippinensis Ames

****Eria polyura*** Lindl.

****Eria profusa*** Lindl.

Eria propinqua Ames

Eria puguahaanensis Ames

Eria ramosa Ames

****Eria rhodoptera*** Rchb. *f.*

Eria ringens Rchb. *f.*

Eria senilis Ames

Eria taylorii Ames

Eria tomentosiflora Hay.

****Eria tridens*** Ames

****Eria vanoverberghii*** Ames

Eria ventricosa Leavitt

Eria whitfordii Leavitt

Eria woodiana Ames

****Eria xanthocheila*** Ridl.

Eria zamboangensis Ames

Erythrodes Bl.

Erythrodes boettcheri Ames

Erythrodes vrydagzynoides Ames

Erythrodes weberi Ames

Erythrodes wenzelii Ames

Erythrorchis Bl.

Erythrorchis atlissima (Bl.) Bl.

Euanthe Schltr.

****Euanthe sanderiana*** (Rchb. *f.*) Schltr.

****Euanthe sanderiana*** var. ***albata*** Rchb. *f.*

Euanthe sanderiana var. ***froebeliana*** Cogn.

Euanthe sanderiana var. ***labello-viridi*** Linden and Rodigas

Eulophia R. Br. *ex* Lind.

Eulophia andamanensis Rchb. *f.*

Eulophia dentata Ames

Eulophia graminea Naves

Eulophia macrostachya Lindl.

Eulophia merrillii Ames

Eulophia pulchra (Thouars) Lindl.

Eulophia spectabilis (Dennst.) Suresh

Eulophia stricta (Presl) Ames
Eulophia vanoverberghii Ames
Eulophia zollingeri (Rchb. *f.*) J.J. Sm.

Flickingeria Hawkes
Flickingeria bancana (J.J. Sm.) Hawkes
Flickingeria candoonense (Ames) Seid.
Flickingeria chrysographota (Ames) Hawkes
Flickingeria eurora (Ames) Hawkes
****Flickingeria fimbriata*** (Bl.) Hawkes
Flickingeria fimbriatolabellum (Hay.) Hawkes
****Flickingeria interjecta*** (Ames) Hawkes
****Flickingeria junctiloba*** Fessel and Lückel
Flickingeria loheriana (Krzl.)
****Flickingeria luxurians*** (J.J. Sm.) Hawkes
Flickingeria pseudoconvexa (Ames) Hawkes
Flickingeria purpureostelidia (Ames) Hawkes
****Flickingeria scopa*** (Lindl.) Brieger
Flickingeria unicornis (Ames) Hawkes
****Flickingeria xantholeuca*** (Rchb. *f.*) Hawkes

Galeola Lour.
Galeola altissima (Bl.) Rchb. *f.*
Galeola hydra Rchb. *f.*
Galeola philippinensis Ames

Gastrochilus D. Don
Gastrochilus calceolaris (Buch-Ham *ex* J.E. Sm.) D. Don.

Gastrodia R. Brown
Gastrodia javanica (Bl.) Lindl.

Geodorum Jackson
Geodorum densiflorum (Lam.) Schltr.

Glomera Bl.
Glomera merrillii Ames

Goodyera R. Brown
****Goodyera clausa*** (A.A. Eaton *ex* Ames) Schltr.
Goodyera elmeri (Ames) Ames
Goodyera fumata Thw.
Goodyera luzonensis Ames
Goodyera philippinensis (Ames) Schltr.
Goodyera procera (Ker.-Gawl.) Hook.
Goodyera ramosii Ames
Goodyera rubicunda (Bl.) Lindl.
Goodyera viridiflora Bl.

Grammatophyllum Bl.
****Grammatophyllum elegans*** Rchb. *f.*
****Grammatophyllum martae*** Quis. *ex* Val. and Tiu
****Grammatophyllum measuresianum*** Weathers
Grammatophyllum multiflorum Lindl.
****Grammatophyllum multiflorum*** fma. ***citrinum*** Hort.
****Grammatophyllum multiflorum*** var. ***tigrinum*** Lindl.
****Grammatophyllum speciosum*** Bl.
****Grammatophyllum stapeliiflorum*** (Teij. and Binn.) J.J. Sm.
Grammatophyllum wallisii Rchb. *f.*

Grastidium Bl.
Grastidium acuminatissimum Bl.
Grastidium agusanense (Ames) M.A. Clem. and D.L. Jones
Grastidium alagense (Ames) M.A. Clem. and D.L. Jones
Grastidium aliciae (Ames and Quis.) M.A. Clem. and D.L. Jones
Grastidium leytense (Ames) M.A. Clem. and D.L. Jones
****Grastidium luzonense*** (Lindl.) M.A. Clem. and D.L. Jones
****Grastidium ornithoflorum*** (Ames) M.A. Clem. and D.L. Jones
Grastidium pergracile (Ames) M.A. Clem. and D.L. Jones
****Grastidium verruculosum*** (Ames) M.A. Clem. and D.L. Jones

Grosourdya Rchb. *f.*
****Grosourdya appendiculata*** (Bl.) Rchb. *f.*
Grosourdya pustulata (Senghas)
Grosourdya tripercus (Ames) Garay

Habenaria Willdenow
Habenaria alagensis Ames
Habenaria aristulifera Rchb. *f.*
Habenaria articulata Lindl.
Habenaria boadanensis Ames
Habenaria congesta Ames
Habenaria curranii Ames
Habenaria delessertiana Krzl.
Habenaria dentata (Sw.) Schltr.
Habenaria diphylla (Nim.) Dalz.
Habenaria falcigera Rchb. *f.*
Habenaria hystrix Ames
Habenaria intrudens Ames
Habenaria leibergii Ames
Habenaria lingulosa Ames
Habenaria malintana (Blco.) Merr.
Habenaria mearnsii Ames
Habenaria militaris var. ***philippinensis*** Ames
Habenaria ponerostachys Rchb. *f.*
Habenaria reticulata Ames

Habenaria rhodocheila var. ***philippinensis*** (Ames) E.A. Christenson
Habenaria robinsonii Ames
Habenaria rosulata Ames
Habenaria vanoverberghii Ames
Habenaria warburgiana Krzl.
Habenaria zephyrica Ames
Herminium Linnaeus
Herminium lanceum (Thunb. *ex* Sw.) Vuijk.
Herpysma Lindl.
Herpysma merrillii Ames
Hetaeria Bl.
Hetaeria cristata Bl.
Hetaeria lancifolia Ames
Hetaeria leytensis Ames
Hetaeria mindanaensis Ames
Hetaeria oblongifolia Bl.
Hetaeria wenzelii Ames
Hippeophyllum Schltr.
****Hippeophyllum wenzelii*** Ames
Hylophila Lindl.
Hylophila lanceolata (Bl.) Miq.
Hymenorchis Schltr.
Hymenorchis vanoverberghii (Ames) Garay
Kingidium P.F. Hunt
****Kingidium philippinense*** (Ames) Gruß and Röllke
Kuhlhasseltia J.J. Sm.
Kuhlhasseltia whiteheadii (Rendle) Ames
Lecanorchis Bl.
Lecanorchis javanica Bl.
Lepidogyne Bl.
Lepidogyne longifolia Bl.
Liparis L.C. Richard
Liparis benguetensis Ames
Liparis bontocensis Ames
Liparis bootanensis Griff.
****Liparis caespitosa*** (Thouars) Lindl.
Liparis clemensiae Ames
Liparis compressa Ames
****Liparis condylobulbon*** Rchb. *f.*
Liparis cumingii Ames
Liparis distans Clarke
Liparis dumaguetensis Ames
****Liparis elegans*** Lindl.
Liparis elmeri Ames
Liparis fragilis Ames
Liparis halconensis Ames
Liparis jarensis Ames
****Liparis latifolia*** (Bl.) Lindl.
Liparis leytensis Ames
Liparis linearifolia Ames
Liparis lyonii Ames
Liparis magnicallosa Ames
****Liparis merrillii*** (Ames) Schltr.
Liparis monophylla Ames
Liparis negrosiana Ames
Liparis nutans Ames
Liparis palawanensis Ames
Liparis pallida (Bl.) Lindl.
****Liparis parviflora*** (Bl.) Lindl.
Liparis pectinata Ridl.
****Liparis philippinensis*** (Ames) Schltr.
Liparis prava Ames
Liparis propinqua Ames
Liparis rizalensis Ames
Liparis tricallosa Rchb. *f.*
Liparis trichoglottis Ames
****Liparis viridiflora*** (Bl.) Lindl.
****Liparis wenzelii*** Ames
Ludisia A.C. Rich.
Ludisia discolor A. Rich.
Luisia Gaud.
Luisia cordatilabia Ames and Quis.
Luisia foxworthii Ames
Luisia ramosii Ames
Luisia teretifolia Gaud.
Macodes Lindl.
Macodes petola (Bl.) Lindl.
Macropodanthus L.O. Wms.
****Macropodanthus philippinensis*** L.O. Wms.
Malaxis Solander *ex* Swartz
Malaxis arietina Ames
Malaxis atrosanguinea Ames
Malaxis bulusanensis Ames
Malaxis epiphytica Ames
Malaxis monophyllos (Linn.) Sw.
Malleola J.J. Sm. and Schltr.
Malleola cf. ***ligulata*** J.J. Sm.
****Malleola constricta*** Ames
Malleola lyonii Ames
Mastigion Garay et al.
Mastigion putidum (Teij. and Binn.) Garay et al.
Megalotus Garay
****Megalotus bifidus*** (Lindl.) Garay
Micropera Lindl.
Micropera loheri (L.O. Wms.) Garay
Micropera pallida (Roxb.) Lindl.
****Micropera philippinensis*** (Lindl.) Garay
****Micropera utriculosa*** (Ames) Garay

Microsaccus Bl.
- ***Microsaccus brevifolius*** J.J. Sm.
- ***Microsaccus wenzelii*** Ames

Microtatorchis Schltr.
- ***Microtatorchis aristata*** Garay
- ***Microtatorchis compacta*** (Ames) Schltr.

Microtis R. Brown
- ***Microtis unifolia*** (Forst. *f.*) Rchb. *f.*

Myrmechis Bl.
- ***Myrmechis gracilis*** Bl.
- ***Myrmechis perpusilla*** Ames
- ***Myrmechis philippinensis*** Ames

Nephelaphyllum Bl.
- ***Nephelaphyllum mindorensis*** Ames
- ****Nephelaphyllum pulchrum*** Bl.

Nervilia Comm. *ex* Gaud.
- ***Nervilia aragoana*** Gaud.
- ***Nervilia crocoformis*** (Zoll. and Mor.) Seid.
- ***Nervilia dilatata*** (Bl.) Schltr.
- ***Nervilia discolor*** (Bl.) Schltr.
- ***Nervilia plicata***

Neuwiedia Bl.
- ***Neuwiedia veratrifolia*** Bl.

Oberonia Lindl.
- ***Oberonia anceps*** Lindl.
- ***Oberonia aporophylla*** Rchb. *f.*
- ***Oberonia basilanensis*** Ames
- ***Oberonia benguetensis*** Ames
- ***Oberonia cylindrica*** Lindl.
- ***Oberonia elmeri*** Ames
- ***Oberonia hispidula*** Ames
- ***Oberonia iridifolia*** (Roxb.) Lindl.
- ***Oberonia leytensis*** Ames
- ***Oberonia lipensis*** Ames
- ***Oberonia luzonensis*** Ames
- ***Oberonia mcgregorii*** Ames
- ***Oberonia merrillii*** Ames
- ***Oberonia mindorensis*** Ames
- ***Oberonia minima*** Ames
- ***Oberonia minutissima*** Ames
- ***Oberonia monstruosa*** (Bl.) Lindl.
- ***Oberonia obesa*** Ames
- ***Oberonia reilloi*** Ames
- ****Oberonia setigera*** Ames
- ***Oberonia surigaensis*** Ames
- ***Oberonia thisbe*** Rchb. *f.*
- ****Oberonia toppingii*** Ames
- ***Oberonia wenzelii*** Ames

Octarrhena Thwaites
- ***Octarrhena amesiana*** Schltr.
- ***Octarrhena caulescens*** Ames
- ***Octarrhena elmeri*** Ames
- ***Octarrhena ensifolia*** (Ames) Schltr.
- ***Octarrhena gemmifera*** Ames

Omoea Bl.
- ****Omoea philippinensis*** Ames

Osyricera Blume
- ***Osyricera erosipetala*** (C. Schweinf.) Garay et al.

Pachystoma Bl.
- ***Pachystoma pubescens*** Bl.

Paphiopedilum Pfitzer
- ****Paphiopedilum acmodontum*** Schoser *ex* Wood
- ****Paphiopedilum adductum*** Asher
- ***Paphiopedilum anitum*** Golamco
- ****Paphiopedilum argus*** (Rchb. *f.*) Stein
- ***Paphiopedilum argus*** var. ***moensii*** (Linden) Stein
- ***Paphiopedilum argus*** var. ***nigricans*** Hort *ex* Stein
- ****Paphiopedilum ciliolare*** (Rchb. *f.*) Stein
- ****Paphiopedilum fowliei*** Birk
- ****Paphiopedilum fowliei*** fma. ***christianae*** Braem
- ****Paphiopedilum haynaldianum*** fma. ***album*** (Hort. *ex* Asher) Braem
- ****Paphiopedilum haynaldianum*** (Rchb. *f.*) Stein
- ****Paphiopedilum hennisianum*** fma. ***christiansenii*** (Gruß and Röth) Gruß and Röth
- ***Paphiopedilum hennisianum*** fma. ***purpureum*** Hort.
- ****Paphiopedilum hennisianum*** (Wood) Fowlie
- ***Paphiopedilum laevigatum*** (Batem.) Pfitz.
- ***Paphiopedilum*** x***mindanaense*** Golamco
- ****Paphiopedilum parnatanum*** Cavestro
- ****Paphiopedilum philippinense*** (Rchb. *f.*) Stein
- ****Paphiopedilum philippinense*** fma. ***alboflavum*** Gruß
- ****Paphiopedilum randsii*** Fowlie
- ****Paphiopedilum roebbelenii*** (Rchb. *f.*) Pfitz.
- ***Paphiopedilum superbiens*** (Rchb. *f.*) Stein
- ***Paphiopedilum urbanianum*** fma. ***alboviride*** Braem
- ****Paphiopedilum urbanianum*** Fowlie

Pennilabium J.J. Sm.
- ***Pennilabium confusum*** (Ames) Garay
- ***Pennilabium luzonense*** (Ames) Garay

Peristylus Bl.
- ***Peristylus bilobus*** Rolfe

Peristylus copelandii (Ames) Carr.
Peristylus goodyeroides (D. Don.) Lindl.
Peristylus lacertifer J.J. Sm.
Phaius Lour.
Phaius amboinensis Bl.
Phaius borneensis J.J. Sm.
***Phaius flavus** (Bl.) Lindl.
Phaius fragilis L.O. Wms.
Phaius linearifolius Ames
Phaius lyonii Ames
Phaius mishmensis (Lindl.) Rchb. *f.*
Phaius philippinensis N.E. Br.
Phaius ramosii Ames
***Phaius tancarvilleae** (Banks *ex* l'Heritier) Bl.
Phalaenopsis Bl.
***Phalaenopsis amabilis** (Linn.) Bl.
Phalaenopsis amabilis var. **aurea** (Hort.) Rolfe
***Phalaenopsis aphrodite** Rchb. *f.*
Phalaenopsis bastianii Gruß and Röllke
***Phalaenopsis cornucervi** (Breda) Bl. and Rchb. *f.*
***Phalaenopsis equestris** (Schauer) Rchb. *f.*
***Phalaenopsis equestris** var. **alba** (Hort.) Sweet
Phalaenopsis equestris var. **leucaspis** Rchb. *f.*
Phalaenopsis equestris var. **leucotanthe** Rchb. *f. ex* Godefroy-Lebauf
Phalaenopsis equestris var. **rosea** Val. and Tiu
***Phalaenopsis fasciata** Rchb. *f.*
Phalaenopsis fasciata var. **flava** Val. and Tiu
Phalaenopsis fuscata Rchb. *f.*
***Phalaenopsis heiroglyphica** (Rchb. *f.*) Sweet
Phalaenopsis x**intermedia** Lindl.
Phalaenopsis x**intermedia** var. **diezii** (Corvera) Quis.
***Phalaenopsis lindenii** Loher
***Phalaenopsis lindenii** var. **alba** Hort.
***Phalaenopsis lueddemanniana** Rchb. *f.*
Phalaenopsis lueddemanniana var. **delicata** Rchb. *f.*
Phalaenopsis lueddemanniana var. **ochracea** Rchb. *f.*
Phalaenopsis x**leucorrhoda** Rchb. *f.*
***Phalaenopsis mariae** Burb. *ex* Warner and Wms.
***Phalaenopsis micholitzii** Rolfe
***Phalaenopsis pallens** (Lindl.) Rchb. *f.*
Phalaenopsis pallens var. **alba** (Ames and Quis.) Sweet
Phalaenopsis pallens var. **denticulata** (Rchb. *f.*) Sweet
Phalaenopsis pallens var. **trullifera** Sweet
***Phalaenopsis philippinensis** Golamco *ex* Fowlie and C.Z. Tang
***Phalaenopsis pulchra** (Rchb. *f.*) Sweet
Phalaenopsis reichenbachiana Rchb. *f.* and Sander
***Phalaenopsis sanderiana** Rchb. *f.*
Phalaenopsis sanderiana var. **alba** (Veitch) Stein
Phalaenopsis sanderiana var. **marmorata** Rchb. *f.*
***Phalaenopsis schilleriana** Rchb. *f.*
Phalaenopsis schilleriana var. **immaculata** Rchb. *f.*
Phalaenopsis schilleriana var. **pallida** Val. and Tiu
Phalaenopsis schilleriana var. **purpurea** O'Brien
Phalaenopsis schilleriana var. **splendens** Warner
***Phalaenopsis stuartiana** Rchb. *f.*
Phalaenopsis stuartiana var. **bella** Rchb. *f.*
Phalaenopsis stuartiana var. **punctatissima** Rchb. *f.*
***Phalaenopsis sumatrana** Korthals and Rchb. *f.*
Phalaenopsis x**veitchiana** Rchb. *f.*
Philippinaea Schltr. and Ames
Philippinaea wenzelii (Ames) Ames and Schltr.
Pholidota Lindl. *ex* Hook.
Pholidota articulata Lindl.
Pholidota carnea (Bl.) Lindl.
Pholidota imbricata Lindl.
***Pholidota ventricosa** (Bl.) Rchb. *f.*
Phragmorchis L.O. Wms.
Phragmorchis teretifolia L.O. Wms.
Phreatia Lindl.
Phreatia amesii Krzl.
Phreatia aristulifera Ames
Phreatia caulescens Ames
Phreatia densiflora (Bl.) Lindl.
Phreatia infundibuliformis Ames
Phreatia luzoniensis Rolfe *ex* Ames
Phreatia mearnsii Ames
Phreatia negrosiana Ames
Phreatia prorepens Rchb. *f.*
Phreatia ramosii Ames
***Phreatia secunda** (Bl.) Lindl.
***Phreatia sulcata** (Bl.) J.J. Sm.

Phreatia vanoverberghii Ames
Phreatia wenzelii Ames
Phreatia xantholeuca Krzl.
Pilophyllum Schltr.
Pilophyllum villosum (Bl.) Schltr.
Platanthera L.C. Richard
Platanthera angustata Lindl.
Platanthera elmeri (Ames) Schltr.
Platanthera halconensis (Ames) Schltr.
Plocoglottis Bl.
***Plocoglottis acuminata** Bl.
Plocoglottis bicallosa Ames
Plocoglottis bicomata L.O. Wms.
Plocoglottis copelandii Ames
Plocoglottis lucbanensis Ames
Plocoglottis mindorensis Ames
Poaephyllum Ridl.
Poaephyllum grandiflorum Quis.
Poaephyllum pauciflorum Ridl.
Podochilus Bl.
Podochilus bicaudatus Schltr.
Podochilus cumingii Schltr.
Podochilus hystricinus Ames
****Podochilus intricatus*** Ames
Podochilus longilabris Ames
Podochilus lucescens Bl.
Podochilus plumosus Ames
Podochilus ramosii Ames
Podochilus sciuroides Rchb. *f.*
Podochilus strictus Ames
Polystachya Hooker
****Polystachya concreta*** (Jacquin) Garay and Sweet
Pomatocalpa Breda
Pomatocalpa andamanica (Hook. *f.*) J.J. Sm.
****Pomatocalpa bicolor*** (Lindl.) J.J. Sm.
Pomatocalpa kunstleri (Hook. *f.*) J.J. Sm.
Pomatocalpa latifolia (Lindl.) J.J. Sm.
****Pomatocalpa spicata*** Breda
Pristiglottis Cretz. and J.J. Sm.
Pristiglottis mindanaensis (Ames) Cretz. and J.J. Sm.
Pteroceras Hasskarl
Pteroceras cladostachyum (Hook. *f.*) Pedersen
****Pteroceras leopardinum*** (Par. and Rchb. *f.*) Seid. and Smit.
****Pteroceras longicalcarum*** (Ames and Rolfe) Garay
Pteroceras philippinense (Ames) Garay
Pteroceras teres (Bl.) Holtt.
****Pteroceras unguiculatum*** (Lindl.) H.Æ. Pedersen
Renanthera Lour.
Renanthera elongata Lindl.
Renanthera matutina Lindl.
****Renanthera monachica*** Ames
****Renanthera philippinensis*** Ames and Quis.
****Renanthera storiei*** Rchb. *f.*
Renanthera storiei var. ***citrina*** Val. and Tiu
Rhomboda Lindl.
Rhomboda blackii (Ames) Ormd.
Rhomboda lanceolata (Lindl.) Ormd.
Rhomboda weberi (Ames) Ormd.
Rhynchostylis Bl.
****Rhynchostylis praemorsa*** (Willd.) Bl.
Rhynchostylis violacea Rchb. *f.*
Rhytionanthus Garay et al.
Rhytionanthus plumatum (Ames) Garay et al.
Robiquetia Gaud.
****Robiquetia cerina*** (Rchb. *f.*) Garay
****Robiquetia compressa*** (Lindl.) Schltr.
Robiquetia discolor (Rchb. *f.*) Seid. and Garay
Robiquetia pantherina (Krzl.) Ames
Robiquetia vanoverberghii Ames
Saccolabiopsis J.J. Sm.
Saccolabiopsis tenella (Ames) Garay
Sarcophyton Garay
Sarcophyton crassifolium (Lindl.) Garay
****Sarcophyton pachyphyllus*** (Ames) Garay
Schoenorchis Bl.
Schoenorchis micrantha Bl.
Schoenorchis vanoverberghii Ames
Spathoglottis Bl.
****Spathoglottis chrysantha*** Ames
Spathoglottis elmeri Ames
****Spathoglottis kimballiana*** var. ***angustifolia*** Ames
Spathoglottis x***parsonii*** Ames and Quis.
****Spathoglottis plicata*** Bl.
Spathoglottis tomentosa Lindl.
****Spathoglottis vanoverberghii*** Ames
Spiranthes L.C. Richard
Spiranthes sinensis (Pers.) Ames
Staurochilus Ridl. *ex* Pfitzer
Staurochilus agusanensis (Ames and Quis.) Fessel and Lückel
****Staurochilus fasciatus*** Rchb. *f.*
Staurochilus guibertii (Linden and Rchb. *f.*) E.A. Christenson

Staurochilus intermedius (L.O. Wms.) Fessel and Lückel
*****Staurochilus ionosma*** (Lindl.) Schltr.
*****Staurochilus loheriana*** (Krzl.) E.A. Christenson
Staurochilus luchuensis (Rolfe) Fukuyama
*****Staurochilus luzonensis*** Ames
Staurochilus tamesisii (Quis. and Schweinf.) Fessel and Lückel

Stereosandra Bl.
Stereosandra javanica Bl.

Synarmosepalum Garay et al.
Synarmosepalum kittredgei Garay et al.

Taeniophyllum Bl.
Taeniophyllum copelandii Ames
Taeniophyllum elmeri Ames
Taeniophyllum leytensis Ames
Taeniophyllum merrillii Ames
Taeniophyllum philippinensis Rchb. *f.*
Taeniophyllum saccatum L.O. Wms.

Tainia Bl.
Tainia elmeri Ames

Thecostele Rchb. *f.*
*****Thecostele alata*** (Roxb.) Par. and Rchb. *f.*

Thelasis Bl.
Thelasis capitata Bl.
Thelasis carinata Bl.
Thelasis micrantha (Brongn.) J.J. Sm.
Thelasis obtusa Bl.
Thelasis pygmaea Hance
Thelasis triptera Rchb. *f.*

Thelymitra Forster
Thelymitra javanica Bl.

Thrixspermum Lour.
Thrixspermum acuminatissimum (Bl.) Rchb. *f.*
Thrixspermum agusanense Ames
Thrixspermum amesianum L.O. Wms.
Thrixspermum amplexicaule (Bl.) Rchb. *f.*
Thrixspermum angustatum L.O. Wms.
Thrixspermum elmeri L.O. Wms.
Thrixspermum elongatum Ames
Thrixspermum eximium L.O. Wms.
Thrixspermum fantasticum L.O. Wms.
Thrixspermum hystrix (Bl.) Rchb. *f.*
Thrixspermum integrum L.O. Wms.
Thrixspermum ligulatum L.O. Wms.
Thrixspermum linearifolium Ames
Thrixspermum merguense (Hook. *f.*) Ktze.
Thrixspermum pensile Schltr.
Thrixspermum quinquelobum Ames
*****Thrixspermum robinsonii*** Ames
Thrixspermum rostratum Ames
*****Thrixspermum subulatum*** (Bl.) Rchb. *f.*
Thrixspermum vanoverberghii Ames
Thrixspermum weberi Ames
Thrixspermum wenzelii Ames

Trachoma Garay
Trachoma brevirhachis (L.O. Wms.)
Trachoma rhopalorhachis (Rchb. *f.*) Garay

Trichoglottis Bl.
*****Trichoglottis amesiana*** L.O. Wms.
*****Trichoglottis atropurpurea*** Rchb. *f.*
*****Trichoglottis bataanensis*** Ames
Trichoglottis brachiata Ames
Trichoglottis calochila L.O. Wms.
*****Trichoglottis geminata*** J.J. Sm.
*****Trichoglottis latisepala*** Ames
Trichoglottis mindanaensis Ames
*****Trichoglottis philippinensis*** Lindl.
Trichoglottis rosea (Lindl.) Ames
Trichoglottis solerederi Krzl.

Trichotosia Bl.
Trichotosia binabayensis Krzl.
*****Trichotosia fastigiatifolia*** (Ames) Krzl.
Trichotosia fusca Teij. and Binn.
Trichotosia halconensis Krzl.
Trichotosia lagunensis Ames
Trichotosia leytensis Ames
Trichotosia odorifera Krzl.
Trichotosia ramosii Krzl.
*****Trichotosia vulpina*** (Rchb. *f.*) Krzl.

Tropidia Lindl.
Tropidia calcarata Ames
Tropidia mindanaensis Ames
Tropidia mindorensis Ames
Tropidia robinsonii Ames
Tropidia septemnervis (Schauer) Rchb. *f.*

Tuberolabium Yamamoto
Tuberolabium escritorii (Ames) Garay
*****Tuberolabium quisumbingii*** (L.O. Wms.) E.A. Christenson
*****Tuberolabium rhopalorrachis*** (Rchb. *f.*) J.J. Wood
Tuberolabium sarcochiloides (Schltr.) Garay

Vanda Jones *ex* R. Brown
*****Vanda javierae*** Tiu *ex* Fessel and Lückel
*****Vanda lamellata*** Lindl.
*****Vanda lamellata*** var. ***boxalli*** Rchb. *f.*
Vanda lamellata var. ***calayana*** Val. and Tiu
*****Vanda lamellata*** var. ***flava*** Val. and Tiu

*****Vanda lamellata*** var. ***remediosae*** Ames and Quis.
Vanda limbata Bl.
*****Vanda luzonica*** Loher *ex* Rolfe
*****Vanda merrillii*** Ames and Quis.
Vanda merrillii var. ***immaculata*** Ames and Quis.
Vanda merrillii var. ***rotorii*** Ames and Quis.
*****Vanda roeblingiana*** Rolfe
*****Vanda scandens*** Holtt.
*****Vanda ustii*** Golamco, Claustro and de Mesa
Vandopsis Pfitzer
Vandopsis lissochiloides (Gaud.) Pfitz.
Vanilla Swartz
Vanilla calopogon Rchb. *f.*
Vanilla ovalis Blco.
Vexillabium Maekawa
Vexillabium yakushimense (Yam.) Maekawa
Vrydagzynea Bl.
Vrydagzynea albida Bl.
Vrydagzynea weberi Ames
Zeuxine Lindl.
Zeuxine affinis (Lindl.) Benth. *ex* Hook. *f.*
Zeuxine elmeri Ames
*****Zeuxine flava*** (Wall. *ex* Lindl.) Benth. *ex* Hook. *f.*
Zeuxine leytensis Ames
Zeuxine luzonensis Ames
Zeuxine marivelensis Ames
Zeuxine nervosa (Lindl.) Benth. *ex* Hook. *f.*
Zeuxine parvifolia (Ridl.) Seid.
Zeuxine philippinensis Ames
Zeuxine strateumatica (Linn.) Schltr.
Zeuxine weberi Ames

INDEX

Page numbers for illustrations are in bold italics. Orchid names in upright text and not in bold type are synonyms.

THE AUTHOR

Jim Cootes started seriously growing orchids about 1975. He first visited the Philippines in 1977 and has been there at least once a year since 1977. He lived in the Philippines for an extended period between 1997 and 2000. On countless trips throughout the islands and their mountains, he studied the orchids in their natural habitat, and discovered previously undescribed species, two of which were named in his honour—*Bulbophyllum cootesii* and *Dendrochilum cootesii*. He has named one new species, *Amesiella monticola*, in conjunction with David P. Banks, and is currently working on numerous undescribed species. An albino form of *Dendrobium fairchildiae* will be described by this author in the near future. There are a number of species from near his home in Mindoro (where he plans to live upon his retirement) which are most likely undescribed and he is currently researching these plants.

For Jim Cootes, an interest in Philippine orchids that began in the early 1980s grew into an obsession about 1990 when he decided to write this book on the orchids of the Philippines. He began collecting Philippine orchid species about 1985, after a trip to the mountains in central Luzon. Many of the plants obtained were grown and flowered, and identification was attempted. About 15 percent of these plants could not be identified. He started to gather all available literature on Philippine orchids and still a number of these plants defied identification. Most have since been described as new species.

An enormous amount of work remains to be done on the orchids of the Philippines and there are undoubtedly a multitude of new species awaiting description. Cootes hopes this book is a start in that direction. The author is willing to attempt to identify Philippines orchid species. Send a slide, or JPEG image via e-mail, including the dimensions of the flowers and plant. The author can be contacted at jecootes@ozemail.com.au.

Published in North America in 2001 by
Timber Press, Inc.
The Haseltine Building
133 S.W. Second Avenue, Suite 450
Portland, Oregon 97204, U.S.A.

by special arrangement with
Times Media Private Limited
Times Centre
1 New Industrial Road, Singapore 536196

ISBN 0-88192-516-0

A catalog record for this book is available from the Library of Congress.

Printed in Singapore